# STATICS+++

## NEVADA EDITION

**JAMES W. DALLY**

**ROBERT J. BONENBERGER, JR.**

University of Maryland, College Park

## Collaborating with Ann-Marie Vollstedt

University of Nevada @ Reno

College House Enterprises, LLC
Knoxville, Tennessee

The manuscript was prepared with 11 point Times New Roman font. The Word files were then converted to pdf files using Adobe Acrobat DC Pro.

**College House Enterprises, LLC.**
**5713 Glen Cove Drive**
**Knoxville, TN 37919, U. S. A.**
**Phone    (865) 558 6111**
**FAX    (865) 558 6111**
**email    jwd@www.collegehousebooks.com**
**Visit our web site    http://www.collegehousebooks.com**

**ISBN 978-1-935673-41-5**

This book is for Anne
who has shared Jim's dreams for so many years


and to the memory of Ann and Bob
who had encouraged and supported Bob until they were called.

# ABOUT THE AUTHORS

**James W. Dally** obtained a Bachelor of Science and a Master of Science degree, both in Mechanical Engineering from the Carnegie Institute of Technology. He obtained a Doctoral degree in mechanics from the Illinois Institute of Technology. He has taught at Cornell University, Illinois Institute of Technology, the U. S. Air Force Academy and served as Dean of Engineering at the University of Rhode Island. He is currently a Glenn L. Martin Institute Professor of Engineering (Emeritus) at the University of Maryland, College Park.

Dr. Dally has also held positions at the Mesta Machine Co., IIT Research Institute and IBM. He is a fellow of the American Society for Mechanical Engineers, Society for Experimental Mechanics, and the American Academy of Mechanics. He was appointed as an honorary member of the Society for Experimental Mechanics in 1983 and elected to the National Academy of Engineering in 1984. Professor Dally was selected by his peers to receive the Senior Faculty Outstanding Teaching Award in the College of Engineering and the Distinguish Scholar Teacher Award from the University. He was also a member of the University of Maryland team receiving the 1996 Outstanding Educator Award sponsored by the Boeing Co.

Professor Dally has co-authored several other books: *Experimental Stress Analysis, Experimental Solid Mechanics, Photoelastic Coatings, Instrumentation for Engineering Measurements, Packaging of Electronic Systems, Production Engineering and Manufacturing, and Introduction to Engineering Design, Books 1, 2, 3, 4, 5, 6, 7, 8, 9, 10 and 11.* He has authored or coauthored about 200 scientific papers and holds five patents.

**Robert J. Bonenberger, Jr.** obtained B.S.E., M.S. and Ph.D. degrees in Mechanical Engineering from the University of Maryland, Baltimore County and College Park campuses. He has taught undergraduate and graduate students in mechanics, strength of materials, and experimental stress analysis. Currently, he is an Assistant Research Scientist and Lecturer in the Clark School of Engineering and a member of the Keystone Program at the University of Maryland, College Park.

Previously, Dr. Bonenberger worked in the Fracture Mechanics Section at the Naval Research Laboratory, both as a postdoctoral fellow and as a contract employee. He is a member of the American Society for Mechanical Engineers, Society for Experimental Mechanics, American Society for Materials, and American Society for Engineering Education. His research interests include material behavior at high strain rates, experimental stress analysis, and fracture mechanics. He has authored or co-authored 20 scientific papers.

# PREFACE

This textbook has been prepared to support a course offering for Statics at the University of Nevada at Reno. Statics provides the first exposure of engineering students to the study of mechanics. While Statics is a relatively simple subject, many students find it difficult, and they often perform far below our expectations. In an effort to improve the curriculum, several members of the faculty at the University of Maryland have been working to enhance the student's learning experience when studying the first two courses in mechanics. This textbook indicates some of the changes in the philosophy adopted by the faculty when presenting the subject matter traditionally offered in the frist two introductory mechanics courses.

The changes in the philosophy were based on five premises:

1. Present the fundamental concepts in a more interesting manner.
   - Change the approach to make it more realistic and less abstract.
   - Couple the mechanics content tightly to design.
2. Provide a smooth transition from Introduction to Engineering Design to an integrated treatment that encompasses both Statics and Mechanics of Materials.
   - Provide the analysis methods and the scaling relations for verifying the safety of the design of this model.
3. Emphasize modeling structural components by stressing throughout the text the importance of preparing a complete free body diagram (FBD).
   - Show the method for constructing complete FBDs.
   - Integrate the FBD with the application of the equilibrium equations.
   - Approach the solution of equilibrium problems with equilibrium relations based on force and moment components.
   - Introduce vectors in conjunction with forces and moments, but only employ vector analysis after both FBDs and equilibrium concepts have been firmly established.
4. Emphasize the design of structural components for safety.
   - Stress is compared to strength to give safety factors for components.
   - Behavior of engineering materials is introduced and physical properties such as strength and modulus of elasticity are described.
   - Sizing of structural components for safety and cost are demonstrated.

We began developing notes for this book with a pilot offering in the spring semester of 1999. Many revisions were made before a limited first edition of the textbook was published in the summer of 1999. The second edition was published in 2000 and was used by about a thousand students. Five different versions of the Statics textbook have been published over the past 17 years. Over 10,000 students have studied Statics using these books. This edition includes homework problems following each of its 9 chapters, with a sufficient number of problems for student assignments for several semesters. Also the errors discovered during the extensive usage have been corrected; however, errors always occur even with careful proof reading by many diligent people. We would greatly appreciate students and instructors calling any errors they find to our attention. The e-mail address of one of the authors is given on the copyright page.

# ACKNOWLEDGEMENTS

As is always the case when major changes are made to the curriculum, the College administrators must lead the way. We are indeed fortunate to have several administrators who not only supported this effort, but also insisted that the mechanics offerings be markedly improved. They seek not only a much more favorable educational experience for the students, but also much better understanding and retention of the course content by the students. We thank Dr. William L. Fourney, Associate Dean of Students and Lead Professor in the Keystone Program.

Special thanks are due to Dr. Ann-Marie Vollstedt for her assessment of the content, its sequencing and the need for emphasis on vectors and vector algebra.

Many thanks are due to several individuals for their significant contributions to the development of this textbook. Dr. Bill Fourney led our efforts to change the curriculum to provide a more effective approach for presenting Statics and Mechanics of Materials so students would better understand the material and retain this knowledge. Bill has stayed with the project, taught sections of students every semester, and provided the leadership needed to keep others involved and interested. Dr. Hugh Bruck also made major contributions. He also taught an early pilot section and made many excellent suggestions for changes in the sequence of the content. Dr. Gary Pertmer, Assistant Dean, has assumed the leadership role in organizing the many sections of this course taught each year.

Many instructors teaching the course made valuable suggestions for improvements to the textbook. These include: Dr. Mary Bowden, Dr. Hugh Bruck, Dr. James Duncan, Dr. Bongtae Han, Dr. Kwan-Nan Yeh, Mr. Christopher Baldwin, and Mr. Thomas Beigel of the University of Maryland, College Park; Dr. Abhijit Nagchaudhuri of the University of Maryland, Eastern Shore; and Dr. Asif Shakur, of Salisbury State University.

James W. Dally
Robert J. Bonenberger
College Park, MD

August, 2017

DEDICATION
ABOUT THE AUTHORS
PREFACE

# CONTENTS

## CHAPTER 4  AXIALLY LOADED STRUCTURAL MEMBERS

## CHAPTER 5  MATERIAL PROPERTIES

## CHAPTER 6  TRUSSES

## CHAPTER 7 SPACE STRUCTURES AND 3-D EQUILIBRIUM

## CHAPTER 8 FRAMES AND MACHINES

## CHAPTER 9  FRICTION

# APPENDICES

## APPENDIX C  PROPERTIES OF AREAS

# LIST OF SYMBOLS

| | | | |
|---|---|---|---|
| A | area | SF | safety factor |
| %A | percent reduction in area | T | torque |
| **a** | acceleration vector | t | time |
| a, b, c, ….dimensions | | **u** | unit vector |
| C | constant, center dimension | V | shear force |
| **D** | vector difference | **v** | velocity vector |
| D | diameter | $\mathrm{V}$ | volume |
| d | diameter or distance | W | weight, watt |
| %e | percent elongation | $\mathrm{W}$ | work |
| E | elastic modulus | w | width dimension |
| $\mathrm{E}$ | modulus scale factor | x, y z | Cartesian coordinates |
| F | force magnitude | | |
| **F** | force as a vector | | |
| $F_f$ | friction force | $\alpha$, $\beta$, $\gamma$ | direction cosines |
| FBD | free body diagram | $\Delta$ | delta |
| g | gravitational constant | $\delta$ | deflection or displacement |
| G | universal gravitational constant | $\varepsilon$ | strain |
| G | shear modulus | $\varepsilon_T$ | true strain |
| h | height | $\phi$ | angle of friction, angle of twist |
| **i, j, k** | unit vectors | $\gamma$ | shear strain |
| I | moment of inertia | $\pi$ | 3.1416 radians |
| J | polar moment of inertia | $\mu$ | coefficient of friction |
| k | number, spring rate | $\nu$ | Poisson's ratio |
| L | length dimension | $\Sigma$ | Summation sign |
| $\mathrm{L}$ | load scale factor | $\sigma$ | stress |
| ln | natural logarithm | $\sigma_{design}$ | design stress |
| M | moment magnitude | $\sigma_f$ | failure stress |
| **M** | moment as a vector | $\sigma_T$ | true stress |
| $\mathrm{M}$ | multiplier | $\theta$ | angle |
| **MA** | mechanical advantage | $\theta_s$ | angle of repose |
| MOS | margin of safety | $\tau$ | shear stress |
| m | mass, subscript for mode | $\omega$ | angular velocity |
| N | normal force or number | | |
| n | number | | |
| P | internal force | | |
| p | pressure, subscript for prototype | | |
| Q | first moment of the area | | |
| q | distributed loading | | |
| r | radius, radius of gyration, distance | | |
| **r** | position vector | | |
| R | reaction force, radius, resistance | | |
| $R_e$ | radius of the earth | | |
| **S** | vector sum | | |
| $\mathrm{S}$ | geometric scale factor | | |
| s | distance or dimension | | |
| $S_{design}$ | design strength | | |
| $S_y$ | yield strength | | |
| $S_{ys}$ | yield strength in shear | | |
| $S_u$ | ultimate tensile strength | | |
| $S_{us}$ | ultimate tensile strength in shear | | |

# CHAPTER 1

## BASIC CONCEPTS IN MECHANICS

### 1.1 INTRODUCTION

The subject of mechanics is usually divided into four different courses, which include:

1. Statics
2. Dynamics
3. Mechanics of Materials
4. Fluid Mechanics

Statics and dynamics both deal with rigid bodies that are subjected to a system of forces. In the traditional study of statics, we are concerned with determining either internal and/or external forces acting on a structural element that is in a state of equilibrium (usually at rest). In dynamics, the forces acting on the body produce motion and the body accelerates or decelerates. The analysis in dynamics deals with determining position, velocity (angular or linear) and acceleration as some function of time. Newton's laws guide our study of mechanics[1]. Consider Newton's second law:

$$\sum \mathbf{F} = \frac{d}{dt}(m\mathbf{v}) \qquad (1.1)$$

where $\sum \mathbf{F}$ is the sum of all of the forces acting on the rigid body; m is the mass of the body; $\mathbf{v}$ is the velocity and d/dt is the derivative operator

In dividing the study of mechanics into its four subjects, scholars have considered the special situation where the velocity is constant (often zero) and developed **statics** based on this simplification of Newton's second law. In this special situation:

$$\sum \mathbf{F} = 0 \qquad (1.2)$$

Of course, we study motion in dynamics—the velocity of the rigid body is changing ($\mathbf{v} \neq 0$) and the general form of Eq. (1.1) applies. However, in most situations the mass of the rigid body is constant ($dm/dt = 0$) and if this is the case, Eq. (1.1) reduces to:

$$\sum \mathbf{F} = m\frac{d\mathbf{v}}{dt} = m\mathbf{a} \qquad (1.3)$$

where $\mathbf{a} = d\mathbf{v}/dt$ is the acceleration of the rigid body.

---

[1] Sir Isaac Newton (1642-1727) formulated three laws of motion and the law of universal gravitational attraction.

In the study of both statics and dynamics, the material from which the body is manufactured is of no concern providing the body remains essentially rigid under the action of the imposed forces.

However, in the course mechanics of materials[2], the deformation of the body is an essential consideration in the analysis. In this course, small deformations of the body are assumed (plane sections remain plane). This assumption enables us to determine the distribution of internal forces and stresses in the body. The material from which the body is fabricated is of critical importance in mechanics of materials for two reasons. First, the deformations of the body due to the forces are markedly affected by the rigidity of the material (its elastic modulus). Second, the behavior of the body—whether it fails or not—depends on its strength. Of course, strength is a physical property of the material.

In fluid mechanics the situation is entirely different. The body is either a gas or a liquid. The deformations are sufficiently large to be considered as flows. The flow can be compressible (gasses under higher pressures) or incompressible (liquids or gasses at low pressures). The flow can occur in closed channels or open channels. It may be internal to a conduit or external to some surface. The flow may be stable (laminar) or unstable (turbulent). The phase of the material may change during a process and the resulting flow consists of two-phases (some liquid and some gas). Because of the complexities inherent in fluid mechanics, it is studied after a student has established a thorough understanding of the other three branches of mechanics.

Before beginning the design analysis of any component, vehicle, structure or even a model of a structure, it is essential to thoroughly understand several basic concepts in mechanics and the behavior of materials. The first two principal subject areas in mechanics of materials are **Statics** and **Mechanics of Materials**. In this chapter, we introduce several of the basic concepts and physical laws included in these two subjects. These concepts and/or laws provide a foundation upon which the design analysis of machine components and structures is based. We also briefly describe the history of mechanics to give you a sense of the age of the principles used in analyzing modern machine elements and structural components.

## 1.2 STATICS AND MECHANICS OF MATERIALS

This textbook integrates two closely related subjects in mechanics, namely **statics** and **mechanics of materials**. In studying **statics**, we assume the body under consideration is perfectly rigid. As such it does not deform under the action of applied forces. You will solve many different types of problems determining **external forces** acting on some structures and the **internal forces** developed in others by using only **free body diagrams** and the **equilibrium equations**. The equilibrium equations are developed from the laws of motion proposed by Sir Isaac Newton to describe the balance of forces acting on a stationary body or on a body moving with a constant velocity. Solutions for forces acting on bodies in equilibrium employ only three basic steps:

- Construct a complete set of free body diagrams.
- Apply the appropriate equations of equilibrium.
- Execute the mathematics required to solve one or more equations of equilibrium.

Statics is a relatively easy subject that you can quickly master. You will find it easy to solve many different problems that are approached with the procedure outlined above.

---

[2] The course mechanics of materials is also known as mechanics of deformable bodies or strength of materials.

**Mechanics of materials** is a relatively simple extension of statics, which accounts for the effect that **material deformations**[3] have on the **internal stresses** generated in a body by applied forces. To solve the problems that arise in mechanics of materials, we begin with the same three steps, and then add two more. The additional steps are to accommodate the effect of the deformations when the body is loaded:

- Construct a complete set of free body diagrams.
- Apply the appropriate equations of equilibrium.
- Assume the geometry of the deformations (usually plane sections remain plane).
- Employ the appropriate relations between stress and strain.
- Execute the mathematics required in solving the equations.

By comparing these lists, it is evident that statics and mechanics of materials are closely related. Indeed, we must use the equilibrium equations in statics before we can begin to solve typical problems in mechanics of materials. However, the solutions to statics problems are obtained on a global scale, whereas the solutions to mechanics of materials problems are obtained on a local scale.

In studying mechanics of materials, we quickly encounter the concepts of stress and strain. Mathematically these concepts are somewhat complex because they are tensor quantities, but physically they are simple and easily understood. Stress is a concept based on the equilibrium of a portion of a body. Consider a part of a body produced by sectioning. The internal force acting on the section cut is developed by some distribution of stresses over the area exposed by the cut.

Strain, on the other hand, is a geometric concept. We determine strain by the change in geometry that occurs when a body deforms under load. When considered individually, both stress and strain are independent of the material from which the body is made. It is only when we write a relation for strain in terms of stress, or vice versa, that we must consider the properties of the material used to fabricate the body.

Materials used in constructing various structures are largely ignored in the study of statics. The equilibrium equations are the same for all materials, and the internal and external forces for statically determinant structures[4] do not depend upon the materials employed to fabricate the structure. Materials are much more important in studies of the mechanics of materials. We usually relate stress and strain in our solutions, and must employ the elastic constants that describe the rigidity of the materials. In addition, we may be required to predict the margin of safety for a structure subjected to specific loading. In the solution to problems dealing with structural failure, we employ the appropriate **"strength"** of the material.

In this textbook, we stress the physical aspects of both statics and mechanics of materials. We consider it essential that you construct complete free body diagrams to model the structure and to define the unknown forces. We are more interested in your understanding of the equations of equilibrium than your use of **vector algebra** to solve them. Of course, it is important that you be able to correctly manipulate the equations resulting from the application of the principle of equilibrium. Whether these equations are written using **trigonometric** functions or in **vector** format is of lesser concern providing they are correct.

---

[3] In mechanics of materials, we assume the deformations are so small that they do not significantly affect the magnitude or the direction of the internal and external forces acting on the body.

[4] Structures are classified as statically determinant if the reactions at their supports may be determined using only the equations of equilibrium.

The general subject of mechanics and materials is both interesting and important. Let's try to enjoy the experience of learning this fundamental subject. In writing this textbook, we have used many examples to:

1. Illustrate drawing free body diagrams.
2. Show the correct procedures for applying the equilibrium equations.
3. Show the geometric changes due to deformations of the body under load.
4. Illustrate distributions of stress over areas exposed by section cuts.
5. Demonstrate the procedures and techniques involved in solving different types of problems.

## 1.3 HISTORY OF MECHANICS

As a student of the first course in mechanics, you are probably in your late teens or early twenties. It might be hard for you to imagine that mechanics has been under development for more than 2000 years. Early pioneers in mechanics, about two or three hundred years ago, understood most of the material included in this book. Although research is still conducted in mechanics, this first course covers the classical content that is considered the foundation for mechanics. To illustrate that mechanics is a classical subject, let's note the contributions of a few of the pioneers who established the foundations of mechanics. For a much more extended treatment of the history of mechanics see references [1, 2].

One of the earliest contributors to our knowledge of mechanics was Archimedes (287 – 212 BC) who discovered that we could use a lever to increase the weight that a person could lift. The concept of the lever with its fulcrum used in lifting a weight W with a force W/4 is illustrated in Fig. 1.1. Archimedes also described the use of pulleys and inclined planes for moving materials at construction sites. On a different application of engineering in ancient times, Archimedes is credited with discovering the law of buoyancy where the upward force on a submerged body is equal to the weight of the water displaced by that body.

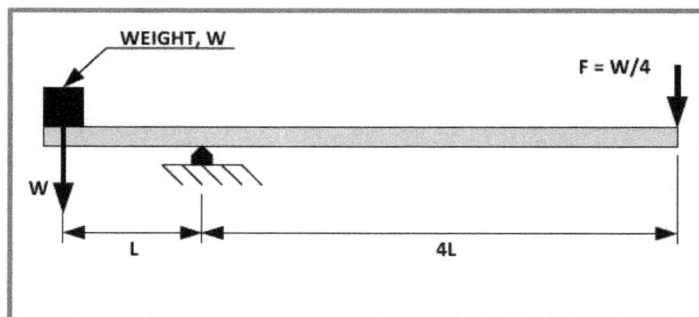

Fig. 1.1 The lever with a fulcrum permits an applied force to be increased with the lever ratio.

Leonardo da Vinci (1452 – 1519) was a man with incredible talent and one of the great Renaissance masters. He was a painter, sculptor, musician, architect and engineer. At the age of about 30, he served as the principal engineer for the Duke of Milan supervising the construction of bridges and war machines for the Duke's military ventures [3]. Leonardo is not known for his mathematical discoveries, but rather for his engineering innovations that implied his thorough understanding of mechanics.

The 17th century was the height of the Renaissance period for science and probably the most productive for developing the foundations of mechanics. During this century, Robert Hooke (1635 – 1703) introduced the concept of elasticity by noting that a body deformed in proportion to the applied forces. This concept leads to the well-known Hooke's law:

$$\sigma = E\,\varepsilon \tag{1.4}$$

where $\sigma$ is the stress, $\varepsilon$ is the strain, and E is the material property known as the modulus of elasticity.

The linearity of Hooke's law is illustrated in Fig. 1.2. Robert Hooke also developed springs and used them in watches and clocks to improve their accuracy and to store energy. He discussed planetary motion, but did not develop the mathematics necessary to describe the observed motion of the planets. It was Sir Isaac Newton who extended Hooke's early ideas about planetary motion and wrote the gravitational law that governs the motion of the planets.

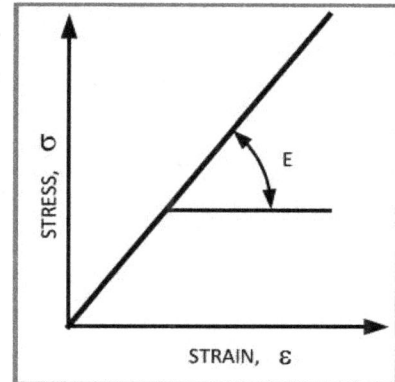

Fig. 1.2 The linear elastic relation between stress and strain.

Sir Isaac Newton (1642 – 1727) was a giant in both mathematics and mechanics. He was only 24 years old when he developed the mathematics known today as differential calculus. Later Newton turned his attention to planetary motion, and formulated his famous laws of motion. We will cover these three laws in more detail in the next section. However, these laws are the foundation of much of the content contained in mechanics courses on statics and dynamics. He is best known for his universal law of gravity [4] that explains the attractive forces between two bodies. The gravitational force is an internal force because it acts on each elemental volume of mass within a body. The attractive forces **F** between two masses, illustrated in Fig. 1.3, are given by:

$$F = G\, m_a\, m_b\, /r^2 \qquad\qquad (1.5)$$

where F is the magnitude of the gravitational force; $G = 6.673 \times 10^{-11}$ m$^3$/(kg - s$^2$) is the universal gravitational constant; $m_a$ and $m_b$ are the masses of bodies A and B and r is the distance between the centers of the two bodies.

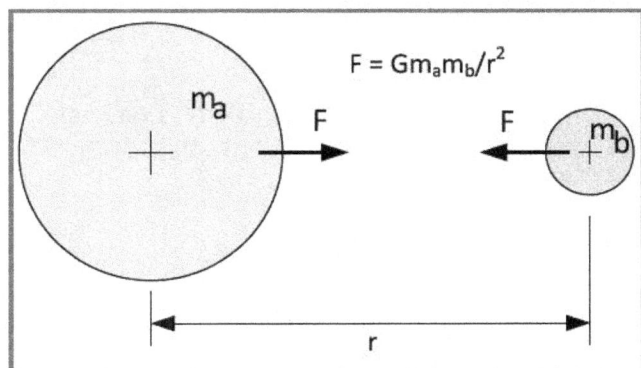

Fig. 1.3 Illustration of the attractive forces on both bodies developed by gravity.

Because Newton's law of gravity is extremely important, we will describe it in more detail later in this Chapter.

Jacob Bernoulli[5] (1654 – 1705) made two contributions to mechanics of materials; one was correct and the other was in error. We cite his contributions to indicate that even the great and famous make mistakes from time to time. Jacob was interested in determining the elastic curve representing the deflection of a beam. He correctly showed that the curvature of the beam was proportional to the bending moment at each point along the length of the beam. He then made the error of assuming that the elastic curve was positioned along the bottom edge of the beam. We will establish in a later course titled Mechanics of Materials that the elastic curve must coincide with the neutral axis of the beam.

The history of mechanics is rich with accomplishments of many mathematicians. As the subject evolved, engineers designing safe efficient structures and machines such as airplanes, automobiles, bridges, machines, skyscrapers, and space satellites reduced the mathematical formulations to practice.

## 1.4 NEWTON'S LAWS OF MOTION

Sir Isaac Newton wrote three laws of motion that form the foundation for both statics and dynamics. They are:

(1) If the sum of all of the forces acting on a body is zero (i. e. $\Sigma \mathbf{F} = 0$), the body will:

    a.   Remain at rest.
    b.   Move at a constant velocity.

(2) If the sum of all of the forces $\mathbf{F}$ acting on a body is not zero, the body will undergo a time rate of change of the linear momentum (mv) given by:

$$\sum \mathbf{F} = \frac{d}{dt}(m\mathbf{v}) \tag{1.1}$$

When the mass m of the body remains constant with respect to time dm/dt = 0, and Eq. (1.1) reduces to:

$$\sum \mathbf{F} = m\frac{d\mathbf{v}}{dt} = m\mathbf{a} \tag{1.3}$$

(3) The force exerted by body A on body B is equal in magnitude but opposite in direction to the force that body B exerts on body A. This third law is sometimes called the law of action and reaction.

### 1.4.1 Newton's First Law

The first of Newton's laws is written as:

$$\Sigma \mathbf{F} = 0 \tag{1.2}$$

Another way of representing Eq. (1.2) is:

---

[5] There were three famous Bernoulli's, namely Jacob, John and Daniel. John, the younger brother of Jacob, was a splendid mathematician. Daniel, the son of John, was famous because of his work in fluid mechanics, his outstanding book Hydrodynamica, and for his suggestion to Euler for deriving the equations for elastic curves.

$$F_1 + F_2 + \ldots\ldots + F_n = 0 \qquad\qquad (1.2a)$$

where n forces are acting on the body.

Equation (1.2), in one form or the other, is used extensively in both statics and mechanics of materials. It is used every time that we write the **equilibrium** equations. To illustrate the meaning of the mathematical symbol $\Sigma F$, let's examine the drawing of the potato like body presented in Fig. 1.4. We have four forces acting on this body. They are pointed in four different directions and have four different magnitudes. The forces are **vector** quantities that must be characterized by specifying both a magnitude and a direction[6]. When the **vector sum** of these forces is zero, the body is in equilibrium.

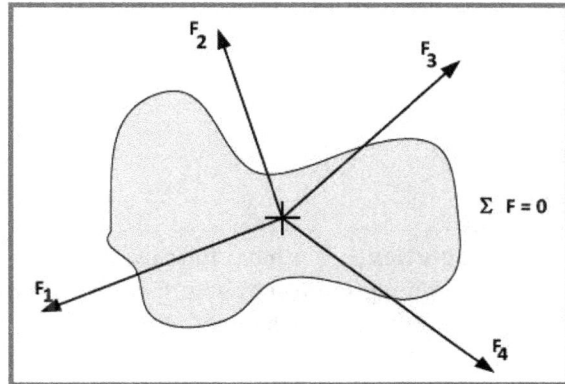

Fig. 1.4 Four forces acting on a body that is in equilibrium if $\Sigma\mathbf{F} = 0$.

We may recast the vector representation of Eq. (1.2) by writing the equivalent **scalar equations** as:

$$\Sigma F_x = 0 \qquad \text{and} \qquad \Sigma F_y = 0 \qquad\qquad (1.2b)$$

$$F_{1x} + F_{2x} + \ldots.. + F_{nx} = 0 \qquad \text{and} \qquad F_{1y} + F_{2y} + \ldots.. + F_{ny} = 0 \qquad\qquad (1.2c)$$

We have accounted for the direction of the forces in this equation by considering only those forces in either the x or y directions. When the forces are constrained to the x and y directions, we employ the **scalar form of the equilibrium equations**. However, when the forces act in directions other than the x and y, the vectors must first be decomposed into their **components** acting in the x and y direction before they can be used in the scalar equilibrium equations.

Let's consider an example to demonstrate the application of the equilibrium equation. We will first consider only forces acting in either the x or y directions.

## EXAMPLE 1.1

Consider the forces acting on the body shown in Fig. E1.1, and determine if it is in equilibrium.

---

[6] We use the bold font to represent forces as vector quantities when both magnitude and direction are to be specified. Sometimes we consider only the magnitude of the force, and in this case normal fonts are used to represent this scalar magnitude.

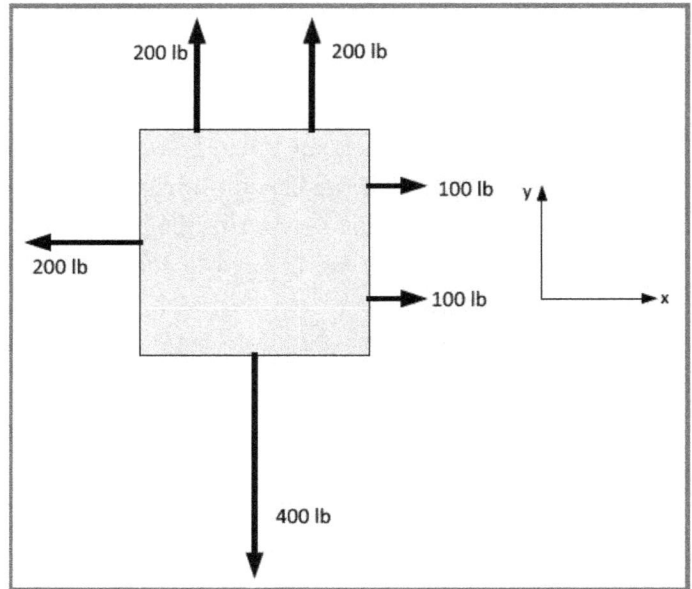

Fig. E1.1

**Solution:** We have inserted a x-y coordinate system in Fig. E1.1 as a reference to employ in summing the forces acting on the square block. We apply Eq. 1.2 to determine if this block is in equilibrium. When we write $\Sigma \mathbf{F} = 0$, it is important to note that $\mathbf{F}$ is a **vector** quantity and the summation of the forces is the **vector form of the equilibrium equation**. Let's take this fact into account by considering first all of the forces in the x direction, and then all of the forces in the y direction. With this approach, we are accounting for both the directions and the magnitude of each of the six forces acting on the block. We write:

$$\Sigma F_x = 0 \qquad \text{and} \qquad \Sigma F_y = 0$$

$$+ 100 + 100 - 200 = 0 \qquad \text{and} \qquad + 200 + 200 - 400 = 0$$

These results show that the equilibrium conditions are satisfied in both the x and y directions. The forces in the positive directions of both x and y cancel with those in the negative directions. Hence, the body is in equilibrium.

## 1.4.2 Newton's Second and Third Laws

Newton's second law $\Sigma \mathbf{F} = d/dt(m\mathbf{v}) = m\mathbf{a}$ is the equation used most frequently in the study of dynamics. When the sum of the forces is not zero, the body of mass m is subjected to an acceleration $\mathbf{a}$. Depending on your choice of disciplines, you may pursue Newton's second law later in the curriculum.

The third law is often called the law of action and reaction. We illustrate the concept of active and reactive forces in Fig. 1.5.

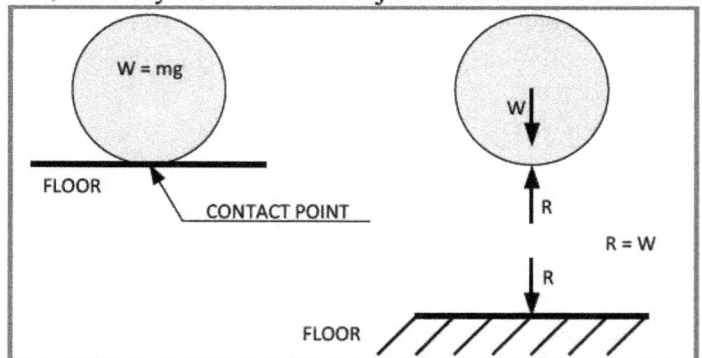

Fig. 1.5 The active force W due to gravity produces a reaction force R at the contact point.

In this illustration, a spherical shaped mass m with a weight W rests on the floor at a contact point. The sphere is in equilibrium under the action of two forces. The first force is the weight W due to gravity that acts downward. The second is the reaction force developed at the contact point. The reaction force R is equal in magnitude to the weight W, but opposite in direction.

When any two bodies are in contact (i.e. the sphere and the floor), two forces develop at the contact point. These forces are equal in magnitude and opposite in direction.

## EXAMPLE 1.2

Suppose that a professional woman weighing 116 lb is approaching an elevator. Let's determine the reaction forces that develop between the floor and her feet.

**Solution:** We start by drawing a free body diagram of the woman. In the free body diagram shown in Fig. E 1.2, the woman is separated from the floor. She is suspended in space and maintained in equilibrium by vertical (y direction) forces applied to her feet.

At the two contact points with the floor, we have drawn arrows downward representing the forces applied to the floor (W/2). The reaction forces R = W/2 are drawn upward on each foot. Clearly the applied and reaction forces are equal in magnitude and opposite in direction. Is the woman in equilibrium?

Fig. E 1.2 Free body diagram of a woman standing on a floor.

## 1.5 FORCES

Statics involves a study of forces that act on and within members of a structure. We seek to determine the external forces acting on bodies, and forces developed by stresses within structural members (internal forces). Some forces occurring under static (steady state) conditions include:

- Gravitational
- Pressure acting over a defined area
- Friction
- Magnetic
- Electrostatic

In addition, forces developed under dynamic conditions are referred to as inertial forces and include:

- Centrifugal
- Centripetal
- Coriolis

First, let's examine the forces due to gravity because they are by far the most important. We continuously work and expend huge amounts of energy to overcome gravitational forces. Gravitational forces are the primary concern when we design building and bridge structures against failure by collapse or rupture. Even in vehicle design where other dynamic forces are significant, gravitational forces are critical in the design of both the structure and the power train.

Weight is a force produced by the Earth's gravitational pull on the mass of our body as illustrated in Fig. 1.6. Suppose we examine the force on a body due to gravity by modifying Eq. 1.5, and letting:

$m_a = m_e$ the mass of the Earth.
$m_b = m_b$ the mass of our body.
$r = R_e$ the radius of the Earth.

Then we rewrite Eq. 1.5 to give:

$$F = G\, m_e\, m_b\, /R_e^2 \qquad\qquad (1.6)$$

In setting $r = R_e$, we assumed that Earth bound bodies, either those of people or objects, are very small compared to the radius of the Earth which is 3960 mi. or $6.37 \times 10^6$ m.

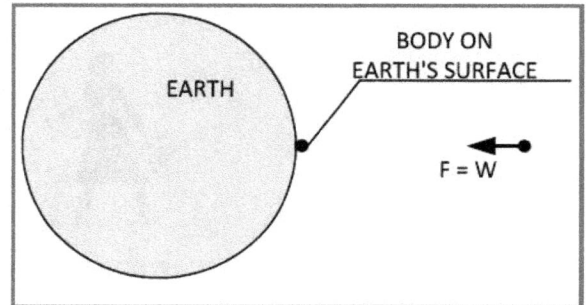

Fig. 1.6 Bodies on Earth's surface are small relative to the Earth's radius.

Next, we collect together all of the quantities in Eq. 1.6 that have anything to do with the Earth, and set them equal to $g_e$.

$$g_e = Gm_e\, /R_e^2 \qquad\qquad (1.7)$$

Note that $g_e$ is the **gravitational constant** equal to 32.17 ft/s² or 9.807 m/s². We will drop the subscript in subsequent discussion of the gravitational constant with the understanding that g is to be applied to Earth bound bodies.

Strictly speaking g is not constant since it varies a bit as we move from one location to another. The Earth is not a perfect sphere, and R does not remain constant as we move from Pikes' Peak to Death Valley. However, the variations are so small that we neglect them without introducing significant error in our design analyses.

Combining Eqs. 1.6 and 1.7 gives:

$$F = m_b\, g = W \qquad\qquad (1.8)$$

The force F in Eq. 1.8 is the weight W of a body on Earth having a mass $m_b$.

From our definition, it clear that the units of g are ft/s$^2$ or m/s$^2$; hence, g is an acceleration. Indeed, if we jump off of a ladder, our body accelerates with a = g until we hit the ground. Clearly, this relationship is consistent with Newton's second law.

If we travel from Washington, D. C. to Denver, CO, we observe that our weight remains essentially the same. So we get confused and begin to think that the constant in Eq. 1.8 is our weight. It is a reasonable thought, but erroneous. When measuring weight, it is essentially constant if we remain Earth bound. However, the constant quantity in Eq. 1.8 is not the weight W, but the mass $m_b$. To prove it, go to the moon, and measure your weight. It is known that we weigh much less on the moon—about one sixth as much as here on Earth. Since our mass $m_b$ is constant, we weigh less because the gravitational constant for the moon is only about g/6. The smaller gravitational constant for the moon is due to its much smaller mass when compared to the mass of the Earth. The mass of our body is the same whether we are on the moon, Mars the Earth or anywhere in space.

## EXAMPLE 1.3

(a) Determine the weight of an object with a mass of 22 kg (a kilogram is the unit for mass in the International System).

**Solution:**

We use Eq. 1.8 to calculate the weight as:

$$F = W = mg = (22 \text{ kg})(9.807 \text{ m/s}^2) = 215.8 \text{ N}$$

where N is the symbol for **newton**, which is the unit for force in the SI system. One newton is also equivalent to 1kg-m/s$^2$. We will discuss units used in the SI system in more detail in the next section.

(b) Determine the weight of an object with a mass of 14 slugs (a slug is the unit for mass in the U.S. Customary system).

**Solution:**

Using Eq. 1.8 again gives:

$$F = W = mg = (14 \text{ slug})(32.17 \text{ ft/s}^2) = 450.4 \text{ lb}$$

The slug has units of (lb-s$^2$)/ft. An object weighing 32.17 lb on Earth has a mass of 1 slug.

Did you drive your car to the university today? If so, the pressure developed by the combustion of gasoline within the cylinders of your car's engine provided the force to propel it along the roads. Pressure, p acts over an area, A of some surface to create a force.

$$F = p A \qquad\qquad (1.9)$$

where A is the cross sectional area of the piston.

We illustrate the force F produced by the action of the pressure p on the piston shown in Fig 1.7. The magnitude of the force is determined by using Eq. 1.9; its direction is normal to the surface of the piston.

Fig. 1.7 The pressure acting on the piston produces a force F = pA.

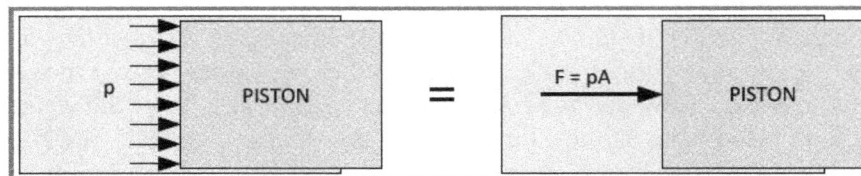

## 1.6 BASIC QUANTITIES AND UNITS

In the study of mechanics, we will encounter four basic quantities—length, time, force and mass. These quantities are shown with their respective units for the SI and the U. S. Customary systems of units in Table 1.1.

**Table 1.1**
**Basic Quantities and Units**

| System of Units | Length | Time | Mass | Force | g |
|---|---|---|---|---|---|
| International System of Units (SI) | meter (m) | second (s) | kilogram (kg) | newton (N), $(kg\text{-}m)/s^2$ | 9.807 $m/s^2$ |
| U. S. Customary (FPS) | foot (ft) | second (s) | slug $(lb\text{-}s^2)/ft$ | pound (lb) | 32.17 $ft/s^2$ |

The basic units are not independent because Eq. 1.3 requires that the units be **dimensionally homogenous**. To maintain the dimension homogeneity of Eq. 1.3, we define the units for length, time and mass in the **SI** system, and then derive the remaining basic unit for the force, the **newton**, in terms of those units. For the **U. S. Customary** system, the basic unit for mass, the **slug**, is derived in terms of the units for length, time, and force.

In the International System of Units (SI), the length is given in meters (m), the time in seconds (s), and the mass in kilograms (kg). The unit for force is called a newton (N) in honor of Sir Isaac. The newton is derived from Eq. 1.3 so that a force of 1N will impart an acceleration of 1 $m/s^2$ to a mass of 1 kg [i. e. 1 N = (1 kg) (1 $m/s^2$)]. For dimension homogeneity, it is clear that N is equivalent to $(kg\text{-}m)/s^2$. In the SI system the gravitation constant g = 9.807 $m/s^2$. With this value of the acceleration due to gravity on Earth, the weight of a mass of 1 kg is:

$$W = mg = (1kg)(9.807 \text{ m/s}^2) = 9.807 \text{ N}$$

In the U. S. Customary System, the length is given in feet (ft), the force in pounds (lb), and the time in seconds (s). The unit for mass is called a slug, which is derived from Eq. 1.3 so that a force of 1 lb will impart an acceleration of 1 $ft/s^2$ to a mass of 1 slug [i. e. 1 lb = (1 slug)(1 $ft/s^2$)]. For dimension homogeneity, it is clear that a slug is equivalent to $(lb\text{-}s^2)/ft$. In the U. S. Customary System the gravitation constant g = 32.17 $ft/s^2$. With this value of the acceleration due to gravity on Earth, the mass of a body weighing 32.17 lb is:

$$m = W/g = 32.17 \text{ lb}/(32.17 \text{ ft/s}^2) = 1 \text{ slug}$$

In addition to the quantities presented in Table 1.1, we deal with several other quantities in our studies of statics and mechanics of materials. These quantities are listed in Table 1.2.

**Table 1.2**
**Units of Other Frequently Used Quantities**

| System of Units | Moment M | Stress σ | Strain ε |
|---|---|---|---|
| SI | newton-meter, (N-m) | pascal, (Pa) = (N/m$^2$) <br> mega pascal, (MPa) = (N/mm$^2$) | dimensionless |
| U. S. Customary, (FPS) | foot-pound, (ft-lb) | pound/square foot, (lb/ft$^2$) <br> pound/square in., (lb/in.$^2$) = psi | dimensionless |

We have shown the quantities in Table 1.2 in terms of the basic units for length, time and force. However, in practice other units are often employed. For instance, a moment may be expressed as in-lb instead of ft-lb, and a stress expressed as MPa instead of Pa or psi (lb/in$^2$) instead of (lb/ft$^2$). We are fortunate that strain is a dimensionless quantity; we will not have to convert from one system of units to another, or to convert from one unit to another within the same system. A technique to convert from one system of units to another is described in the next section.

Often the metric prefixes are employed in expressing numerical results. Indeed, we used a metric prefix M with the unit for stress (Pa) in the previous paragraph. The metric prefixes are useful in dealing with either very large or very small numbers. Accordingly, it is common practice to employ them together with the symbol for the units (e.g. MPa).

**Table 1.3**
**SI Prefixes**

| Multiplication Factor | Prefix Name | Prefix Symbol |
|---|---|---|
| $10^{18}$ | exa | E |
| $10^{15}$ | peta | P |
| $10^{12}$ | tera | T |
| $10^{9}$ | giga | G |
| $10^{6}$ | mega | M |
| $10^{3}$ | kilo | k |
| $10^{2}$ | hecto* | h |
| $10^{1}$ | deka* | da |
| $10^{-1}$ | deci* | d |
| $10^{-2}$ | centi* | c |
| $10^{-3}$ | milli | m |
| $10^{-6}$ | micro | μ |
| $10^{-9}$ | nano | n |
| $10^{-12}$ | pico | p |
| $10^{-15}$ | femto | f |
| $10^{-18}$ | atto | a |

*To be avoided when possible.

## 1.7 CONVERSION OF UNITS

We show two different techniques for the conversion of units from one system to another. The first technique utilizes the conversion factors listed in Table 1.4.

**Table 1.4**
**Unit Conversion Factors**

| Quantity | U. S. Customary | SI Equivalent |
|---|---|---|
| Acceleration | $ft/s^2$<br>$in/s^2$ | $0.3048 \text{ m/s}^2$<br>$0.0254 \text{ m/s}^2$ |
| Area | $ft^2$<br>$in^2$ | $0.0929 \text{ m}^2$<br>$645.2 \text{ mm}^2$ |
| Distributed Load | lb/ft<br>lb/in. | 14.59 N/m<br>0.1751 N/mm |
| Energy | ft-lb | 1.356J |
| Force | kip = 1000 lb<br>lb | 4.448 kN<br>4.448N |
| Impulse | lb-s | 4.448 N-s |
| Length | ft<br>in<br>mi | 0.3048 m<br>25.40 mm<br>1.609 km |
| Mass | lb mass<br>slug<br>ton mass | 0.4536 kg<br>14.59 kg<br>907.2 kg |
| Moments or Torque | ft-lb<br>in-lb | 1.356 N-m<br>0.1130 N-m |
| Area Moment of Inertia | $in^4$ | $0.4162 \times 10^6$<br>$mm^4$ |
| Power | ft-lb/s<br>hp | 1.356 W<br>745.7 W |
| Stress and Pressure | $lb/ft^2$<br>$lb/in^2$ (psi)<br>ksi = 1000 psi | 47.88 Pa<br>6.895 kPa<br>6.895 MPa |
| Velocity | ft/s<br>in/s<br>mi/h (mph) | 0.3048 m/s<br>0.0254 m/s<br>0.4470 m/s |
| Volume | $ft^3$<br>$in^3$<br>gal | $0.02832 \text{ m}^3$<br>$16.39 \text{ cm}^3$<br>3.785 L |
| Work | ft-lb | 1.356 J |

Consider the following examples to illustrate the use of this conversion table.

## EXAMPLE 1.4

(a) Determine the SI equivalent of a force of 200 kip.

> **Solution:** Using the data listed in Table 1.4 found in Section 1.7, we write:
>
> F = (200 kip) (4.448 kN/kip) = 889.6 kN

(b) Determine the SI equivalent of the mass of 4 slugs.

**Solution:** Using the data listed in Table 1.4 in the row for mass, we write:

$$m = (4 \text{ slug}) (14.59 \text{ kg/slug}) = 58.36 \text{ kg}$$

## EXAMPLE 1.5

(a) Determine the U. S. Customary equivalent of a force of 350 N.

**Solution:** Using the data listed in Table 1.3 in the row for force, we write:

$$F = (350 \text{ N}) (\text{lb}/4.448 \text{ N}) = 78.69 \text{ lb}$$

(b) Determine the U. S. Customary equivalent of a mass of 84 kg.

**Solution:** Using the data listed in Table 1.3 in the row for mass, we write:

$$m = (84 \text{ kg})/(\text{slug}/14.59 \text{ kg}) = 5.757 \text{ slug}$$

In solving these exercises, we demonstrated the technique employing data from Table 1.4.
A second technique, for converting units from one system to another, is based on multiplying the term to be converted by a combination of units, which equals one. Let's consider a simple conversion from say 16 MPa to its equivalent value expressed in terms of psi.

$$16 \text{ MPa} \times \overset{1^{st}}{\frac{10^6 \text{ Pa}}{\text{MPa}}} \times \overset{2^{nd}}{\frac{\text{N}/\text{m}^2}{\text{Pa}}} \times \overset{3^{rd}}{\frac{\text{m}^2}{(39.37)^2 \text{in}^2}} \times \overset{4^{th}}{\frac{\text{lb}}{4.448 \text{ N}}} \times \overset{5^{th}}{\frac{\text{psi}}{\text{lb}/\text{in}^2}} = 2321 \text{ psi}$$

Let's examine this relation term by term:

1. The first term in the expression is a stress of 16 MPa. Recall that MPa is the unit for stress in the SI system.
2. The next unit term converts the metric prefix M to $10^6$. Examine Table 1.3 for definitions of the metric prefixes.
3. The next unit term converts Pa to $\text{N}/\text{m}^2$.
4. The next unit term converts m to in.—note both the conversion factor and units are squared.
5. The next unit term converts N to lb.
6. The last term before the equal sign recognizes the symbol psi for $\text{lb}/\text{in}^2$.

If you find that a conversion table is not convenient and you are able to remember a few basic conversions, multiplication by a number of unit terms is a good method for converting units from one system to another.

## EXAMPLE 1.6

Using the unit multiplication method convert the stress of 20 ksi to its equivalent in the SI system of units.

**Solution:**

$$20 \text{ ksi} \times \frac{10^3 \text{ psi}}{\text{ksi}} \times \frac{\text{lb}/\text{in}^2}{\text{psi}} \times \frac{4.448 \text{ N}}{\text{lb}} \times \frac{(39.37)^2 \text{in}^2}{\text{m}^2} \times \frac{\text{Pa}}{\text{N}/\text{m}^2} \times \frac{\text{M}}{10^6} = 137.9 \text{MPa}$$

This approach is longer than the more direct conversion method. However, if you do not remember that a stress of 1 ksi is equivalent to 6.895 MPa, it provides a means of performing the conversion using only the most fundamental of the conversion factors.

## 1.8 SIGNIFICANT FIGURES

With the advent of the hand held calculator, we usually obtain solutions with ten or more figures. These results are misleading because the data used in the formulas to calculate these ten figure results are rarely accurate to more than 0.2% or one part in 500. To avoid the implication of fictitious accuracies, we recommend that the results from your calculator be written with four significant figures. This practice yields a computational accuracy of 1/1000 or 0.1% that is consistent with the accuracy of the physical data and the analytical model upon which the formula is based.

Let's consider several numerical results and convert them to results acceptable in engineering practice.

### EXAMPLE 1.7

Convert the following ten digit numerical results shown in the table below to a format with four significant figures.

**Solution:**

**Table E1.7**
**Examples for Writing Numbers with Four Significant Figures**

| Ten Digit Number | Four Significant Figures | Ten Digit Number | Four Significant Figures |
|---|---|---|---|
| (a) 6.142857143 | 6.143 | (f) 0.182926829 | 0.1829 |
| (b) 0.000304136 | $3.041 \times 10^{-4}$ | (g) 1226308544 | $1.226 \times 10^{9}$ |
| (c) 0.026941846 | 0.02694 | (h) 1628720.675 | $1.629 \times 10^{6}$ |
| (d) 308.5225000 | 308.5 | (i) 0.0000005984 | $5.984 \times 10^{-7}$ |
| (e) 22.76923077 | 22.77 | (j) 88.62508918 | 88.63 |

(a) In converting we rounded the fourth digit up since the fifth digit was larger than 5.
(b) We retained the first four numbers after the leading zeros. The zeros were accommodated by using $10^{-4}$.
(c) We retained the first four numbers after the leading zero. Since only one zero was involved, we retained it in the representation.
(d) We maintained the fourth digit since the fifth digit was less than 5.
(e) We rounded the fourth digit up since the fifth digit was larger than 5.
(f) We maintained the fourth digit since the fifth digit was less than 5.
(g) We retained the first four numbers since the fifth digit was less than 5. The very large number was represented by using $10^{9}$.
(h) We rounded the fourth digit up since the fifth digit was larger than 5. The very large number was represented by using $10^{6}$.
(i) We retained the first four numbers after the leading zeros. The zeros were accommodated by using $10^{-7}$.
(j) We rounded the fourth digit up since the fifth digit was equal to 5.

## 1.9 SCALARS, VECTORS AND TENSORS

In mechanics we deal with three types of quantities, namely **scalars, vectors** and **tensors**. In normal every day life it is not essential to recognize the difference among them. If we have true friends, food, shelter and some money, all is well. Incidentally, true friends, food, shelter and money are all scalar quantities. We need only to count these items in describing them. For example, we might have $74 and be content after eating two big Macs. Other quantities encountered in engineering that are scalar and listed in Table 1.4 include:

- Area
- Energy
- Length
- Mass
- Moment of inertia
- Power
- Volume
- Work

Scalar quantities are the easiest of the three with which to work. We can use simple arithmetic to add, subtract, multiply and divide when we use them. No additional operations are required to manipulate scalar quantities. This not the case with vectors and tensors. Additional mathematical operations and descriptions must be introduced to manipulate equations containing vectors or tensors.

Unfortunately we must deal with both vector and tensor quantities in mechanics. They are not **that** difficult, but we will need to recognize they are not scalar quantities. We will also need mathematics higher than arithmetic to deal with them. Let's start with quantities that must be described with vectors. Vector quantities require two descriptors—magnitude and direction. The magnitude indicates the size of the quantity, and the direction gives its orientation. Quantities that we cited in Table 1.4 which require vector representation for their complete description include:

- Acceleration
- Force
- Impulse
- Moments
- Velocity

We have already discussed gravitational forces. Our weight is the magnitude of the force due to the pull of gravity on the mass of our body. The direction of the gravitational force relative to an Earth bound coordinate system is toward the center of the Earth.

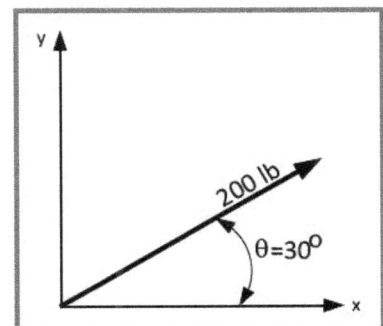

Fig. 1.8 A force vector with a magnitude—200 lb and direction—30° relative to the x-axis.

We represent a typical vector quantity, such as a force, with an arrow as shown in Fig. 1.8. The length of the arrow is proportional to the magnitude of the force, and its inclination relative to the x-y coordinate system gives its direction. We will use vectors extensively in our studies of statics, learning to accommodate both their magnitude and directions in the solution of many different types of equilibrium problems. Vectors are independent of the orientation chosen for the x-y coordinate system. However, the scalar value of the vector's direction and the components of the vector in the x and y directions are dependent on the orientation of the reference coordinate system. This property of vectors will be of importance when coordinate systems are chosen to solve equilibrium problems, because choices can be made that simplify the solution to the problem without changing the physical description of the problem.

Tensor quantities are even more difficult to describe than vector quantities because their complete description requires information about three characteristics. Like vectors, we must specify the magnitude and direction. In addition we must specify the orientation of the plane upon which a tensor quantity acts. Stresses, $\sigma$ are tensor quantities. We demonstrate the three characteristics—magnitude, direction, and the orientation of the plane upon which the stresses act in Fig. 1.9.

The round bar is loaded with forces in the axial direction. A section cut exposes an internal surface (plane) of the bar normal to its axis. Stresses with a magnitude of 140 psi act in the axial direction.

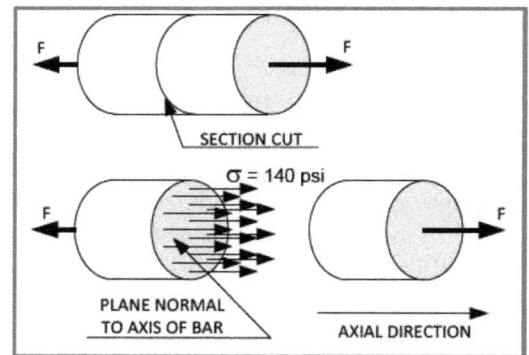

Fig. 1.9 Stresses of 140 psi act in the direction of the axis of the bar. The stresses are acting on the plane normal to the bar's axis.

## 1.10 SUMMARY

In statics we assume that the body is perfectly rigid—it does not deform under load. With this assumption, a free body diagram, and the equations of equilibrium, we are able to solve the many different types of problems arising in statics. Mechanics of materials is an extension of statics where very small changes in the geometry of the body due to deformations are used to determine the stresses and strains. Simple procedures are outlined for the solution of typical problems arising in both subjects.

A very brief review of the history of mechanics is given to show that both subjects have been well understood for many years. The contributions of Archimedes, Da Vinci, Hooke, Newton, and Jacob Bernoulli are mentioned.

Newton's three laws of motion that form the foundation for mechanics have been described. Newton's laws lead to Eqs. 1.2 and 1.3 which are so important that they are repeated here:

$$\Sigma\,\mathbf{F} = 0 \qquad\qquad (1.2)$$

$$\Sigma\,\mathbf{F} = \mathbf{ma} \qquad\qquad (1.3)$$

A simple example is provided to illustrate the use of the equilibrium equations.

Forces are described in considerable detail. Since forces due to gravity and pressure are so common, we provided the equations used to determine their magnitudes. The force or weight due to gravity is:

$$F = m_b\,g = W \qquad\qquad (1.8)$$

The force due to pressure is:

$$F = p\,A \qquad\qquad (1.9)$$

The basic quantities and their units that are employed in statics and mechanics of materials have been given. A table listing the conversion factors for the SI and U. S. Customary systems has been included. Examples showing different techniques for converting units from one system to the other are presented.

## REFERENCES

1  Timoshenko, S. P., History of Strength of Materials, Dover edition, Dover Publications, New York, NY 1983.
2  Todhunter, I. and K. Pearson, History of the Theory of Elasticity and Strength of Materials, Cambridge University Press, Cambridge, U. K., 1886.
3  Uccelli, A., Leonardo da Vinci, Reynal and Co., New York, NY 1956.
4  Newton, Sir Isaac, Philosophiae Naturalis Principia Mathematica, 1687.

## PROBLEMS

1.1  Design a lever that will permit you to lift a weight of 1,000 N a distance of 100 mm. In the design, prepare a dimensioned sketch of the lever, state the force that you apply to the lever, and show the analysis proving that you will be able to lift the weight and move it upward by the specified distance.

1.2  The *Sojourner*, a mechanical rover that explored Mars during a mission in 1997, had a mass of 10.5 kg. Determine the weight of *Sojourner* (in newtons) on (a) Earth, (b) Mars. Assume Mars has a mass of $0.64 \times 10^{24}$ kg and a radius of 3390 km.

1.3  Weather satellites are often placed in geosynchronous orbit around Earth, so that they appear stationary in the sky when viewed from a point on the equator. The orbital altitude for such a satellite is 35,800 km. If a satellite weighs 500 N on the Earth's surface, calculate its weight in orbit. Assume Earth has a mass of $5.976 \times 10^{24}$ kg and a radius of 6370 km.

1.4  An astronaut weighs 150 lb on the Earth's surface. When orbiting Earth in the space shuttle *Atlantis*, however, the astronaut's weight is measured as 130 lb. Determine the altitude of the shuttle's orbit. Assume Earth has a mass of $5.976 \times 10^{24}$ kg and a radius of 6370 km.

1.5  Suppose that you are in a space ship traveling in a circular orbit about the Earth, as shown in the figure to the right. Using a spreadsheet, prepare a graph showing your weight as a function of your position relative to the surface of the Earth. Consider radii from $1.0(R_e)$, on the Earth's surface, to $15(R_e)$.

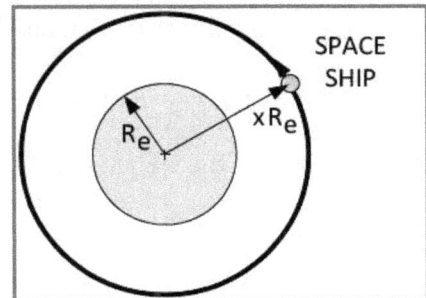

1.6    Determine the gravitational force exerted on the Earth by (a) the sun, (b) the moon.  Also find the ratio of the forces ($F_{sun}/F_{moon}$).  Assume masses of $5.976 \times 10^{24}$ kg for the Earth, $1.990 \times 10^{30}$ kg for the sun and $7.350 \times 10^{22}$ kg for the moon.  The mean distance (center-to-center) between the sun and the Earth is $149.6 \times 10^{6}$ km and between the moon and the Earth is $384 \times 10^{3}$ km.

1.7    Prove whether or not the blocks, shown in the figure (a) and (b) below, are in equilibrium.

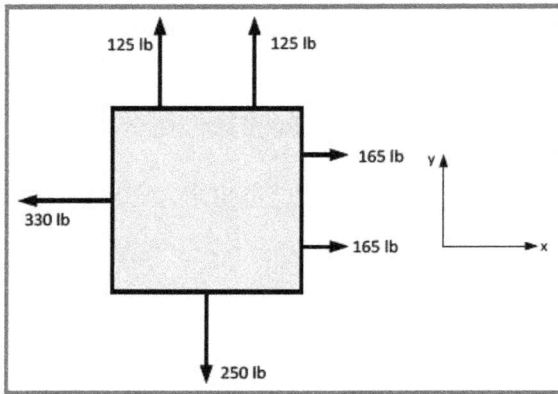

(a)

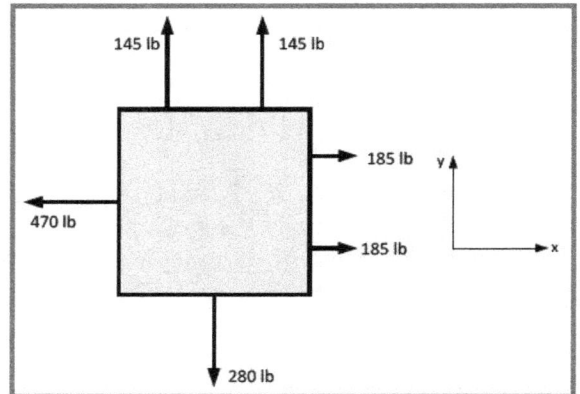

(b)

1.8    Prove whether or not the blocks, shown in the figures (a) and (b) below, are in equilibrium.

1.9    Draw a free body diagram of an automobile parked on a level street showing the reaction forces on the tires.  Also show the active forces on the pavement.  Assume that the auto weighs 3,200 lb. State the assumption made in distributing the weight of the auto among the four tires.

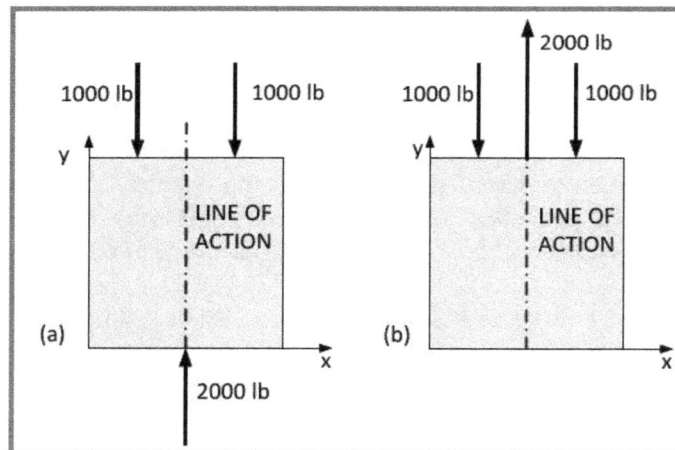

1.10   Determine the weight of a body with the following masses in SI units:

(a) 65 kg        (b) 19 kg        (c) 58 kg        (d) 17.3 kg        (e) 34.7 kg        (f) 81.2 kg

1.11   Determine the weight of a body with the following masses in U. S. Customary units:

(a) 7.5 slug        (b) 12 slug        (c) 65 slug        (d) 4.2 slug        (e) 33.7 slug        (f) 49.4 slug

1.12 If a pressure of 3,000 psi acts on a piston with a diameter of 1.75 in., determine the force required to maintain the piston in equilibrium.

1.13 If a pressure of 300 MPa acts on a piston with a diameter of 150 mm, determine the force required to maintain the piston in equilibrium.

1.14 If a force of 3,000 lb acts on a piston with a diameter of 5.0 in., determine the pressure required to maintain the piston in equilibrium.

1.15 If a force of 5,000 N acts on a piston with a diameter of 100 mm, determine the pressure required to maintain the piston in equilibrium.

1.16 If a piston is in equilibrium under a pressure of 1.5 ksi and a force of 10.5 kip, determine the diameter of the piston.

1.17 If a piston is in equilibrium under a pressure of 55 MPa and a force of 675 kN, determine the diameter of the piston.

1.18 List the four basic quantities and give their units in both the SI and U. S. Customary systems.

1.19 Determine the SI equivalent of the masses listed below.

(a) 7.5 slug    (b) 12 slug    (c) 65 slug    (d) 4.2 slug    (e) 33.7 slug    (f) 49.4 slug

1.20 Determine the U. S. Customary equivalent of the masses listed below.

(a) 65 kg  (b) 19 kg    (c) 58 kg    (d) 17.3 kg    (e) 34.7 kg    (f) 81.2 kg

1.21 Determine the SI equivalent of the following forces:

(a) 2,400 lb    (b) 425 lb    (c) 688 lb    (d) 18 tons    (e) 52 kip    (f) 95 kip

1.22 Determine the U. S. Customary equivalents for the following forces:

(a) 2,234 N    (b) 89.8 N    (c) 13.2 kN    (d) 96.9 mN    (e) 142.3 $\mu$N    (f) 44.8 MN

1.23 Determine the SI equivalent of the following lengths:

(a) 13 ft
(b) 153 in.
(c) 41.3 yards
(d) 16.4 miles
(e) 6 ft - 2 in.
(f) 1.8 miles - 662 yards

1.24 Determine the U. S. Customary equivalents for the following lengths:

(a) 7.4 m
(b) 500 mm
(c) 12.9 cm
(d) 13.2 km
(e) 18.8 mm
(f) 64.3 cm

1.25 Convert the following speeds into SI equivalent units:

(a) 60 mph
(b) 45 mph
(c) 30 mph
(d) 22 ft/s
(e) 90 ft/s
(f) 580 mph
(g) 16 ft/s
(h) 224 ft/s
(i) 187 mph

1.26    Convert the following speeds into U. S. Customary equivalent units:

    (a) 120 km/h             (d) 123 m/s             (g) 312 m/h
    (b) 61 km/h               (e) 510 m/s             (h) 16.7 m/s
    (c) 88 km/h               (f) 17.5 km/s          (i) 43.5 km/s

1.27    Convert the following stresses into SI equivalent units:

    (a) 12,000 psi           (d) 132.5 ksi           (g) 194,200 psi
    (b) 2,320 psi            (e) $30 \times 10^6$ psi     (h) 32 psi
    (c) 1,980 psi            (f) $10.5 \times 10^6$ ksi    (i) 96,230 psi

1.28    Convert the following stresses into U. S. Customary equivalent units:

    (a) 1,000 MPa          (d) 1,400 kPa        (g) 207 GPa
    (b) 121 MPa             (e) 144 kPa          (h) 16,430 MPa
    (c) 86.4 MPa            (f) 9,642 kPa        (i) 42.3 kPa

1.29  Prepare a short written description of the differences among scalars, vectors, and tensors.  Also prepare a sketch illustrating a quantity of each type.

1.30  Consider each quantity listed in the table below and classify it as a scalar, vector or tensor.

## Table
## Unit Conversion Factors

| Quantity | U. S. Customary | SI Equivalent |
|---|---|---|
| Acceleration | ft/s$^2$ | 0.3048 m/s$^2$ |
| Area | ft$^2$ | 0.0929 m$^2$ |
| Distributed Load | lb/ft | 14.59 N/m |
|  | lb/in. | 0.1751 N/mm |
| Energy | ft-lb | 1.356J |
| Force | kip = 1000 lb | 4.448 kN |
|  | lb | 4.448N |
| Impulse | lb-s | 4.448 N-s |
| Length | ft | 0.3048 m |
|  | in | 25.40 mm |
|  | mi | 1.609 km |
| Mass | lb mass | 0.4536 kg |
|  | slug | 14.59 kg |
|  | ton mass | 907.2 kg |
| Moments or Torque | ft-lb | 1.356 N-m |
|  | in-lb | 0.1130 N-m |
| Area Moment of Inertia | in$^4$ | 0.4162 x 10$^6$ mm$^4$ |
| Power | ft-lb/s | 1.356 W |
|  | hp | 745.7 W |
| Stress and Pressure | lb/ft$^2$ | 47.88 Pa |
| Velocity | ft/s | 0.3048 m/s |
| Volume | ft$^3$ | 0.02832 m$^3$ |
| Work | ft-lb | 1.356 J |

# CHAPTER 2

# VECTORS — FORCES AND MOMENTS

## 2.1 INTRODUCTION

Complex structures, vehicles and machines are constructed from many different structural elements such as rods, columns, struts, beams, cables, etc. To properly size these elements so they will perform their function safely for the duration of their design life, we must determine the forces and the moments acting at each point on each element. These forces and moments are vector quantities that are defined by specifying both their magnitude and direction. In this chapter, two different approaches for dealing with vector quantities are described:

1. A relatively simple approach based on trigonometry, which is easy to use when analyzing two-dimensional (plane) structures.
2. A more elegant technique utilizing vector algebra, which is very useful in analyzing three-dimensional structures. Unfortunately, this technique requires an understanding of a mathematical topic that is new to many engineering students.

We will solve two-dimensional examples employing the trigonometric techniques. We will also introduce vector algebra and demonstrate the vector dot and cross products to show their application in determining the unknown forces and moments in structural elements.

## 2.2 INTERNAL AND EXTERNAL FORCES

In dealing with forces, we distinguish between those that are applied to the structure (**external**), and those that develop within a structural element (**internal**). The external forces include the active loads applied to the structure such as those shown in Fig. 2.1a-c, and the reaction forces, shown in Fig. 2.1 d, that develop at the supports to maintain the structure in equilibrium.

In Fig. 2.1a, the simply supported beam is loaded with a **concentrated force** at a local point near its center. The concentrated force, applied at a point, is an idealization. Forces are always distributed over some area; however, with concentrated forces, we assume that the area is so small that it approaches a point. The symbol F will be used to designate the magnitude of concentrated forces.

In Fig. 2.1b, the beam is loaded with a **uniformly distributed load** that is applied over most of its length. Uniformly distributed loads along beams are specified in terms of force/unit length (i. e. lb/ft or N/m). The symbol q is used to designate the magnitude of the distributed forces applied to a beam.

A distributed load that is increasing as we move from the left end of the beam to the right is illustrated in Fig. 2.1c. Again, the symbol q is used to designate the magnitude of the distributed forces; however, in this case we must recognize that q is a function of position x along the length of the beam, designated by $q(x)$.

The last example of external forces is shown in Fig. 2.1d. The beam and its loading are identical with that shown in Fig. 2.1b, but the supports have been removed. The reaction forces developed by the supports to maintain the beam in equilibrium are shown as concentrated forces. The symbol R will be used to designate the magnitude of the reaction forces.

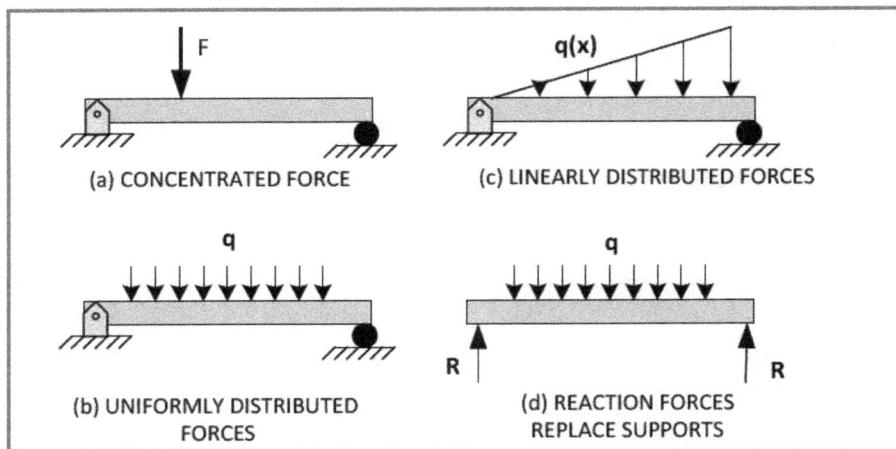

Fig. 2.1 Examples of different types of external forces applied to the structure (beam).

Internal forces develop within a structural member due to the action of the applied external forces. These internal forces are not visible—although we try to visualize them by making imaginary cuts through a structural member. Let's examine a bar subjected to two external forces F applied at each end as shown in Fig. 2.2. We make a section cut perpendicular to the axis in the central region of the bar. This is an imaginary cut, not a real one, but it permits us to visualize either segment of the bar. We examine the segment on the left, and find a **normal stress**, $\sigma$, which is uniformly distributed over the area exposed by the section cut. When this stress is integrated over the area of the bar, an internal force $P_{int}$ is generated that acts along the axis of the bar. The magnitude of $P_{int}$ is given by:

$$P_{int} = \int \sigma \, dA \qquad (2.1)$$

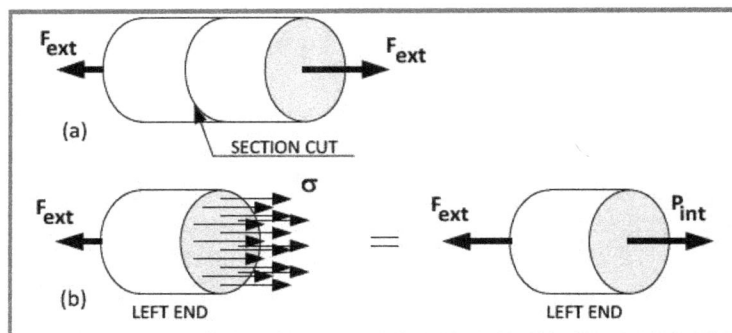

Fig. 2.2 (a) A tension bar with external loads and a section cut dividing the bar into two parts. (b) Two different representations of the left side of the bar.

The left end of the bar must be in equilibrium, which implies that:

$$\Sigma F_x = 0 \qquad (1.2 \text{ bis})$$

Summing the forces in the x direction gives:

$$P_{int} = F_{ext} \qquad (2.2)$$

In this elementary example of a rod in tension, we have found the relation between the internal force P within the bar and the external force F applied to its ends. We will use the same three-step approach throughout the book in solving much more complex problems. In your solutions to the exercises remember to:

1. Make an appropriate section cut.
2. Use the equations of equilibrium.
3. Solve for internal forces in structural members.

## 2.3 FORCE VECTORS

Forces are vector quantities, and as such it is necessary to specify both their magnitude and direction. Thus far, we have focused on the magnitude to illustrate the physical aspects of different types of forces. Let's now move to a more mathematical discussion of forces and vectors. A typical force vector is presented with an arrow as shown in Fig. 2.2.

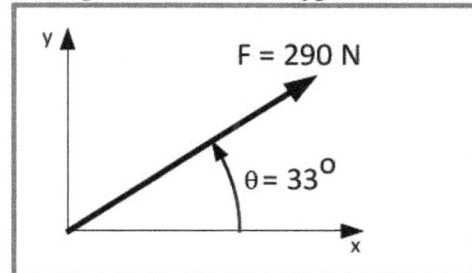

Fig. 2.3 A force vector is represented by an arrow with a length proportional to its magnitude (290 N). The direction of the vector ($\theta = 33°$) is specified relative to the x-axis.

The length of the arrow indicates the magnitude, and the orientation of the arrow relative to the x-axis gives the direction. The tail end of the arrow is positioned at the origin of the coordinate system, and the arrowhead indicates the tip.

### EXAMPLE 2.1

Construct a graph showing a force vector with a magnitude of 3700 N and a direction of 150° relative to an x-y coordinate system.

**Solution:**

First, define an x-y coordinate system making certain to include the negative side of the x-axis. Next, layout a line from the origin inclined by 180° − 150° = 30° relative to the negative x-axis. Select a scale for the magnitude; we suggest 1 mm = 100 N. Measure a length of 37 mm along this line, and place an arrowhead at this point. Place arrowheads on the coordinate system, and include the measure of the angle and the magnitude of the force vector. The result obtained is illustrated in Fig. E2.1:

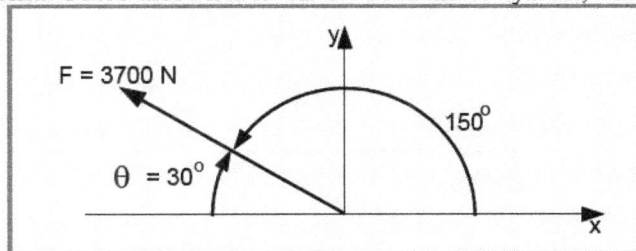

Fig. E2.1

Force vectors have a line of action, as indicated in Fig. 2.4, which is collinear with the direction of the vector.

Fig. 2.4 The line of action is collinear with the force vector.

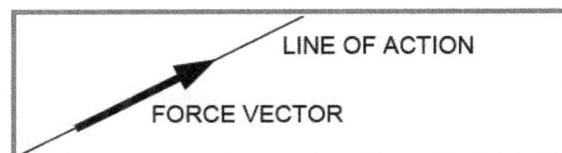

We can move the force vector along the line of action if it is convenient to do so. Shifting the force along the line of action does not change the state of equilibrium of the body. To illustrate this concept of the **transmissibility** of forces, suppose the body shown in Fig. 2.5a is in equilibrium with the force $F_1$ applied at point A. The body remains in equilibrium as we move the force $F_1$ along the line of action from point A to point B as indicated in Fig. 2.5b. In the first instance, we are pushing on the body with $F_1$, and in the second, we are pulling on the body. In both cases the force $F_1$ acts on the body with the same magnitude and direction; hence the body remains in equilibrium.

Fig. 2.5 The body remains in equilibrium as we shift $F_1$ along the line of action from A to B.

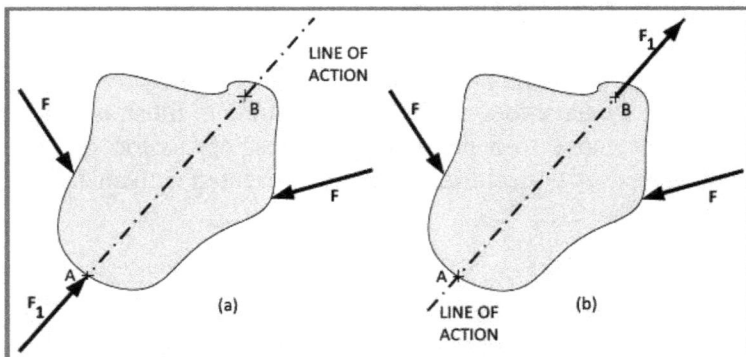

## EXAMPLE 2.2

Let's consider the block with three applied forces as shown below. Are both blocks in equilibrium after we shift the 200 N force along the line of action from the bottom edge to the top edge?

Fig. E2.3

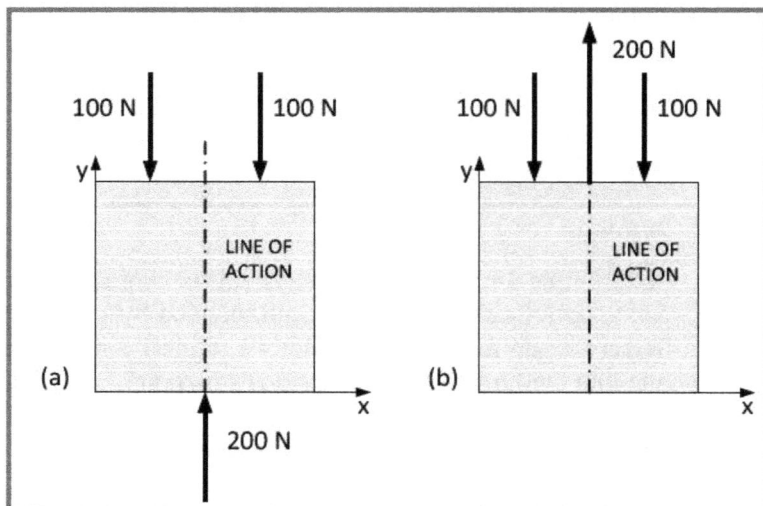

**Solution:**

The block in Fig. (a) is in equilibrium with a 200 N force applied to the bottom edge. From Eq. 2.4b, we write the equation for equilibrium in the y direction to obtain:

$$\Sigma F_y = 0$$

$$200 - 100 - 100 = 0$$

We can write the same equations for the block shown in Fig. (b). The fact that the 200 N force has been shifted along the line of action does not change the equilibrium relation.

We have already made a distinction between applied (active) forces F and reactive forces R. Reactive forces develop at structural supports and grow in magnitude as the active loads are applied to maintain the body in equilibrium. Reactive forces are shown in free body diagrams (FBDs)[1] when the supports are removed from a structural member. We illustrate a typical FBD in Fig. 2.6 showing three reactive forces that maintain a uniformly loaded, simply supported beam in equilibrium. In constructing the FBD, we remove the supports and apply the reactive forces at their location. The magnitude of R is determined from the equilibrium relations as will be discussed later in Chapter 3. The directions of both $R_{yL}$ and $R_{yR}$ are perpendicular to the surface of the beam. The direction of $R_x$ is taken in the positive x direction. The two reactive forces $R_{yL}$ and $R_x$ at the left support replace the effect of the pinned joint. It is emphasized that you must specify directions for the reactive forces, which are consistent with the type of supports that are removed in constructing the FBD.

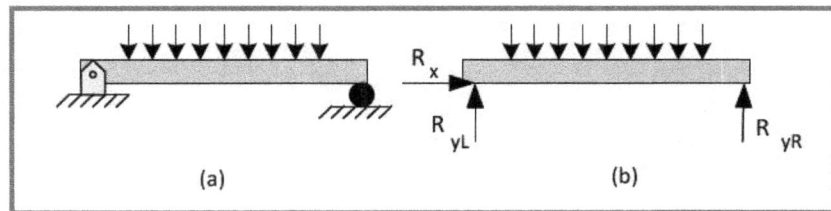

Fig. 2.6 Reactive forces replace the supports when drawing the FBD of the beam.

## EXAMPLE 2.3

To illustrate the direction of reactive forces, let's consider a circular pipe supported in a right-angled rack as shown below. Draw a FBD of the pipe and show the angle that the reaction forces make with a horizontal line.

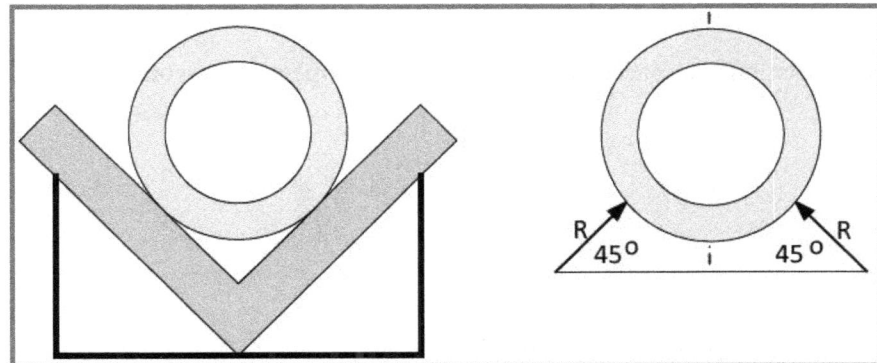

Fig. E2.3

**Solution:**

We remove the rack and support the pipe with a pair of reaction forces R. These forces are perpendicular to the surface of the pipe at the contact points. The lines of action of the reaction forces are radial and pass through the center of the pipe[2]. Since the rack is symmetric with a 90° included angle, it is easy to visualize the reaction forces at a 45° angle relative to a horizontal line. The third force W represents the weight of the pipe, which acts downward along the pipe's centerline.

---

[1] We will use the abbreviation FBD to represent free body diagram throughout the remainder of this textbook.
[2] Friction forces at the contact points are assumed to be negligible.

## 2.4 ADDING AND SUBTRACTING VECTORS

Suppose that we have two force vectors **A** and **B**, shown in Fig. 2.7a, that are to be added to give the vector sum $S_v = A + B$. There are several approaches that we can follow to determine this vector sum; however, to begin let's visualize the process of adding vectors by examining Fig. 2.7b. In this illustration, we have shifted vector **B** while retaining its direction and magnitude. Its new position places its tail at the tip of vector **A**. We then construct a line from the tail of vector **A** to the tip of vector **B** to obtain both the magnitude and direction of the vector sum $S_v$.

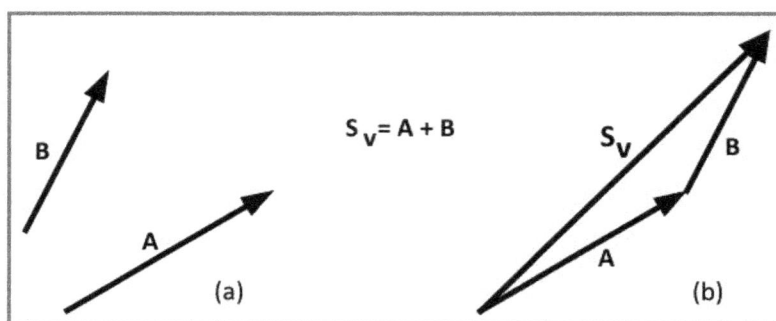

Fig. 2.7 An illustration of the physical interpretation of vector addition $S_v = A + B$.

Now that we can visualize adding **A** + **B**, let's add **B** + **A**. Will reversing the order of adding vectors **A** and **B** give the same result? Let's repeat the process of moving vectors to obtain the results shown in Fig. 2.8. In this case, we have shifted vector **A**, maintaining its magnitude and direction, so that its tail coincides with the tip of vector **B**. The line from the tail of **B** to the tip of **A** is the vector $S_v$. From these illustrations, it is evident that:

$$S_v = A + B = B + A \qquad (2.3)$$

This relation is known as the **commutative property of vector addition**.

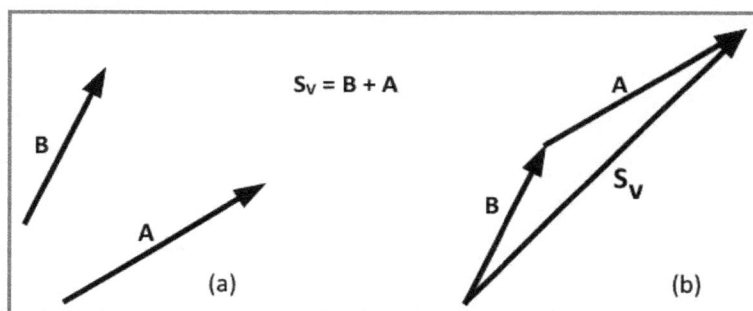

Fig. 2.8 Illustration of the physical interpretation of vector addition. $S_v = B + A$.

With a graphical approach, it is possible to add vectors and estimate the magnitude and direction of $S_v$ using only a scale and a protractor. We will not pursue the graphical technique in this text because there is a much more accurate trigonometric method discussed later.

Let's consider vector subtraction where we seek $D_v$: the vector formed by $A - B$. The subtraction process is illustrated in Fig. 2.9. In Fig. 2.9a, we show the same two vectors **A** and **B**. In Fig. 2.9b, we rotate **B** though 180° to reverse its direction and obtain a new vector − **B**. Next, we add − **B** to **A** following the same procedure used for vector addition to determine $D_v$ as shown in Fig. 2.9c.

From Figs. 2.7 − 2.9, which illustrate the addition or subtraction of two vectors, it is clear that both $S_v$ and $D_v$ are sides of a triangle. The trigonometric properties of a triangle will be valuable in determining the unknown magnitudes and directions of either $S_v$ or $D_v$ in vector addition or subtraction. Let's define the sides and the included angles of a triangle as indicated in Fig. 2.10.

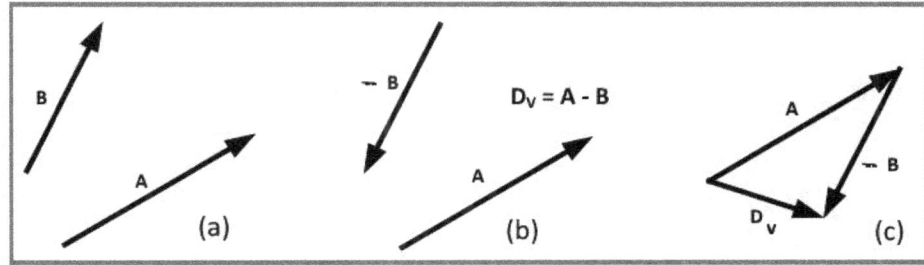

Fig. 2.9 An illustration of the process of vector subtraction.

In analyzing triangles, we make use of the **sine law:**

$$\frac{A}{\sin a} = \frac{B}{\sin b} = \frac{C}{\sin c} \qquad (2.4)$$

And the **cosine law:**

$$C = \sqrt{A^2 + B^2 - 2AB \cos (c)} \qquad (2.5)$$

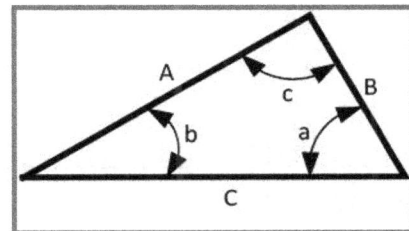

Fig. 2.10 Triangle with sides A, B, and C and included angles of a, b, and c.

The examples shown below illustrate the use of the sine and cosine laws in adding two vectors.

## EXAMPLE 2.4

Add the vectors **A** and **B** defined in Fig. E2.4a to obtain the vector sum $S_v$.

Fig. E2.4

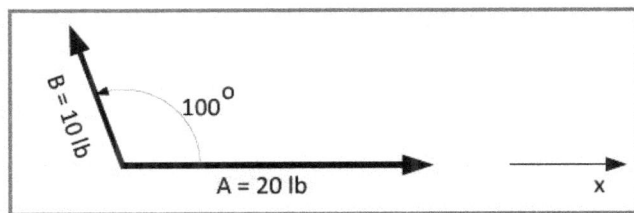

**Solution:**

We begin by constructing an A, B and C triangle where the magnitude of $S_v$ is the same as the side C. The construction of the triangle is presented in Fig. E2.4b.

Fig. E2.4b

We solve for the magnitude of C using the cosine law, which is given in Eq. (2.5).

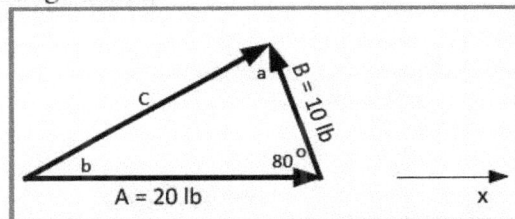

$$C = [A^2 + B^2 - 2AB \cos (c)]^{1/2}$$

$$C - [(20)^2 + (10)^2 - 2(20)(10) \cos(80)]^{1/2}$$

$$C = [ -(500) - 400(0.1736)]^{1/2} = [430.4]^{1/2}$$

$$C = 20.75 \text{ lb}$$

Now that we have the magnitude of $\mathbf{S_v}$, the sine law given in Eq. (2.4) may be used to determine its direction. The direction of $\mathbf{S_v}$ relative to the x-axis is given by the angle b.

$$\frac{B}{\sin (b)} = \frac{C}{\sin (c)}$$

$$\sin (b) = \frac{B \sin (c)}{C} = \frac{(10)\sin (80)}{20.75} = 0.4746$$

$$b = 28.33°$$

The direction of $\mathbf{S_v}$ is at angle of 28.33° relative to the x-axis. The solution is complete because we have determined both the magnitude and the direction of the vector sum $\mathbf{S_v}$. Note, the results are given with four significant figures, and the units for both the magnitude and the direction are stated.

## EXAMPLE 2.5

Two forces **A** and **B** are applied to lift the weight shown in Fig. E2.5. If the weight W is to move vertically upward without swinging when it clears the floor, determine the angle θ to specify for the direction of the force **B**.

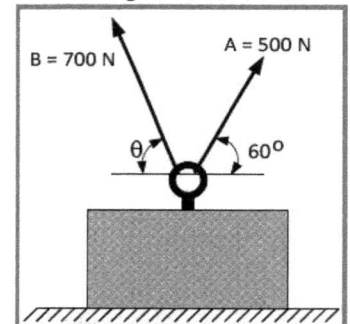

Fig. E2.5

**Solution:**

To begin the solution of this problem, draw the force triangle as shown in Fig. E2.5a. Note that the direction of $\mathbf{S_v}$, the resultant of the vector addition **A** + **B**, must be vertical if the weight is to be lifted without swinging.

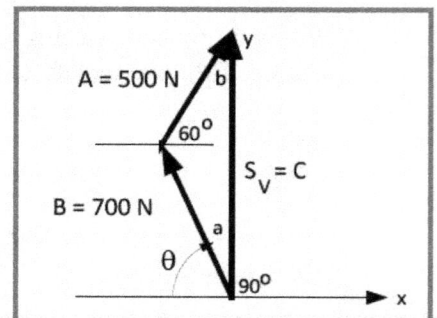

Fig. E2.5a

We will use the force triangle shown in Fig. E2.5b to solve for the angle θ. First, recall from your studies of plane geometry that the sum of the included angles within a triangle is 180°:

$$a + b + c = 180°$$

$$(90° - θ) + b + (θ + 60°) = 180°$$

$$b = 30°$$

Next use the law of sines given by Eq. (2.4) to solve for θ.

$$\frac{A}{\sin a} = \frac{B}{\sin b}$$

$$\frac{500}{\sin (90 - θ)} = \frac{700}{\sin 30}$$

$$\sin (90 - θ) = (5/7)(1/2) = 0.3571$$

$$90 - θ = 20.92° \implies θ = 69.08°$$

We must apply the force B at an angle θ = 69.08°, if the weight is to lift vertically from the floor without swinging.

## EXAMPLE 2.5 (Continued)

What weight can be lifted from the floor with the forces **A** and **B** applied as shown in Fig. E2.5a?

**Solution:**

The magnitude of the vector sum Sv or the side C of our force triangle gives the weight that can be lifted. Let's use the law of cosines given by Eq. (2.5) to determine C.

$$C^2 = A^2 + B^2 - 2AB \cos (c)$$

$$C^2 = (500)^2 + (700)^2 - 2(500)(700) \cos(129.1)$$

$$C^2 = 118.1 \times 10^4 \implies C = 1087\,N$$

These results indicate that only 1087 N can be lifted even though forces of 500 N and 700 N (a total of 1200 N) were applied in the lifting process. Only the vertical components of the forces A and B contributed to lifting. The horizontal components cancelled each other, and did not contribute to lifting the weight. We will explore the components of a force in the next section.

## 2.5 COMPONENTS OF A FORCE VECTOR

In section 2.4, we discussed the techniques for combining two force vectors into a vector sum $S_v$. Let's now reverse the process, and consider a technique for converting a single force vector into two equivalent vectors. Suppose we have a force vector **C** as shown in Fig. 2.11a. We can resolve this vector into a set of two vectors **A** and **B** that is equivalent to **C**. We show such a set of vectors in Fig. 2.11b. Clearly, an infinite set of selections of **A** and **B** can be made that are equivalent to **C**. The only requirement on the choice of **A** and **B** is that they satisfy the equation for vector addition:

$$A + B = C \tag{2.6}$$

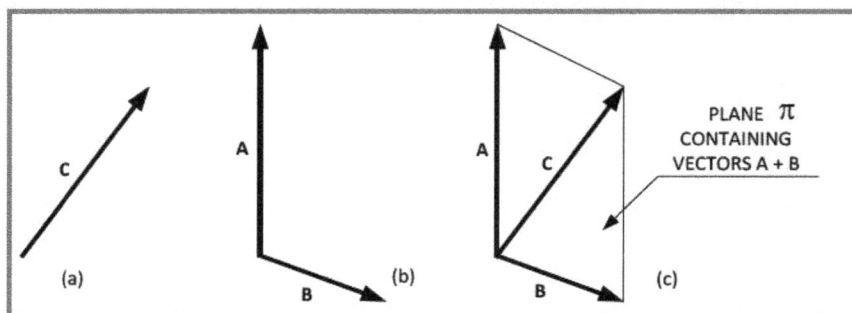

Fig. 2.11 Resolution of a vector **C** into an equivalent vector pair **A** and **B**.

We illustrate the vector addition involved in resolving a single vector into two equivalent vectors with the parallelogram presented in Fig. 2.11c. Note that the vector **C** and its equivalents **A** and **B** are coplanar—they all lie in the same plane $\pi$.

Let's next consider a special case of resolving a vector into an equivalent two vector pair oriented along the Cartesian axes. Suppose a force vector **F** lies in the x-y plane as illustrated in Fig. 2.12a.

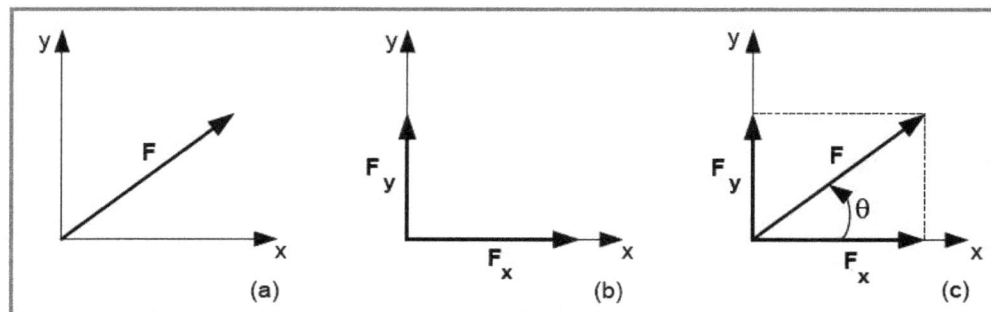

Fig. 2.12 Resolution of vector **F**, lying in the x-y plane, into two Cartesian force vectors $F_x$ and $F_y$.

The equivalent Cartesian force vectors $F_x$ and $F_y$ are shown in Fig. 2.12b where

$$F = F_x + F_y \tag{2.7}$$

The magnitudes of the Cartesian force vectors are given by:

$$F_x = F \cos \theta \quad \text{and} \quad F_y = F \sin \theta \tag{2.8}$$

The relation between F and the Cartesian components $F_x$ and $F_y$ is given by:

$$F = [F_x^2 + F_y^2]^{1/2} \tag{2.9}$$

Resolution of a vector into its Cartesian components is extremely important, as this concept will be used repeatedly in applying the equations of equilibrium. Let's next consider two examples to demonstrate the procedures followed in determining the Cartesian components of a vector.

# EXAMPLE 2.6

Determine the x and y components of the 600 lb force shown in the Fig. E2.6. Also check the solutions for $F_x$ and $F_y$ by verifying the value of 600 lb assigned to F.

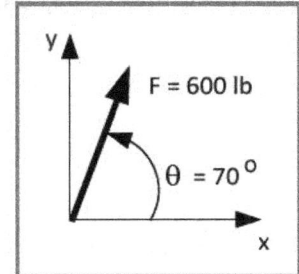

Fig. E2.6

**Solution:**

From Eqs. (2.8), we write:

$$F_x = F \cos \theta = 600 \cos 70° = (600)(0.3420) = 205.2 \text{ lb}$$

$$F_y = F \sin \theta = 600 \sin 70° = (600)(0.9397) = 563.8 \text{ lb}$$

We use Eq. (2.9) to verify the value of 600 lb for the magnitude F.

$$F = [F_x^2 + F_y^2]^{1/2} = [(205.2)^2 + (563.8)^2]^{1/2} = 600 \text{ lb}$$

# EXAMPLE 2.7

Determine the x and y components of the 50 kN force shown in Fig. E2.7. Also check the solutions for $F_x$ and $F_y$ by verifying the value of 50 kN assigned to F.

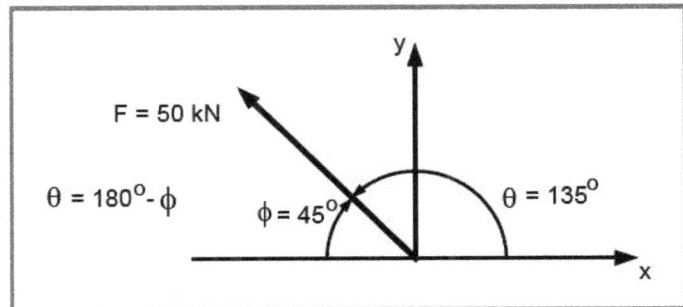

Fig. E2.7

**Solution:**

From Eqs. (2.8), we write:

$$F_x = F \cos \theta = 50 \times 10^3 \cos 135° = (50 \times 10^3)(-0.7071) = -35.36 \text{ kN}$$

$$F_y = F \sin \theta = 50 \times 10^3 \sin 135° = (50 \times 10^3)(0.7071) = +35.36 \text{ kN}$$

We use Eq. (2.9) to verify the value of 50 kN for the magnitude F.

$$F = [F_x^2 + F_y^2]^{1/2} = [(-35.36)^2 + (35.36)^2]^{1/2} = \sqrt{2}\,(35.36) = 50 \text{ kN}$$

### 2.5.1 Adding Three or More Coplanar Vectors

In the previous discussion and examples, we consider ED the addition and subtraction of two vectors, **A** and **B**. A graphical method for performing the addition and subtraction was described and an analytical approach was discussed and demonstrated with examples. The analytical approach involved forming a triangle with sides A, B and C and employing the sine and cosine laws to solve for the unknown vector **C**. This approach is effective if you are required to add two vectors **A** and **B**, but the approach becomes tedious if it is necessary to add three or more vectors together. A much more efficient approach involves the use of adding the Cartesian components of each vector to form a vector sum.

Suppose we have the three vectors, **A**, **B** and **C** that lie in a common plane (coplanar) as shown in Fig. 2.13. Let's form the vector sum $S_v = A + B + C$.

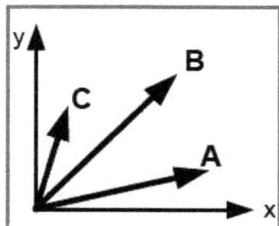

Fig. 2.14 Vector addition of three vectors A, B and C by a graphical technique.

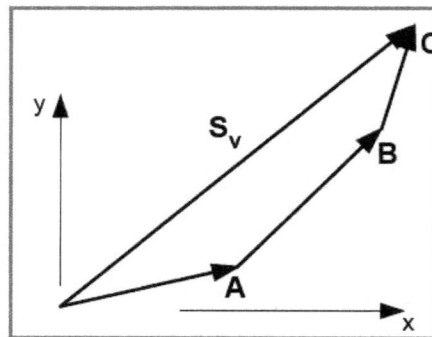

Fig. 2.13 Three vectors A, B and C.

Of course, we may add the three vectors using a graphical technique by connecting them together tip to tail as shown in Fig. 2.14.

Suppose we attempt to perform the addition of the three vectors by using the triangle approach described previously in Section 2.4. Clearly, the geometric shape in Fig. 2.14 is not a triangle because it has four sides (tetragon). To employ the laws of sines and cosines, it is necessary to form two triangles by dividing the tetragon as illustrated in Fig. 2.15.

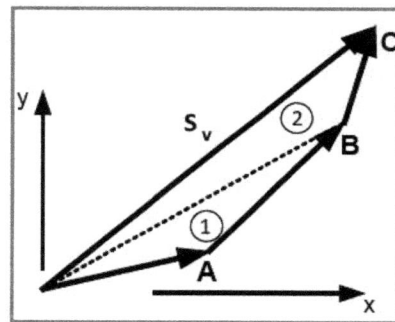

Fig. 2.15 Dividing the vector tetragon into two triangles.

It is possible to solve for the vector sum $S_v$ if we employ the sine and cosine laws twice; however, this approach is time consuming and mathematical errors are common. It is recommended that you resolve the three (or more) vectors into their Cartesian components, and add them in the manner described in the following paragraphs and examples.

Suppose you resolve the vectors **A**, **B** and **C** and determine their Cartesian components using Eq. (2.8). These Cartesian components are illustrated in Fig. 2.16.

The next step is to add the Cartesian components together to obtain:

$$S_{vx} = A_x + B_x + C_x \qquad \text{and} \qquad S_{vy} = A_y + B_y + C_y \qquad (2.10)$$

Substituting Eq. (2.10) into Eq. (2.9) yields the magnitude of the sum of the three vectors as:

$$S_v = [(A_x + B_x + C_x)^2 + (A_y + B_y + C_y)^2]^{1/2} \qquad (2.11)$$

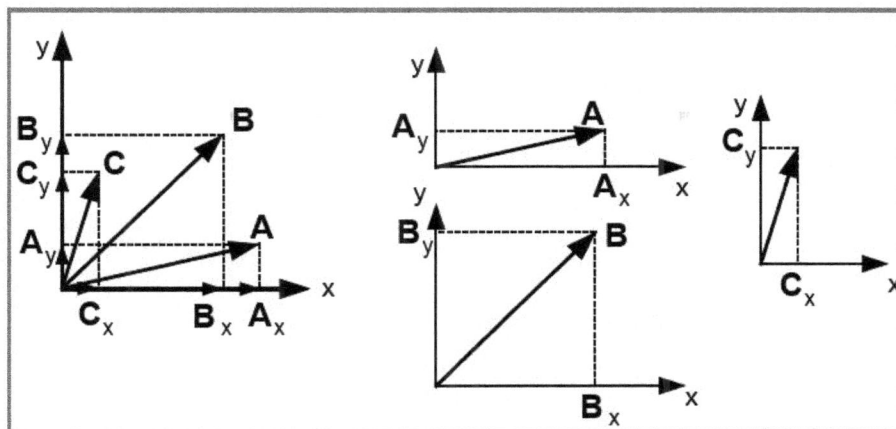

Fig. 2.16 Resolving vectors **A**, **B**, and **C** into Cartesian components $A_x$, $A_y$, $B_x$, $B_y$, $C_x$ and $C_y$.

The vector $S_v$ and its Cartesian components $S_{vx}$ and $S_{vy}$ are presented in Fig. 2.17. Also shown in this figure is the angle $\theta$ that defines the direction of the vector relative to the x-axis.

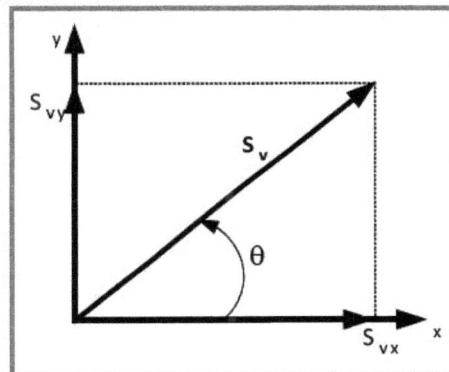

Fig. 2.17 The vector $S_v$ and its Cartesian components $S_{vx}$ and $S_{vy}$.

The direction of the vector $S_v$ is given by $\theta$, which is determined from:

$$\theta = \sin^{-1} (S_{vy}/S_v) = \cos^{-1} (S_{vx}/S_v) = \tan^{-1} (S_{vy}/S_{vx}) \qquad (2.12)$$

## EXAMPLE 2.8

Three vectors $F_1$, $F_2$, and $F_3$ are illustrated in Fig. E2.8. Determine the sum of these three vectors and calculate the angle $\theta$ the vector $S_v$ makes with the x-axis.

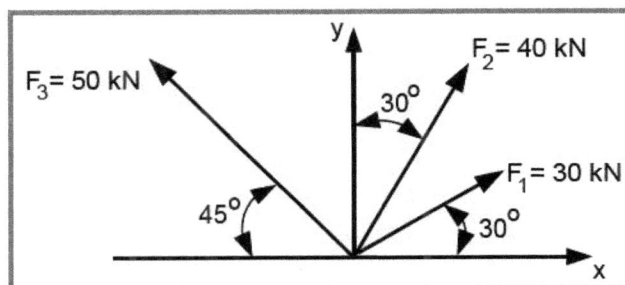

Fig. E2.8

**Solution:** Determine the Cartesian components of the three forces using Eq. (2.8) as:

$F_{1x} = 30 \cos (30°) = 25.98$ kN $\qquad$ $F_{1y} = 30 \sin (30°) = 15.00$ kN
$F_{2x} = 40 \cos (90° - 30°) = 20.00$ kN $\qquad$ $F_{2y} = 40 \sin (60°) = 34.64$ kN
$F_{3x} = 50 \cos (180° - 45°) = -35.36$ kN $\qquad$ $F_{3y} = 50 \sin (135°) = 35.36$ kN

Using Eq. (2.10) determine the Cartesian components of the vector sum $S_v$ as:

$$S_{vx} = F_{1x} + F_{2x} + F_{3x} = 25.98 + 20.00 - 35.36 = 10.62 \text{ kN}$$
$$S_{vy} = F_{1y} + F_{2y} + F_{3y} = 15.00 + 34.64 + 35.36 = 85.00 \text{ kN}$$

The magnitude $S_v$ is determined from Eq. (2.11) as:

$$S_v = [S_{vx}^2 + S_{vy}^2]^{1/2} = [(10.62)^2 + (85.00)^2]^{1/2} = (7338)^{1/2} = 85.66 \text{ kN}$$

Finally, the direction of the vector $S_v$ relative to the x-axis is given by Eq. (2.12) as:

$$\theta = \sin^{-1} (S_{vy} /S_v) = \sin^{-1} (85.00/85.66) = 82.88°$$

## EXAMPLE 2.9

The gusset plate illustrated in Fig. E2.9 is subjected to four forces from the attached uniaxial structural members. If the vector sum of these four forces is zero, determine the forces $F_3$ and $F_4$.

Fig. E2.9

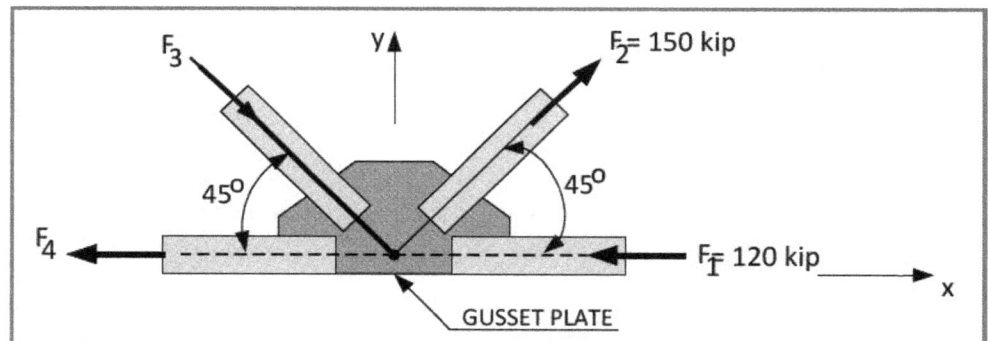

**Solution:**

The fact that the vector sum of the four forces $S_v = 0$, implies that:

$$S_{vx} = S_{vy} = 0 \qquad\qquad (a)$$

Use Eq. (2.8), Eq. (2.10) and Eq. (a) to obtain:

$$S_{vy} = F_1 \sin (0°) + F_2 \sin (45°) - F_3 \sin (45°) + F_4 \sin (0°) = 0$$

$$F_2 = F_3 = 150 \text{ kip} \qquad\qquad (b)$$

Finally, Eq. (2.8), Eq. (2.10), Eq. (a) and Eq. (b) enable us to write:

$$S_{vx} = -F_1 \cos(0°) + F_2 \cos(45°) + F_3 \cos(45°) - F_4 \cos(0°) = 0$$

$$F_4 = \cos(45°)[F_2 + F_3] - F_1 = (0.7071)(300) - 120 = 92.13 \text{ kip}$$

## EXAMPLE 2.10

Determine the magnitude and the direction of the force **F** in Fig. E2.10 if the box is lifted from the floor without swinging. The weight of the box is 5600 N.

Fig. E2.10

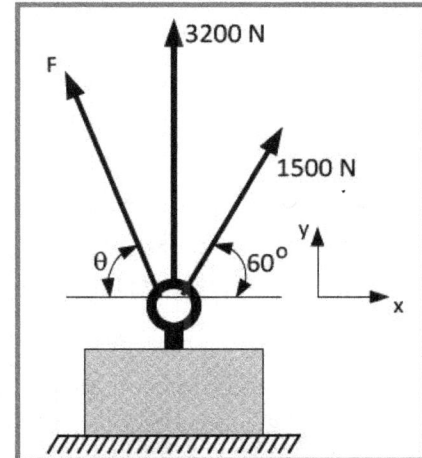

**Solution:**

If the box is lifted without swinging the component of the vector sum of the three forces in the x direction must be zero, or $S_{vx} = 0$

From Eq. (2.8) and Eq. (2.10), we may write:

$$S_{vx} = 1500 \cos(60°) - F \cos\theta = 0 \qquad \text{(a)}$$

$$F \cos\theta = 750 \text{ N} \qquad \text{(b)}$$

For the box to lift from the floor the vertical sum of the three forces must equal the box's weight of 5600 N. This fact permits us to write:

$$S = W = 5600 = 1500 \sin(60°) + 3200 + F \sin\theta \qquad \text{(c)}$$

$$F \sin\theta = 1101 \text{ N} \qquad \text{(d)}$$

Dividing the results of Eq. (d) by Eq. (b) yields:

$$(\sin\theta/\cos\theta) = \tan\theta = 1101/750 = 1.4679 \qquad \text{(e)}$$

$$\theta = 55.74° \qquad \text{and} \qquad F = 1101/\sin(55.74°) = 1332 \text{ N} \qquad \text{(f)}$$

## 2.6 CONCURRENT AND COPLANAR FORCES

**Concurrent** and **coplanar** are descriptive words identifying certain vector systems. Let's start with a concurrent vector system. When two or more forces (or vectors) act at a point, as illustrated in Fig. 2.18, they are concurrent. When a body is subjected to a concurrent set of forces, it does not exhibit a tendency to rotate (i.e. the moments about the point Q are zero). Equilibrium of the body is determined by using only Eq. (1.2).

$$\Sigma \, \mathbf{F} = 0 \qquad\qquad (1.2 \text{ bis})$$

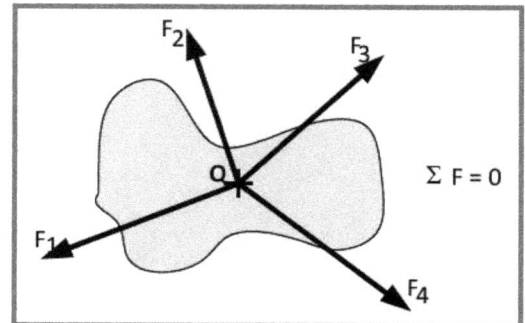

Fig. 2.18 Forces $\mathbf{F}_1$, $\mathbf{F}_2$, $\mathbf{F}_3$, and $\mathbf{F}_4$ are concurrent because they all are applied to point Q.

While the equilibrium of forces in Eq. (1.2) will prevent translational motion of a body, it does not guarantee that the body will not rotate. To understand what makes bodies rotate, the concept of **moments** must be introduced. Moment vectors, **M**, can be produced by force systems that are not concurrent. The equilibrium condition for moments is given by:

$$\Sigma \, \mathbf{M} = 0 \qquad\qquad (2.13)$$

This relation is satisfied automatically for concurrent force systems. We will discuss moments produced by forces in much more detail in Sections 2.8 and 2.9.

Coplanar forces (or vectors) all lie in the same plane. If we have any two vectors, it is easy to show that they are coplanar regardless of their orientation. However, if we deal with three or more vectors, the vector system may or may not be coplanar. Examples of planar and non-planar vector systems are shown in Figs. 2.19 and 2.20 respectively.

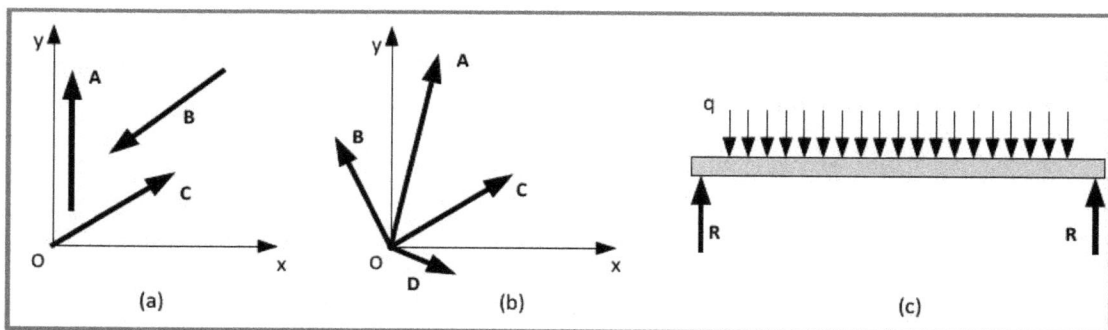

Fig. 2.19 Examples of coplanar forces.

In Fig. 2.19a, the vectors **A**, **B**, and **C** all lie in the x-y plane, and the vector system is coplanar. However, this system is not concurrent. Vectors **A** and **C** both can be shifted along their lines of action to produce an intersection of their tail points, but it is impossible to shift vector **B** along its line of action to intersect at this point. Consequently, vector **B** is offset from this intersection point, and produces a moment about it.

In Fig. 2.19b, the vectors **A**, **B**, **C** and **D** all lie in the x-y plane, and the vector system is coplanar. Moreover, this system is concurrent because all four of the vectors intersect at the origin O.

The force vectors in Fig. 2.19c are coplanar because both of the reaction forces **R** and the individual forces constituting the distributed load q all act in the plane of the beam. This system is not concurrent because all of the force vectors are parallel, and cannot be shifted along their lines of action to intersect at a common point.

The three dimensional force systems illustrated in Figs. 2.20a and 2.20b are not coplanar. In Fig. 2.20a, we show a Cartesian force system with vectors $F_x$, $F_y$ and $F_z$ that coincide with the x, y and z axes, respectively. Since these three forces pass though the origin, the system is concurrent. In Fig. 2.20b, the vector $F_1$ lies in the x-z plane, $F_2$ lies in the y-z plane and $F_3$ lies in the x-y plane. Clearly, the system is not coplanar nor is it concurrent.

Classification of force vector systems is important because it indicates the number of equations of equilibrium that provide useful relations in solving for the forces and moments occurring in structures. We will present the equilibrium relations for the four possible classifications shown below in Chapter 3.

- Coplanar and concurrent.
- Coplanar and non-concurrent.
- Non-coplanar and concurrent
- Non-coplanar and non-concurrent.

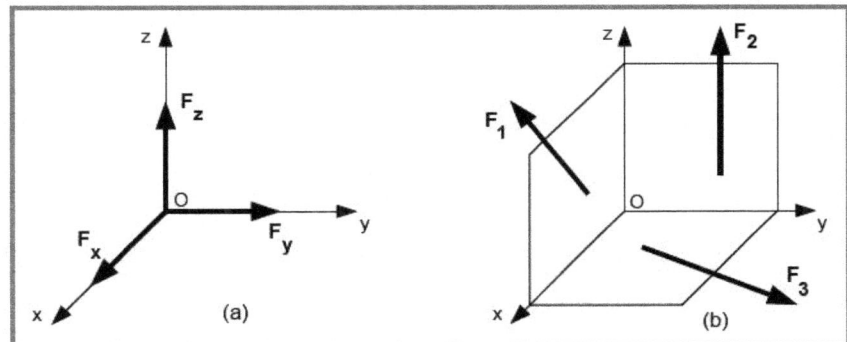

Fig. 2.20 Examples of non-coplanar vectors.

Three-dimensional force systems, like those illustrated in Fig. 2.20, are referred to as **space forces**. They are more complex because of the additional equations that must be satisfied to maintain a body in equilibrium when it is subjected to a three-dimensional system of forces.

## 2.7 SPACE FORCES

When forces are coplanar, we place them on a x-y plane and then apply the equilibrium equations to solve for the unknown forces acting on the structure under analysis. However, in some structures we must consider space forces, which cannot be completely characterized on the x–y plane. Consider the space force **F** shown in Fig. 2.21, which is represented with Cartesian vectors $F_x$, $F_y$ and $F_z$.

The space force **F** has both magnitude and direction. Its magnitude is a scalar quantity F, and its direction relative to the three-dimensional Cartesian coordinate system is given by the angles $\alpha$, $\beta$ and $\gamma$. These quantities, known as the coordinate direction angles, are measured from the line of action of **F** to the positive x, y and z-axes.

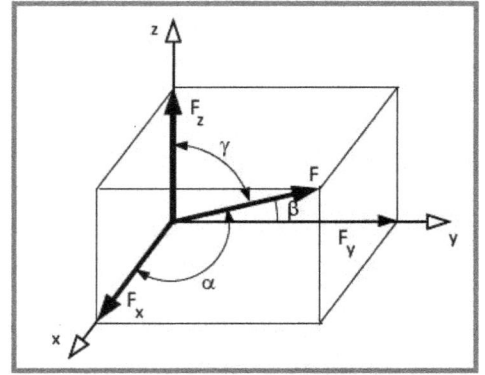

Fig. 2.21 A space force F is shown with its equivalent Cartesian vectors $\mathbf{F_x}$, $\mathbf{F_y}$ and $\mathbf{F_z}$.

The space force $\mathbf{F}$ has both magnitude and direction. Its magnitude is a scalar quantity F, and its direction relative to the three-dimensional Cartesian coordinate system is given by the angles $\alpha$, $\beta$ and $\gamma$. These quantities, known as the coordinate direction angles, are measured from the line of action of $\mathbf{F}$ to the positive x, y and z-axes.

The coordinate direction angles of the space vector F are given by:

$$\cos \alpha = \frac{F_x}{F}$$
$$\cos \beta = \frac{F_y}{F}$$
$$\cos \gamma = \frac{F_z}{F} \tag{2.14}$$

Also, recall the important identity for the coordinate direction angles, which is:

$$\cos^2 \alpha + \cos^2 \beta + \cos^2 \gamma = 1 \tag{2.15}$$

From Eq. (2.14), it is evident that the magnitudes of the Cartesian components of the space force are given by:

$$F_x = F \cos \alpha \qquad F_y = F \cos \beta \qquad F_z = F \cos \gamma \tag{2.16}$$

The terms $\cos \alpha$, $\cos \beta$ and $\cos \gamma$ that appear in Eq. (2.14) and Eq. (2.16) are called direction cosines of the vector F. Finally, the magnitude of F is given by the positive square root of the sum of the squares of its three components.

**EXAMPLE 2.11**

$$F = \sqrt{F_x^2 + F_y^2 + F_z^2} \tag{2.17}$$

A space force with a magnitude of 4,500 N is positioned at the origin of a Cartesian coordinate system. The force is oriented so that the angles $\alpha = 45°$ and $\beta = 60°$. Determine the Cartesian components of the space force and the angle that it makes with the z-axis.

**Solution:**

From Eq. (2.15), we write:

$$\cos^2 \alpha + \cos^2 \beta + \cos^2 \gamma = \cos^2(45°) + \cos^2(60°) + \cos^2 \gamma = 1$$

This relation reduces to:

$$\cos^2 \gamma = 1 - 0.5 - 0.25 = 0.25$$

Solving for the angle $\gamma$ yields:

$$\cos \gamma = 0.5 \quad \Rightarrow \quad \Rightarrow \quad \gamma = 60°$$

Next, let's recall Eq. (2.16) and apply it to determine the Cartesian components of the space force.

$F_x = F \cos \alpha$ ⠀⠀⠀⠀ $F_y = F \cos \beta$ ⠀⠀⠀⠀ $F_z = F \cos \gamma$

$F_x = 4500 \cos 45°$ ⠀⠀ $F_y = 4500 \cos 60°$ ⠀⠀ $F_z = 4500 \cos 60°$

$F_x = 3182$ N ⠀⠀⠀⠀ $F_y = 2250$ N ⠀⠀⠀⠀ $F_z = 2250$ N

We may check the accuracy of the result by employing Eq. (2.17).

$$\mathbf{F} = \sqrt{F_x^2 + F_y^2 + F_z^2}$$

Substituting numerical values into this relation gives:

$$\mathbf{F} = \sqrt{(3182)^2 + (2250)^2 + (2250)^2} = \sqrt{2025 \times 10^4} = 4500 \text{ N}$$

The results confirm the accuracy of the computations.

## EXAMPLE 2.12

A space force is represented by its Cartesian components as:

$F_x = 707$ lb ⠀⠀⠀⠀ $F_y = 883$ lb ⠀⠀⠀⠀ $F_z = 264$ lb

Determine the magnitude and the coordinate direction angles of the space force.

**Solution:** Let's first determine the magnitude of the space force by using Eq. (2.16)

$$\mathbf{F} = \sqrt{F_x^2 + F_y^2 + F_z^2} = \sqrt{(707)^2 + (883)^2 + (264)^2} = 1162 \text{ lb}$$

The coordinate direction angles are calculated from Eq. (2.14) as:

$$\cos \alpha = \frac{F_x}{F} = \frac{707}{1162} = 0.6087$$

$$\cos \beta = \frac{F_y}{F} = \frac{883}{1162} = 0.7602$$

$$\cos \gamma = \frac{F_z}{F} = \frac{264}{1162} = 0.2273$$

Solving for the coordinate direction angles yields:

$$\alpha = 52.51° \qquad \beta = 40.52° \qquad \gamma = 76.86°$$

Again we may check the accuracy of our calculations by using Eq. (2.15). Substituting the results for the coordinate direction angles into this relation yields:

$$\cos^2 \alpha + \cos^2 \beta + \cos^2 \gamma = (0.6087)^2 + (0.7602)^2 + (0.2273)^2 = 1.000$$

Because the identity for the coordinate direction angles is satisfied, we may conclude that the calculation was performed correctly.

## 2.7.1 Unit Vectors

Let's define a force vector **F** in terms of unit vectors as shown below:

$$\mathbf{F} = F_x \mathbf{i} + F_y \mathbf{j} + F_z \mathbf{k} \qquad\qquad (2.18)$$

where $F_x$, $F_y$, $F_z$ are the magnitudes of the Cartesian force components, and **i, j, k** are the Cartesian unit vectors in the x, y and z directions, respectively.

As the name implies, the unit vectors have a magnitude of unity. They provide directions along the x, y and z-axes in Eq. (2.18). The unit vectors **i, j** and **k** are illustrated in Fig. 2.22.

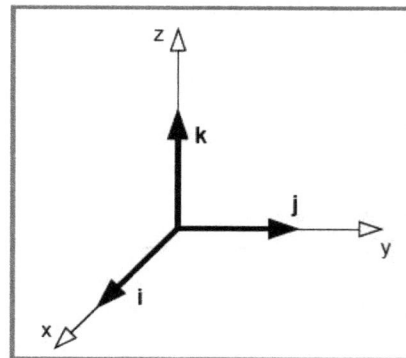

Fig. 2.22 Unit vectors **i, j** and **k** oriented along the x, y and z axes, respectively.

It is also useful to employ unit vectors that are oriented in an arbitrary direction. For example consider a vector **F** as shown in Fig. 2.23.

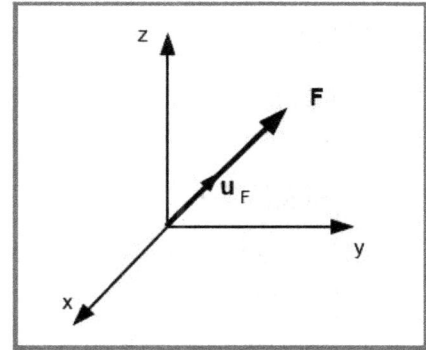

Fig. 2.23 A unit vector $\mathbf{u}_F$ gives the direction of the vector $\mathbf{F}$.

The direction of this unit vector is given by:

$$\mathbf{u}_F = \mathbf{F}/F \qquad (2.19)$$

Hence, the vector $\mathbf{F}$ may be written as:

$$\mathbf{F} = F\mathbf{u}_F \qquad (2.20)$$

A comparison of Eqs. (2.18) and (2.20) indicates that we may write the equation for a force using either the Cartesian unit vectors, which are directed along the x, y and z-axes, or by a single unit vector along the line of action of the force. The direction of the force may also be expressed using either the unit vector $\mathbf{u}_F$ or the coordinate direction angles. This fact implies a relation between these quantities. To show this relation, we substitute Eq. (2.18) into Eq. (2.19) to obtain:

$$\mathbf{u}_F = \mathbf{F}/F = (F_x/F)\mathbf{i} + (F_y/F)\mathbf{j} + (F_z/F)\mathbf{k} \qquad (2.21)$$

Substituting Eq. (2.14) into Eq. (2.21) yields:

$$\mathbf{u}_F = \cos \alpha \, \mathbf{i} + \cos \beta \, \mathbf{j} + \cos \gamma \, \mathbf{k} \qquad (2.22)$$

## EXAMPLE 2.13

A space force is represented by its Cartesian components as:

$$F_x = 633 \text{ N} \qquad\qquad F_y = 216 \text{ N} \qquad\qquad F_z = 521 \text{ N}$$

Write an expression for the force vector $\mathbf{F}$ in terms of the Cartesian unit vectors and determine the magnitude of the force F.

**Solution:**

Using Eq. (2.18), we write:

$$\mathbf{F} = (633 \, \mathbf{i} + 216 \, \mathbf{j} + 521 \, \mathbf{k}) \text{ N}$$

The magnitude F is calculated from Eq. (2.17) as:

$$F = \sqrt{F_x^2 + F_y^2 + F_z^2} = \sqrt{(633)^2 + (216)^2 + (521)^2} = 847.8 \text{ N}$$

## EXAMPLE 2.14

Determine the unit vector describing the direction of the space force specified in Example 2.13. Also, calculate the coordinate direction angles for this force vector.

---

**Solution:**

To determine the unit vector $\mathbf{u_F}$, we substitute the results from Example 2.13 into Eq. (2.21) to obtain:

$$\mathbf{u_F} = \mathbf{F}/F = (F_x/F)\mathbf{i} + (F_y/F)\mathbf{j} + (F_z/F)\mathbf{k} = (633/847.8)\,\mathbf{i} + (216/847.8)\,\mathbf{j} + (521/847.8)\,\mathbf{k}$$

$$\mathbf{u_F} = 0.7466\,\mathbf{i} + 0.2548\,\mathbf{j} + 0.6145\,\mathbf{k}$$

The coordinate direction angles for the space force are determined from Eq. (2.14) as:

$$\cos\alpha = \frac{F_x}{F} = \frac{633}{847.8} = 0.7466$$

$$\cos\beta = \frac{F_y}{F} = \frac{216}{847.8} = 0.2548$$

$$\cos\gamma = \frac{F_z}{F} = \frac{521}{847.8} = 0.6145$$

Solving for the coordinate direction angles gives:

$$\alpha = 41.70° \qquad \beta = 75.24° \qquad \gamma = 52.08°$$

---

## EXAMPLE 2.15

Express the space force **F**, shown in Fig. E2.15, as a Cartesian vector.

Fig. E2.15

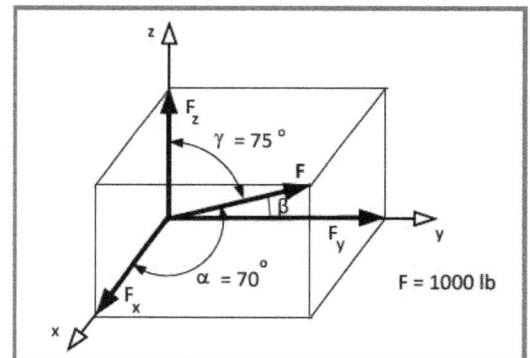

---

**Solution:**

Because only two of the direction cosine angles have been specified, we begin by using Eq. (2.15) to determine the unknown angle γ.

$$\cos^2\beta = 1 - \cos^2\alpha - \cos^2\gamma = 1 - \cos^2(70°) - \cos^2(75°)$$

$$\cos^2\beta = 1 - 0.1170 - 0.06699 = 0.8160$$

$$\cos \beta = \pm\, 0.9033 \qquad \Rightarrow \qquad \Rightarrow \qquad \beta = 25.40° \text{ or } 154.6°$$

By inspection of Fig. E2.15, it is evident that β is an acute angle; therefore:

$$\beta = 25.40°$$

Next, employ Eq. (2.16) to determine the Cartesian components of the space force **F** as:

$$F_x = F \cos \alpha = 1000 \cos (70.0°) = 342.0 \text{ N}$$

$$F_y = F \cos \beta = 1000 \cos (25.4°) = 903.3 \text{ N}$$

$$F_z = F \cos \gamma = 1000 \cos (75.0°) = 258.8 \text{ N}$$

Finally, let's check the accuracy of these results by using Eq. (2.17).

$$F = \sqrt{F_x^2 + F_y^2 + F_z^2} = \sqrt{(342.0)^2 + (903.3)^2 + (258.8)^2} = 999.9 \text{ N}$$

The result of 999.9 N for the magnitude of F is within 1 part in 10,000 for the 1000 N force. The very small difference, of no consequence, is due to rounding errors.

## 2.7.2 Position Vectors

A position vector is a fixed vector in space. We use a position vector to locate a point in space relative to another point. A position vector is similar to a force vector in that it has both magnitude and direction. An example of a position vector is presented in Fig. 2.24. In this case, the position vector **r** locates a point A (x, y, z) relative to the origin O of the coordinate system.

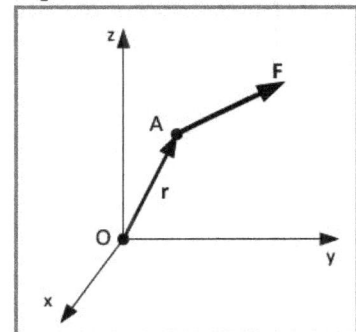

Fig. 2.24 This position vector **r** locates the point of application of **F** relative to the origin.

The position vector in Fig. (2.24) is expressed as:

$$\mathbf{r} = x\mathbf{i} + y\mathbf{j} + z\mathbf{k} \qquad (2.23)$$

The coordinate direction angles of the position vector **r** are:

$$\cos \alpha = x/r$$

$$\cos \beta = y/r \qquad (2.24)$$

$$\cos \gamma = z/r$$

Finally, the magnitude of the position vector r is given by:

$$r = [x^2 + y^2 + z^2]^{1/2} \qquad (2.25)$$

## EXAMPLE 2.16

A force vector **F** is applied to a rectangular space frame at point A as shown in Fig. E2.16. Write the equation for the position vector **r** locating the point of application relative to the origin of the Cartesian coordinates. Also determine its magnitude and direction.

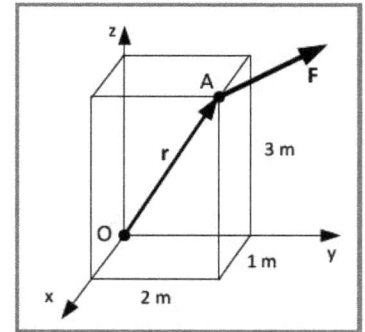

Fig E2.16

**Solution:**

Selecting the dimensions of the rectangular space frame from Fig. E2.16 and using Eq. (2.23) enables us to write:

$$r = (1\ i + 2\ j + 3\ k)\ m$$

The magnitude of the position vector is given by substituting the coordinates into Eq. (2.25):

$$r = \sqrt{x^2 + y^2 + z^2} = \sqrt{(1)^2 + (2)^2 + (3)^2} = 3.742\ m$$

Next, we employ Eq. (2.24) to determine the coordinate direction angles for the vector **r**.

$$\cos \alpha = x/r = 1/3.742 = 0.2673 \quad \Rightarrow \quad \alpha = 74.50°$$

$$\cos \beta = y/r = 2/3.742 = 0.5345 \quad \Rightarrow \quad \beta = 57.69°$$

$$\cos \gamma = z/r = 3/3.742 = 0.8018 \quad \Rightarrow \quad \gamma = 36.70°$$

Let's check the accuracy of the calculations by using Eq. (2.15).

$$\cos^2 \alpha + \cos^2 \beta + \cos^2 \gamma = (0.2673)^2 + (0.5345)^2 + (0.8018)^2 = 1.000$$

Okay! The results for the direction cosine angles check.

Let's consider a more general case for a position vector that locates point B relative to point A. In this case neither point A nor B are located at the origin of the coordinate system. The position vector for this situation is presented in Fig. 2.25.

The **i**, **j** and **k** components of the position vector **r** in Fig. 2.25 are written by taking the coordinates of the tip of the vector B $(x_B, y_B, z_B)$ and subtracting them from the coordinates of the tail A$(x_A, y_A, z_A)$.

$$\mathbf{r} = (x_B - x_A)\mathbf{i} + (y_B - y_A)\mathbf{j} + (z_B - z_A)\mathbf{k} \qquad (2.26)$$

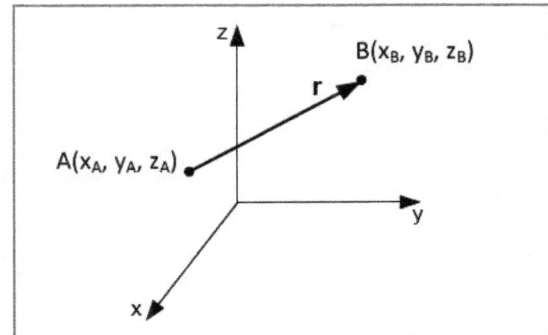

Fig. 2.25 A position vector r that locates point
B relative to point A.

The coordinate direction angles of the position vector **r** are given by:

$$\cos \alpha = (x_B - x_A)/r$$

$$\cos \beta = (y_B - y_A)/r \qquad (2.27)$$

$$\cos \gamma = (z_B - z_A)/r$$

These coordinate direction angles are measured from a local coordinate system positioned at point A (the tail of the position vector **r**).

The magnitude of the position vector in Fig. 2.25 is given by:

$$\mathbf{r} = \sqrt{(x_B - x_A)^2 + (y_B - y_A)^2 + (z_B - z_A)^2} \qquad (2.28)$$

## EXAMPLE 2.17

Determine the magnitude and direction of the position vector that extends from point A to B in Fig. 2.25 if the coordinates of points A and B are given by:

$$A(2, 0, -6) \text{ and } B(-4, 4, 6)$$

The coordinates are expressed in m.

**Solution:**

Let's write the expression for r by using Eq. (2.26).

$$\mathbf{r} = (x_B - x_A)\mathbf{i} + (y_B - y_A)\mathbf{j} + (z_B - z_A)\mathbf{k} = (-4 - 2)\,\mathbf{i} + (4 - 0)\,\mathbf{j} + (6 + 6)\,\mathbf{k}$$

$$\mathbf{r} = (-6\,\mathbf{i} + 4\,\mathbf{j} + 12\,\mathbf{k})\,\text{m}$$

The magnitude of the position vector is given by:

$$r = \sqrt{(-6)^2 + (4)^2 + (12)^2} = 14 \text{ m}$$

The coordinate direction angles are computed from Eq. 2.27.

$$\cos \alpha = (x_B - x_A)/r = -6/14 = -0.4286 \qquad \Rightarrow \qquad \alpha = 115.4°$$

$$\cos \beta = (y_B - y_A)/r = 4/14 = 0.2857 \qquad \Rightarrow \qquad \beta = 72.40°$$

$$\cos \gamma = (z_B - z_A)/r = 12/14 = 0.8571 \qquad \Rightarrow \qquad \gamma = 31.00°$$

We may also ascertain the direction of **r** by writing the expression for the unit vector **u**$_r$. Let's adapt Eq. (2.21) and write:

$$\mathbf{u_r} = \mathbf{r}/r = (-6\,\mathbf{i} + 4\,\mathbf{j} + 12\,\mathbf{k})/14 = -0.4286\,\mathbf{i} + 0.2857\,\mathbf{j} + 0.8571\,\mathbf{k}$$

Note the coefficients of the unit vectors are identical to the results for the coordinate direction cosines.

## EXAMPLE 2.18

A helicopter is attempting to tow a disabled supply truck along a road parallel to the y-axis in Fig. E2.18a. Unfortunately, the pilot is not properly aligned for the task with an initial position given by the coordinates $x = -10$, $y = -15$, and $z = 20$ feet relative to the hitch point on the truck. If a force component $F_y$ of 1250 lb is required to tow the truck, determine the force that the helicopter must apply to the towline.

Fig. E2.18

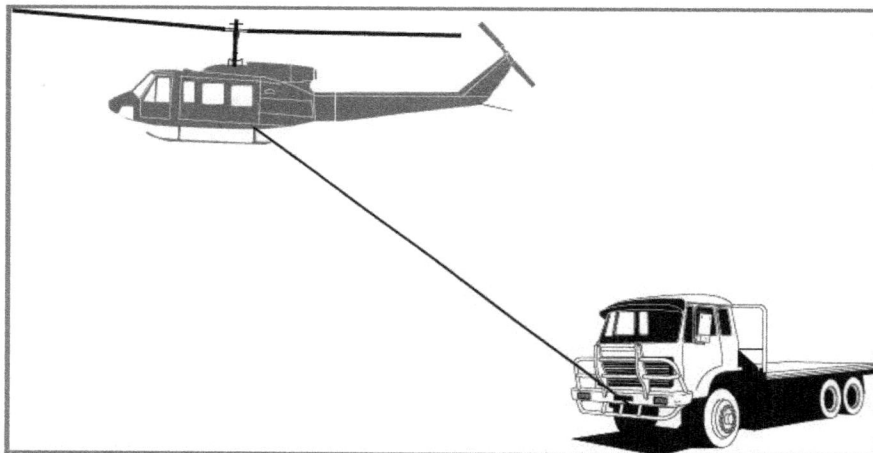

**Solution:**

Prepare a drawing of the truck tow point and the helicopter hitch point relative to a three-dimensional Cartesian coordinate system as shown in Fig. E2.18a. We will use the tow point on the truck as the origin of the Cartesian coordinates.

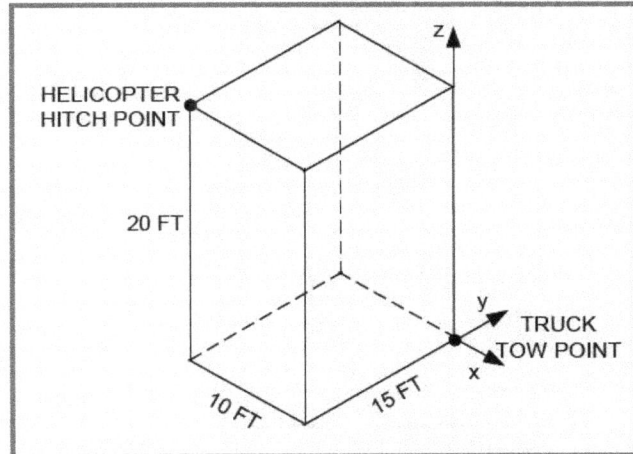

Fig. E2.18 a

Let's begin by determining the coordinate direction angles of the force that the helicopter will exert on the truck. The magnitude of position vector from the truck tow point to the helicopter hitch point is given by Eq. (2.25).

The coordinate direction angles are given by Eq. (2.24) as indicated below:

$$r = \sqrt{x^2 + y^2 + z^2} = \sqrt{(-10)^2 + (-15)^2 + (20)^2} = 26.926 \text{ ft}$$

$$\cos \alpha = \frac{x}{r} = \frac{-10}{26.926} = -0.3714 \implies \alpha = 111.8^\circ$$

$$\cos \beta = \frac{y}{r} = \frac{-15}{26.926} = -0.5571 \implies \beta = 123.9^\circ$$

$$\cos \gamma = \frac{z}{r} = \frac{20}{26.926} = 0.7428 \implies \gamma = 42.03^\circ$$

Because the force and the rope are along the same line of action, the coordinate direction angles are identical for the position vector $r$ and the force $F$. Recognizing this fact permits us to solve for the magnitude of the force necessary to tow the truck by using Eq. (2.14).

$$F = F_y / \cos \beta$$

$$F = -1250/(-0.5571) = +2244 \text{ lb}$$

The negative sign for the force component $F_y$ is used to recognize that the force is being applied in the negative y direction.

## EXAMPLE 2.19

A large pontoon is moored at a dock with a line attached to the cylinder at a mooring cleat as indicated in Fig. E2.19. If the cleat is at a position on the pontoon given by coordinates, $x = 4$ m, $y = -6$ m and $z = -3$ m, determine the tension in the mooring line if the wind and current are producing a drag force of 665 N in the x direction.

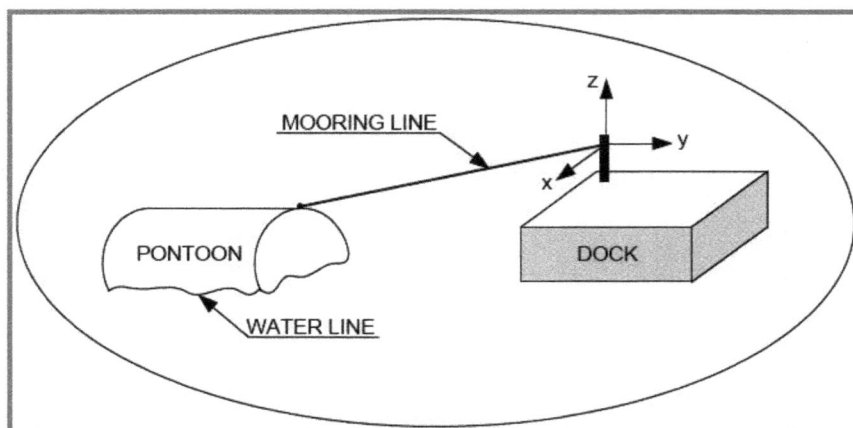

Fig. E2.19

**Solution:**

This situation is similar to the previous example where a rope or line is used to apply a force. In these situations, the position vector describing the orientation of the line and the force coincide. Since we know the coordinates of the rope, we can determine its length and orientation from Eqs. (2.24) and (2.25). The length of the position vector from the dock to the cleat is:

$$r = \sqrt{(4)^2 + (-6)^2 + (-3)^2} = 7.810 \text{ m}$$

The coordinate direction angles are given by modifying Eq. (2.24) as indicated below:

$$\cos \alpha = \frac{r_x}{r} = \frac{4}{7.810} = 0.5121 \Rightarrow \alpha = 59.20°$$

$$\cos \beta = \frac{r_y}{r} = \frac{-6}{7.810} = -0.7682 \Rightarrow \beta = 140.2°$$

$$\cos \gamma = \frac{r_z}{r} = \frac{-3}{7.810} = -0.3841 \Rightarrow \gamma = 121.6°$$

The magnitude of the force on the mooring line is determined from Eq. (2.16) as:

$$F = F_x/\cos \alpha = 665/0.5121 = 1299 \text{ N}$$

## 2.8 MOMENTS

If you have ever used a wrench or screwdriver to tighten a bolt or screw, you have generated a moment. A moment $M_o$ is produced when a force F, as shown in Fig. 2.26, is applied in such a manner that it tends to cause a body to rotate about point O. The magnitude of a moment produced by a force is dependent on the location of the point O, and is given by:

$$M_o = F d \qquad (2.29)$$

where d is the perpendicular distance from the point O to the line of action of F.

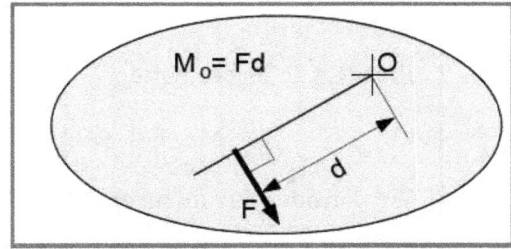

Fig. 2.26 A moment produced by a force depends on the position of the point O.

The units of a moment $M_o$ are given as N-m or ft-lb as was stated in Table 1.2. $M_o$ is a vector quantity that must be specified with both magnitude and direction. The magnitude is given by Eq. 2.29, and the direction of the vector is perpendicular to the plane in which both F and d lie. As shown in Fig. 2.27, the moment $M_o$ has a sense of direction. In this illustration, the moment $M_o$ tends to rotate the body in a **counterclockwise** direction and is **positive**.

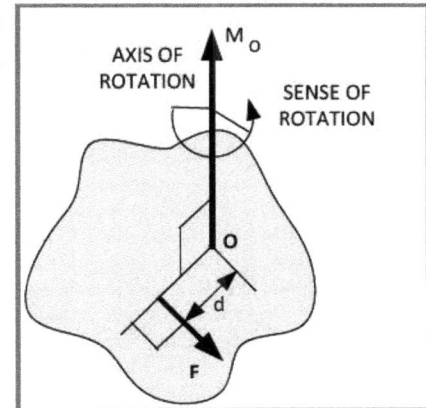

Fig. 2.27 A graphic illustration of the moment $M_o$ as a vector quantity.

To determine the sign of the moment, we use the **right hand rule**. In applying this rule, place the palm of your right hand along the axis of rotation and point your fingers in the direction of the force and rotate your hand. If the direction of the force causes you to rotate **counterclockwise**, with your thumbs pointing up from the plane in which both F and d lie, then the moment is **positive**. However, if you must rotate your hand **clockwise**, with your thumb pointing downward, the moment is **negative**.

## EXAMPLE 2.20

A hexagonal headed bolt is tightened with a wrench as shown in Fig. E2.20. A 35-lb force is applied to the handle of the wrench to produce a moment (torque). If the distance from the centerline of the bolt to the point of application of the force is d = 10 in., find the applied torque.

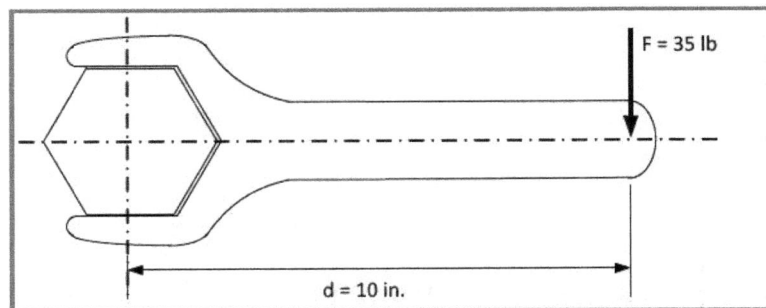

Fig. E2.20

**Solution:**

From Eq. 2.29, we write

$$M = F\,d = (35)(10) = 350 \text{ in-lb} = 29.17 \text{ ft-lb}$$

We consider this moment to be **negative** since it tends to produce a **clockwise** rotation about the head of the bolt.

## EXAMPLE 2.21

Suppose a force of 125 N was applied to the wrench as shown in Fig. E2.21. Determine the torque (moment) applied to the hex headed bolt.

Fig. E2.21

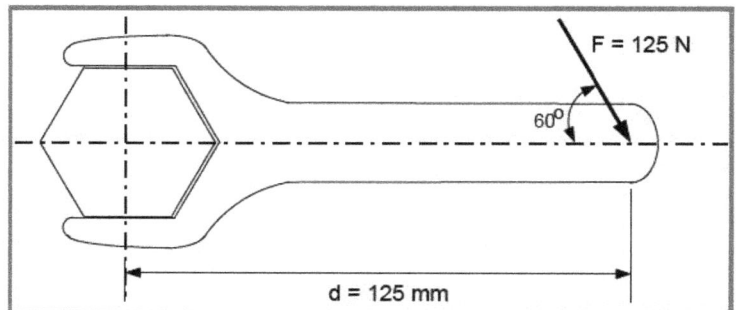

**Solution:**

Let's resolve the 125 N force into two components—one perpendicular to the axis of the wrench and the other parallel.

The component of force parallel to the axis of the wrench is obtained by:

$$F_{=} = 125 \cos (60°) = 62.5 \text{ N}$$

Note that $F_{=}$ does not produce a moment (torque) on the bolt since its moment arm $d = 0$.

The component perpendicular to the axis of the wrench is given by:

$$F_{\perp} = 125 \sin (60°) = 108.3 \text{ N}$$

To determine the moment that this component of force produces on the head of the bolt, we use Eq. 2.29 and write:

$$M = F\,d = (108.3)(0.125) = 13.53 \text{ N-m}$$

Again, this moment is **negative** because it tends to produce a **clockwise** rotation about the head of the bolt.

## 2.9 VECTOR MECHANICS

### 2.9.1 Expressing Forces and Moments as Vectors

When we encounter three-dimensional structures, the analysis becomes more complex often involving six equations of equilibrium and several free body diagrams. Visualization of the structure and force components becomes more difficult. To alleviate the complexity and reduce visualization difficulties, we often use a vector mechanics approach. When we express each force **F** or moment **M** in a complete vector representation, the equilibrium relations reduce to two vector equations:

$$\Sigma\mathbf{F} = 0 \qquad \text{and} \qquad \Sigma\mathbf{M} = 0$$

The $\Sigma$ symbol implies that we sum all of the forces and all of the moments produced by the forces acting on the structure. Hence, we may write these relations as:

$$\Sigma\mathbf{F} = \mathbf{F}_1 + \mathbf{F}_2 + \quad \ldots\ldots\ldots + \mathbf{F}_n = 0 \qquad (2.30)$$

$$\Sigma\mathbf{M} = \mathbf{M}_1 + \mathbf{M}_2 + \ldots\ldots\ldots + \mathbf{M}_n = 0 \qquad (2.31)$$

where forces 1, 2, 3 ……. n act on the structure.

We set the sums to zero assuming the body is in equilibrium. Next let's use unit vectors in expressing the force vectors $\mathbf{F}_1$, $\mathbf{F}_2$, to $\mathbf{F}_n$ as:

$$\mathbf{F}_1 = F_{1x}\,\mathbf{i} + F_{1y}\,\mathbf{j} + F_{1z}\,\mathbf{k}$$

$$\mathbf{F}_2 = F_{2x}\,\mathbf{i} + F_{2y}\,\mathbf{j} + F_{2z}\,\mathbf{k}$$

$$\mathbf{F}_n = F_{nx}\,\mathbf{i} + F_{ny}\,\mathbf{j} + F_{nz}\,\mathbf{k} \qquad (2.32)$$

where $F_{1x}$, $F_{1y}$, $F_{1z}$, etc. are the magnitudes of the Cartesian force components.

Substituting Eq. (2.32) into Eq. (2.30) yields:

$$\Sigma\mathbf{F} = (F_{1x} + F_{2x} + \ldots + F_{nx})\mathbf{i} + (F_{1y} + F_{2y} + \ldots + F_{ny})\mathbf{j} + (F_{1z} + F_{2z} + \ldots + F_{3z})\mathbf{k} = 0 \qquad (2.33)$$

Inspection of this relation shows that:

$$F_{1x} + F_{2x} + \ldots + F_{nx} = \Sigma F_x = 0$$

$$F_{1y} + F_{2y} + \ldots + F_{ny} = \Sigma F_y = 0 \qquad (2.34)$$

$$F_{1z} + F_{2z} + \ldots + F_{nz} = \Sigma F_z = 0$$

These three relations represent the equilibrium equations expressed in terms of scalar force components. We will employ these relations throughout most of this textbook. With the vector mechanics approach, they are combined into a single relation—Eq. (2.33).

## EXAMPLE 2.22

A space force **F** with a magnitude of 800 lb has coordinate direction angles of $\alpha = 45°$ and $\beta = 45°$. Write a vector equation for **F** and prepare a drawing showing the force vector in a three-dimensional Cartesian coordinate system.

**Solution:** The specification of the force **F** in the problem statement is not complete because the coordinate direction angle $\gamma$ is not given. We may determine angle $\gamma$ by recalling Eq. (2.15).

$$\cos^2 \alpha + \cos^2 \beta + \cos^2 \gamma = \cos^2 (45°) + \cos^2 (45°) + \cos^2 \gamma = 1$$

$$½ + ½ + \cos^2 \gamma = 1 \quad \Rightarrow \quad \cos^2 \gamma = 0 \quad \Rightarrow \quad \gamma = 90°$$

The Cartesian components of the force vector are given by Eq. (2.16) as:

$$F_x = F \cos \alpha \qquad\qquad F_y = F \cos \beta \qquad\qquad F_z = F \cos \gamma$$

$$F_x = 800 \cos (45°) = 565.7 \text{ lb} \qquad F_y = 800 \cos (45°) = 565.7 \text{ lb} \qquad F_z = 800 \cos (90°) = 0$$

Next, we may write the vector equation for the 800 lb force as:

$$\mathbf{F} = (565.7 \, \mathbf{i} + 565.7 \, \mathbf{j}) \text{ lb}$$

Note, the multiplier of the **k** unit vector is absent from this result because $F_z = 0$.

Finally, the force vector **F** is shown in the x – y plane in Fig. E2.22.

Fig. E2.22

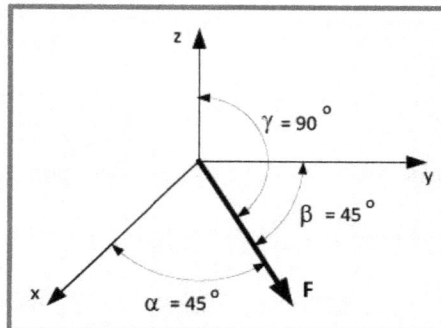

## EXAMPLE 2.23

A space force $\mathbf{F}_1$ with a magnitude of 20 kN has coordinate direction angles of $\alpha = 60°$ and $\beta = 60°$ and $\gamma = 135°$. A second space force $\mathbf{F}_2$ with a magnitude of 12 kN has coordinate direction angles of $\alpha = 90°$ and $\beta = 60°$ and $\gamma = 30°$. Write a vector equation for the summation of these two space forces.

**Solution:** The summation of $\mathbf{F}_1$ and $\mathbf{F}_2$ is given by Eq. (2.33) as:

$$\Sigma\mathbf{F} = (F_{1x} + F_{2x})\mathbf{i} + (F_{1y} + F_{2y})\mathbf{j} + (F_{1z} + F_{2z})\mathbf{k}$$

Let's first determine the Cartesian components of the space forces from Eq. (2.16).

$$F_{x1} = F_1 \cos \alpha_1 \qquad\qquad F_{y1} = F_1 \cos \beta_1 \qquad\qquad F_{z1} = F_1 \cos \gamma_1$$

$F_{x1} = 20 \cos (60°) = 10$ kN     $F_{y1} = 20 \cos (60°) = 10$ kN     $F_{z1} = 20 \cos (135°) = -14.14$ kN

$F_{x2} = F_2 \cos \alpha_2$          $F_{y2} = F_2 \cos \beta_2$          $F_z = F \cos \gamma_2$

$F_{x2} = 12 \cos (90°) = 0$ kN          $F_{y2} = 12 \cos (60°) = 6$ kN          $F_{z2} = 12 \cos (30°) = 10.39$ kN

We then sum the forces to obtain:

$$\Sigma \mathbf{F} = (10 + 0)\mathbf{i} + (10 + 6)\mathbf{j} + (-14.14 + 10.39)\mathbf{k}$$

$$\Sigma \mathbf{F} = \mathbf{F}_1 + \mathbf{F}_2 = (10\,\mathbf{i} + 16\,\mathbf{j} - 3.75\,\mathbf{k})\ \text{kN}$$

While we have summed forces in Example 2.12, the sum has not been set to zero as shown in Eq. (2.33). The sum of the forces acting on a body is zero when the body is at rest or moving at a constant velocity. In this example problem, we have not considered equilibrium because our purpose was to demonstrate the technique for adding together two vectors.

## 2.9.2 The Vector Dot Product

In several examples, we have determined components of force vectors by using Eqs. (2.8) and (2.16). These relations are based on a trigonometric approach where the vector and its components are easy to visualize. However, in some three-dimensional problems, visualization is much more difficult. In these instances, vector algebra methods for determining components of force and moment vectors are useful. To introduce those aspects of vector algebra useful in design analysis two different methods for vector multiplication—the dot product and the cross product are described. The dot product is useful in determining the component of a force along some arbitrary line or the angle between two lines. The cross product is used to determine the moment due to a vector force about some arbitrary point.

Let's define the dot product of vectors **A** and **B** as shown below:

$$\mathbf{A} \bullet \mathbf{B} = AB \cos \theta \qquad\qquad (2.35)$$

where the angle θ is defined in Fig. 2.28.

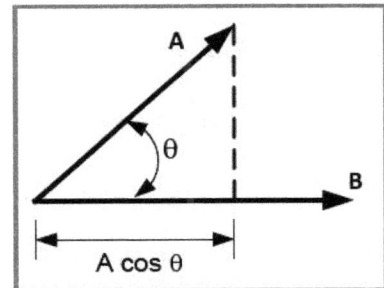

Fig. 2.28 Vectors A and B and their dot product.

In Fig. 2.28, we show the projection of the vector **A** as a dimension measured along vector **B**. The vector dot product **A** • **B** is represented by the projection (A cosθ) multiplied by the magnitude B. The dot product of two vectors is a scalar quantity with unit's dependent on the product of the units of A and B. As a scalar quantity **A** • **B** has a magnitude, but not a direction. We cannot draw an arrow to represent it, because it does not have an orientation.

Next let's consider the dot product of different pairs of unit vectors. It is easy to show that:

$$\mathbf{i} \bullet \mathbf{i} = (1)(1) \cos (0°) = 1 \quad \text{and} \quad \mathbf{i} \bullet \mathbf{j} = (1)(1) \cos (90°) = 0.$$

We list all possible combinations of the dot product of two unit vectors below:

$$i \bullet i = 1 \qquad i \bullet j = 0 \qquad i \bullet k = 0$$
$$j \bullet j = 1 \qquad j \bullet k = 0 \qquad j \bullet i = 0 \qquad (2.36)$$
$$k \bullet k = 1 \qquad k \bullet i = 0 \qquad k \bullet j = 0$$

Suppose that we represent two independent space vectors **A** and **B** each with three components:

$$\mathbf{A} = A_x\, \mathbf{i} + A_y\, \mathbf{j} + A_z\, \mathbf{k}$$

$$\mathbf{B} = B_x\, \mathbf{i} + B_y\, \mathbf{j} + B_z\, \mathbf{k}$$

Next, let's determine the dot product of these two space vectors:

$$\mathbf{A} \bullet \mathbf{B} = (A_x\, \mathbf{i} + A_y\, \mathbf{j} + A_z\, \mathbf{k}) \bullet (B_x\, \mathbf{i} + B_y\, \mathbf{j} + B_z\, \mathbf{k})$$

With the dot product relations for the unit vectors given in Eq. (2.36), it is evident that:

$$\mathbf{A} \bullet \mathbf{B} = A_x B_x + A_y B_y + A_z B_z \qquad (2.37)$$

As expected, the result in Eq. (2.37) is a scalar quantity. When we have two arbitrary vectors represented in vector format, the dot product is the sum of the product of their Cartesian components.

The dot product is a valuable analysis tool that may be used in two different applications:

1. To determine the angle $\theta$ between two space vectors. From Eq. (2.35) it is clear that:

$$\theta = \cos^{-1}\left[(\mathbf{A} \bullet \mathbf{B})/(AB)\right] \qquad (2.38)$$

2. To determine the component of a vector quantity along a line. If we define the direction of this line with a unit vector **u**, then the component of vector **A** along the line defined by **u** is a scalar quantity given by:

$$A_u = \mathbf{A} \bullet \mathbf{u} \qquad (2.39)$$

Let's consider two examples to demonstrate the techniques involved in utilizing these tools.

## EXAMPLE 2.24

Suppose we have two space forces specified in vector format as:

$$\mathbf{F_1} = (12\, \mathbf{i} + 8\, \mathbf{j} - 3\, \mathbf{k})\ \text{lb} \qquad \text{and} \qquad \mathbf{F_2} = (-6\, \mathbf{i} + 4\, \mathbf{j} - 2\, \mathbf{k})\ \text{lb}$$

Determine the angle between the two forces.

**Solution:**

First determine the magnitude of each force from Eq. (2.17)

$$F_1 = [(12)^2 + (8)^2 + (-3)^2]^{1/2} = 14.73\ \text{lb}$$

$$F_2 = [(-6)^2 + (4)^2 + (-2)^2]^{1/2} = 7.483\ \text{lb}$$

Next, use Eq. (2.37) to solve for $\mathbf{F_1} \bullet \mathbf{F_2}$ as:

$$\mathbf{F_1} \bullet \mathbf{F_2} = F_{1x} F_{2x} + F_{1y} F_{2y} + F_{1z} F_{2z} = (12)(-6) + (8)(4) + (-3)(-2) = -34 \text{ lb}^2$$

Finally, apply Eq. (2.38) to determine the angle $\theta$ between the two forces.

$$\theta = \cos^{-1}[(\mathbf{F_1} \bullet \mathbf{F_2})/(F_1 F_2)] = \cos^{-1}\{-34/[(14.73)(7.483)]\} = \cos^{-1}\{-0.3085\} = 108.0°$$

It is a very difficult exercise in solid geometry to determine the angle between two arbitrary lines. With the use of the dot product as a vector tool the task becomes much easier.

## EXAMPLE 2.25

Suppose we have the same two space forces as in Example 2.24 specified in vector format as:

$$\mathbf{F_1} = (12\,\mathbf{i} + 8\,\mathbf{j} - 3\,\mathbf{k}) \text{ lb} \qquad \text{and} \qquad \mathbf{F_2} = (-6\,\mathbf{i} + 4\,\mathbf{j} - 2\,\mathbf{k}) \text{ lb}$$

Determine the magnitude of a single force component produced by these two forces if it is directed along a line of action, which lies in the x – y plane making an angle of 45° with the x-axis.

**Solution:**

It is evident that we will employ Eq. (2.39) to determine the magnitude of this force component. The difficulty we first encounter is to write an expression for the unit vector giving the direction of the line in the x – y plane. It is easy to write a vector describing a line in the x – y plane with a 45° orientation as:

$$\mathbf{v} = \mathbf{i} + \mathbf{j} \tag{a}$$

However, the magnitude of this vector is $(2)^{1/2}$; hence, it is not a unit vector. To convert vector $\mathbf{v}$ to a unit vector $\mathbf{u}$, we must divide $\mathbf{v}$ by its magnitude v as shown below:

$$\mathbf{u} = \mathbf{v}/v \tag{2.40}$$

Applying Eq. (2.40) to Eq. (a) yields:

$$\mathbf{u} = (\mathbf{i} + \mathbf{j})/(2)^{1/2} = 0.7071\,\mathbf{i} + 0.7071\,\mathbf{j} \tag{b}$$

Substituting Eq. (b) into Eq. (2.39) gives:

$$F_{1u} = \mathbf{F_1} \bullet \mathbf{u} \qquad \text{and} \qquad F_{2u} = \mathbf{F_2} \bullet \mathbf{u}$$

$$F_{1u} = \mathbf{F_1} \bullet \mathbf{u} = (12\,\mathbf{i} + 8\,\mathbf{j} - 3\,\mathbf{k}) \bullet (0.7071\,\mathbf{i} + 0.7071\,\mathbf{j}) = 14.14 \text{ lb}$$

$$F_{2u} = \mathbf{F_2} \bullet \mathbf{u} = (-6\,\mathbf{i} + 4\,\mathbf{j} - 2\,\mathbf{k}) \bullet (0.7071\,\mathbf{i} + 0.7071\,\mathbf{j}) = -1.415 \text{ lb}$$

$$F_u = F_{1u} + F_{2u} = 14.14 - 1.415 = 12.73 \text{ lb}$$

Projecting forces and moments, both vector quantities, onto structural components that are oriented in arbitrary directions by using the properties of a dot product is a useful technique.

### 2.9.3 The Vector Cross Product

Let's define the mathematical form of the vector cross product as:

$$\mathbf{C} = \mathbf{A} \times \mathbf{B} = (AB \sin \theta)\mathbf{u} \qquad (2.41)$$

The cross product of vectors **A** and **B** yields a vector quantity **C**. Multiplying the product of the magnitudes of A and B by the sin $\theta$ gives the magnitude of C. The direction of vector **C** is perpendicular to the plane in which **A** and **B** lie. The sense of **C** is determined by the right hand rule—when curling the fingers of the right hand from **A** toward (cross) **B** with the normal to the plane $\pi$ along the palm of your hand, the thumb points in the direction of vector **C**. A graphic representation of **C**, the cross product of vectors **A** and **B** as defined in Eq. (2.41), is presented in Fig. 2.29.

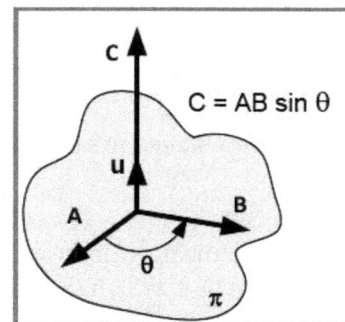

Fig. 2.29 Vector **C** is the cross product **A** × **B**.

Moments are a product of a force times a distance. In our previous discussion, we were always careful to define the distance as the perpendicular distance from the line of action of the force to the point about which the moments were determined. The definition of the moment does not change, but with the vector mechanics approach, we find it helpful to define the moment in terms of a vector cross product as:

$$\mathbf{M_O} = \mathbf{r} \times \mathbf{F} \qquad (2.42)$$

where **r** is the position vector that locates point Q relative to the origin of a Cartesian coordinate system as shown in Fig. 2.30.

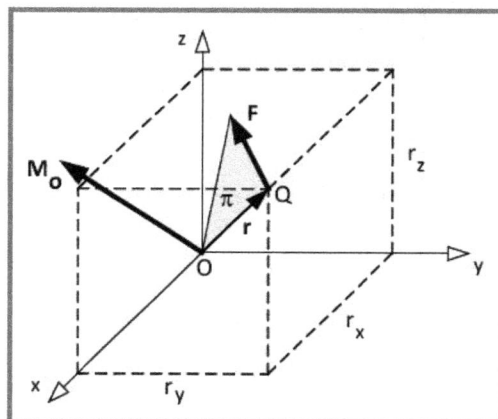

Fig. 2.30 A drawing showing the geometric representation of $\mathbf{M_O} = \mathbf{r} \times \mathbf{F}$.

The moment $\mathbf{M_O}$, relative to the origin O of the coordinate system, is due to the force **F** that is applied at point Q. The location of point Q relative to the origin O is established with a position vector **r**. Vectors **F** and **r** form a plane $\pi$ that is shown in Fig. 2.30; the moment vector $\mathbf{M_O}$ is perpendicular to this plane.

The position vector is determined by the location of point Q as:

$$\mathbf{r} = r_x\,\mathbf{i} + r_y\,\mathbf{j} + r_z\,\mathbf{k} \qquad (2.43)$$

where $r_x$, $r_y$ and $r_z$ are the coordinates of point Q.

Using Eq. (2.18) we write an expression for the force vector as:

$$\mathbf{F} = F_x\,\mathbf{i} + F_y\,\mathbf{j} + F_z\,\mathbf{k}$$

From Eq. (2.41) it is easy to show that the cross vector products of the unit vectors are given by:

$$\begin{aligned}
\mathbf{i}\times\mathbf{i} &= 0, & \mathbf{j}\times\mathbf{j} &= 0, & \mathbf{k}\times\mathbf{k} &= 0 \\
\mathbf{i}\times\mathbf{j} &= \mathbf{k}, & \mathbf{j}\times\mathbf{k} &= \mathbf{i}, & \mathbf{k}\times\mathbf{i} &= \mathbf{j} \\
\mathbf{j}\times\mathbf{i} &= -\mathbf{k} & \mathbf{k}\times\mathbf{j} &= -\mathbf{i} & \mathbf{i}\times\mathbf{k} &= -\mathbf{j}
\end{aligned} \qquad (2.44)$$

Combining Eqs. (2.42), (2.43), (2.18) and (2.45), we obtain:

$$\mathbf{M_O} = \mathbf{r}\times\mathbf{F} = (r_y F_z - r_z F_y)\mathbf{i} + (r_z F_x - r_x F_z)\mathbf{j} + (r_x F_y - r_y F_x)\mathbf{k} \qquad (2.45)$$

We may also express the moment in terms of components relative to the coordinate axes as:

$$\mathbf{M_O} = \mathbf{r}\times\mathbf{F} = M_x\,\mathbf{i} + M_y\,\mathbf{j} + M_z\,\mathbf{k} \qquad (2.46)$$

Comparing the results of Eqs. (2.45) with those of Eq. (2.46) yields the equations for the components of the moments about the coordinate axes as:

$$M_x = (r_y F_z - r_z F_y)$$

$$M_y = (r_z F_x - r_x F_z)$$

$$M_z = (r_x F_y - r_y F_x) \qquad (2.47)$$

When employing the vector cross product $\mathbf{r}\times\mathbf{F}$ to determine the moment $\mathbf{M_O}$, we simultaneously obtain the relations for the moments $M_x$, $M_y$ and $M_z$.

In writing Eq. (2.45), we often employ the determinant given by:

$$\mathbf{M_O} = \begin{vmatrix} \mathbf{i} & \mathbf{j} & \mathbf{k} \\ r_x & r_y & r_z \\ F_x & F_y & F_z \end{vmatrix} \qquad (2.48)$$

Let's apply these results to three example problems to demonstrate the use of the vector cross product in the determination of moments due to forces.

## EXAMPLE 2.26

A force vector $\mathbf{F}$ with the following components ($F_x = 3000$ N, $F_y = 5000$ N, and $F_z = 4500$ N) is applied to a structure at point Q. If point Q is given by the coordinates x = 6 m, y = 5 m, and z = 12 m, determine the moment about the origin O of the coordinate system.

**Solution:**

The moment is given by Eq. (2.42) as $\mathbf{r} \times \mathbf{F}$. To execute this simple cross product, we write both $\mathbf{r}$ and $\mathbf{F}$ in vector format as:

$$\mathbf{r} = r_x \mathbf{i} + r_y \mathbf{j} + r_z \mathbf{k} = (6\,\mathbf{i} + 5\,\mathbf{j} + 12\,\mathbf{k})\ \text{m}$$

$$\mathbf{F} = (3000\,\mathbf{i} + 5000\,\mathbf{j} + 4500\,\mathbf{k})\ \text{N}$$

To execute the cross product, let's employ Eq. (2.45) to obtain:

$$\mathbf{M_O} = \mathbf{r} \times \mathbf{F} = (r_y F_z - r_z F_y)\,\mathbf{i} + (r_z F_x - r_x F_z)\,\mathbf{j} + (r_x F_y - r_y F_x)\,\mathbf{k}$$

$$\mathbf{M_O} = [(5)(4500) - (12)(5000)]\,\mathbf{i} + [(12)(3000) - (6)(4500)]\,\mathbf{j}$$

$$+ [(6)(5000) - (5)(3000)]\,\mathbf{k}$$

$$\mathbf{M_O} = (-37{,}500\,\mathbf{i} + 9{,}000\,\mathbf{j} + 15{,}000\,\mathbf{k})\ \text{N-m}$$

The magnitude of the moment is given by the square root of the sum of the squares of the moment components as:

$$M_O = [M_x^2 + M_y^2 + M_z^2]^{1/2} \tag{2.49}$$

$$M_O = [(-37.5)^2 + (9)^2 + (15)^2]^{1/2} = 41.38\ \text{kN-m}$$

## EXAMPLE 2.27

If a structure is loaded at point Q with a force given by:

$$\mathbf{F} = (150\,\mathbf{i} + 240\,\mathbf{j} + 80\,\mathbf{k})\ \text{lb}$$

Determine the moment $\mathbf{M_O}$ using the determinant format presented in Eq. (2.48). Note that point Q is located at the position $x = 3$, $y = 6$ and $z = 5$ ft.

**Solution:** Let's first write the relation for the position vector $\mathbf{r}$ as:

$$\mathbf{r} = r_x \mathbf{i} + r_y \mathbf{j} + r_z \mathbf{k} = (3\,\mathbf{i} + 6\,\mathbf{j} + 5\,\mathbf{k})\ \text{ft} \tag{a}$$

Recall Eq. (2.48) as:

$$\mathbf{M_O} = \begin{vmatrix} \mathbf{i} & \mathbf{j} & \mathbf{k} \\ r_x & r_y & r_z \\ F_x & F_y & F_z \end{vmatrix}$$

Substituting into the matrix, the coefficients from the relations for the position and force vectors gives:

$$\mathbf{M_O} = \begin{vmatrix} \mathbf{i} & \mathbf{j} & \mathbf{k} \\ 3 & 6 & 5 \\ 150 & 240 & 80 \end{vmatrix}$$

Solving the determinant gives:

$$\mathbf{M_O} = [(48 - 120)\mathbf{i} + (75 - 24)\mathbf{j} + (72 - 90)\mathbf{k}](10)$$

$$\mathbf{M_O} = (-720\,\mathbf{i} + 510\,\mathbf{j} - 180\,\mathbf{k}) \text{ ft-lb}$$

## EXAMPLE 2.28

The boom of a crane extends from its base at point O to its tip at point Q as indicated in Fig. E2.28. A three-dimensional coordinate system has been established in this illustration with the point O at the origin and the point Q defined with coordinates (1,1,5). A force with a magnitude of 12 kN is applied by the boom onto a cable that extends from the tip of the boom to point P. Point P is located on the x – y plane with coordinates (3,6, 0). If the coordinates are expressed in meters, determine the following quantities:

(a) The vector representation of the force **F**.
(b) The components $F_x$, $F_y$ and $F_z$.
(c) The moment $\mathbf{M_O}$.
(d) The unit vector describing the direction of $\mathbf{M_O}$.
(e) The moments about the coordinate axes—$M_x$, $M_y$ and $M_z$.
(f) The magnitude of the moment.
(g) The angle between the force and the boom OQ.
(h) The projection of F along the boom OQ.

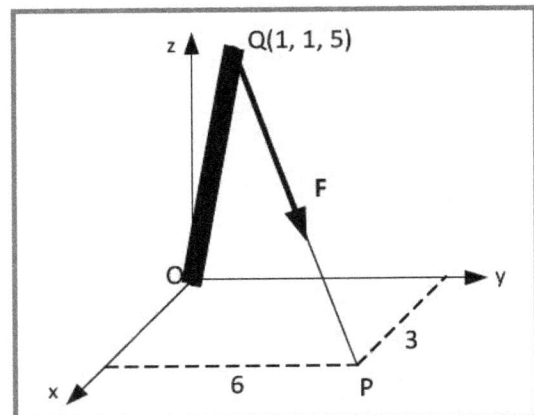

Fig. E2.28

**Solution:**

Let's begin by determining the unit vector that describes the direction of the applied force. Since the force is directed along the cable that originates at point Q and extends through point P, we know its orientation from the coordinates of these two points. The unit vector $\mathbf{u_F}$ is given by:

$$\mathbf{u_F} = (1/u_F)[(x_P - x_Q)\mathbf{i} + (y_P - y_Q)\mathbf{j} + (z_P - z_Q)\mathbf{k}] \qquad (2.50)$$

where u is the magnitude of the length of the line from Q to P given by:

$$u_F = [(x_P - x_Q)^2 + (y_P - y_Q)^2 + (z_P - z_Q)^2]^{1/2} \qquad (2.51)$$

Substituting the coordinates of points P and Q into Eqs. (2.50) and (2.51) gives:

$$u_F = \frac{(3-1)i + (6-1)j + (0-5)k}{\left[(3-1)^2 + (6-1)^2 + (0-5)^2\right]^{1/2}}$$

(a)

$$u_F = 0.2722i + 0.6804j - 0.6804k$$

The vector representation of the force F is then given by:

$$\mathbf{F} = F\,\mathbf{u_F} = (3.266\,i + 8.165\,j - 8.165\,k)\ kN \qquad (b)$$

The components of the force vector are:

$$F_x = 3.266\ kN; \qquad F_y = 8.165\ kN; \qquad F_z = -8.165\ kN \qquad (c)$$

From the coordinates of point Q, it is clear that the position vector **r** is given by:

$$\mathbf{r} = i + j + 5\,k \qquad (d)$$

The moment is determined from Eq. (2.48) as:

$$\mathbf{M_O} = \mathbf{r} \times \mathbf{F} = \begin{vmatrix} i & j & k \\ 1 & 1 & 5 \\ 3.266 & 8.165 & -8.165 \end{vmatrix}$$

Evaluating this determinant yields:

$$\mathbf{M_O} = [(-8.165)(1) - (8.165)(5)]i + [(5)(3.266) + (8.165)(1)]j \\ + [(8.165)(1) - (3.266)(1)]k$$

$$\mathbf{M_O} = (-49.0\,i + 24.5\,j + 4.90\,k)\ kN\text{-}m \qquad (e)$$

The components of this vector relative to the Cartesian coordinates are:

$$M_x = -49.0\ kN\text{-}m; \qquad M_y = 24.5\ kN\text{-}m \qquad M_z = 4.90\ kN\text{-}m \qquad (f)$$

The magnitude of the moment $M_O$ is determined from Eq. (2.49) as:

$$M_O = [(-49.0)^2 + (24.5)^2 + (4.90)^2]^{1/2} = 55.0\ kN\text{-}m \qquad (g)$$

The unit vector describing the direction of the moment vector $M_O$ is:

$$\mathbf{u_{Mo}} = (1/M_O)[\mathbf{M_O}] = (1/55.0)[-49.0\,i + 24.5\,j + 4.90\,k]$$

$$\mathbf{u}_{Mo} = -0.891\ \mathbf{i} + 0.445\ \mathbf{j} + 0.089\ \mathbf{k} \qquad \text{(h)}$$

Let's use Eq. (2.38) to determine the angle θ between the force vector **F** and a vector **QO** that extends from point Q to the origin O.

$$\theta = \cos^{-1}[\mathbf{F} \bullet \mathbf{QO}/(F)(QO)] \qquad \text{(i)}$$

Note that **QO** = – **r**; hence, we use Eqs. (b) and (d) to express Eq. (i) as:

$$\theta = \cos^{-1}\left[\frac{(3.266\mathbf{i} + 8.165\mathbf{j} - 8.165\mathbf{k}) \bullet (-\mathbf{i} - \mathbf{j} - 5\mathbf{k})}{(12)(5.196)}\right] \qquad \text{(j)}$$

$$\theta = \cos^{-1}[0.4714] = 61.87^\circ$$

The projection of the force F along the boom is given by:

$$F_{QO} = F \cos \theta = 12 \cos (61.87^\circ) = 5.657 \text{ kN} \qquad \text{(k)}$$

Another approach to determine $F_{QO}$ involves employing Eq. (2.39). Then we write:

$$F_{QO} = \mathbf{F} \bullet \mathbf{u}_{QO} = \mathbf{F} \bullet (-\mathbf{r}/\mathrm{r}) \qquad \text{(l)}$$

You may wish to verify that Eq. (j) gives the same result as Eq. (k).

## 2.10 SUMMARY

Both external and internal forces have been described. External forces are applied to the structure either by loading or by reactions at the supports. Internal forces develop within structural members due to the action of the external forces. Section cuts on the structural member are made to visualize the stresses that generate the internal forces.

Forces are vector quantities that require the specification of both magnitude and direction for a complete description. We represent forces with arrows where its length is proportional to the magnitude and its orientation relative to a suitable coordinate system gives its direction. The line of action of a force is collinear with the force vector. It is possible to slide a force vector along the line of action to a new position without affecting equilibrium of a body.

The process of vector addition and subtraction has been described. We showed that the vector sum $S_v = \mathbf{A} + \mathbf{B} = \mathbf{B} + \mathbf{A}$ could be represented by a triangle with sides of length A, B and Sv. The sine and cosine laws are employed to determine the magnitude and direction in vector addition or subtraction.

Vector decomposition, the resolution of a single force into two components, was described. Resolution of a force into its two Cartesian components is of particular significance. The equations involved are summarized below.

Classifications of vector systems were discussed with definitions and examples of concurrent and coplanar systems. For concurrent systems, where all of the forces are applied at a common point, it is important to recognize that the equilibrium equation $\Sigma \mathbf{M} = 0$ is satisfied automatically.

A brief discussion was included on moments. Moments, like forces, are vector quantities with both a magnitude and direction. Moments tend to cause bodies upon which they act to rotate. The direction of a moment is perpendicular to the plane in which the force and moment arm lie. Moments are considered **positive** when they tend to produce a **counterclockwise** rotation of the body upon which they act.

Space forces were described with force components in three directions (x, y and z). Coordinate direction angles were defined to provide the orientation of the force vectors in the three-dimensional coordinate system. Unit vectors **i**, **j** and **k** directed along the Cartesian axes were defined. It was shown that the direction of space forces could be expressed in terms of a unit vector $\mathbf{u_F}$. A position vector **r** was introduced to locate the point of application of a force in space.

Finally, vector mechanics was introduced and the three dimensional vector representations of forces, moments and unit vectors were given. These representations permit us to show separately the direction and magnitude of a vector quantity. The vector dot product was described together with two useful applications in analysis of forces acting on structures. The vector cross product was also defined. It is used extensively in determining the moments produced by forces acting on three-dimensional structures. Examples demonstrating the application of the vector dot and cross products were provided.

Key relations you will use on many occasions in the analysis of structures that were introduced in this Chapter are summarized below.

$$P_{int} = \int \sigma \, dA \qquad (2.1)$$

$$\frac{A}{\sin a} = \frac{B}{\sin b} = \frac{C}{\sin c} \qquad (2.4)$$

$$C = \sqrt{A^2 + B^2 - 2AB \cos (c)} \qquad (2.5)$$

For two dimensional forces:

$$\mathbf{F} = \mathbf{F_x} + \mathbf{F_y} \qquad (2.7)$$

$$F_x = F \cos \theta$$
$$F_y = F \sin \theta \qquad (2.8)$$

$$F = [F_x^2 + F_y^2]^{1/2} \qquad (2.9)$$

$$M_o = (F)(d) \qquad (2.29)$$

We indicate the direction of the space forces with coordinate direction angles defined by:

$$\cos \alpha = \frac{F_x}{F} \qquad \cos \beta = \frac{F_y}{F} \qquad \cos \gamma = \frac{F_z}{F} \qquad (2.14)$$

Alternatively the direction may be specified by unit vectors directed along the line of action of the space force.

$$\mathbf{u_F} = \mathbf{F}/F \qquad (2.19)$$

For space forces (three-dimensional) and moments it is often easier to visualize the vectors and more efficient to perform the analysis by employing vector mechanics. The relations frequently used in for three-dimensional analyses include vector representation of forces, moments and position. Also, extensive utilization of the dot and cross products facilitates the computation required.

$$\mathbf{F} = F_x \mathbf{i} + F_y \mathbf{j} + F_z \mathbf{k} \qquad (2.18)$$

$$\mathbf{A} \bullet \mathbf{B} = (A)(B) \cos \theta \qquad (2.35)$$

$$\mathbf{A} \bullet \mathbf{B} = A_x B_x + A_y B_y + A_z B_z \qquad (2.37)$$

$$\theta = \cos^{-1} [(\mathbf{A} \bullet \mathbf{B})/(A)(B)] \qquad (2.38)$$

$$A_u = \mathbf{A} \bullet \mathbf{u} \qquad (2.39)$$

$$\mathbf{C} = \mathbf{A} \times \mathbf{B} = [(A)(B) \sin \theta]\mathbf{u} \qquad (2.41)$$

$$\mathbf{M_O} = \mathbf{r} \times \mathbf{F} \qquad (2.42)$$

$$\mathbf{r} = r_x \mathbf{i} + r_y \mathbf{j} + r_z \mathbf{k} \qquad (2.43)$$

$$\mathbf{M_O} = \mathbf{r} \times \mathbf{F} = (r_y F_z - r_z F_y)\mathbf{i} + (r_z F_x - r_x F_z)\mathbf{j} + (r_x F_y - r_y F_x)\mathbf{k} \qquad (2.45)$$

$$\mathbf{M_O} = \mathbf{r} \times \mathbf{F} = M_x \mathbf{i} + M_y \mathbf{j} + M_z \mathbf{k} \qquad (2.46)$$

$$\mathbf{M_O} = \begin{vmatrix} \mathbf{i} & \mathbf{j} & \mathbf{k} \\ r_x & r_y & r_z \\ F_x & F_y & F_z \end{vmatrix} \qquad (2.48)$$

## PROBLEMS

2.1 Prepare a free body diagram (FBD) showing the following loads applied to a beam:

    (a) concentrated load            (c) non-uniform distributed load
    (b) uniformly distributed load      (d) two reaction loads

2.2 Suppose you are to analyze the rope used in a tug of war. Make a section cut of the rope at a location between the two teams and draw a diagram representing the stresses acting on the area exposed by the section cut. Also draw a diagram representing the internal force acting on the area exposed by the section cut.

2.3    Show a graphic representation of a force vector with a magnitude (F) and an orientation (θ) relative to the positive x-axis for:

(a)  F = 7,500 lb and θ = 105°          (c)  F = 11.2 kN and θ = 48°
(b)  F = 1,400 N and θ = 210°           (d)  F = 1.620 kip and θ = 155°

2.4    Show that the body, illustrated in figures (a) and (b) below, is in equilibrium before and after sliding the 4,000 lb force along its line of action.

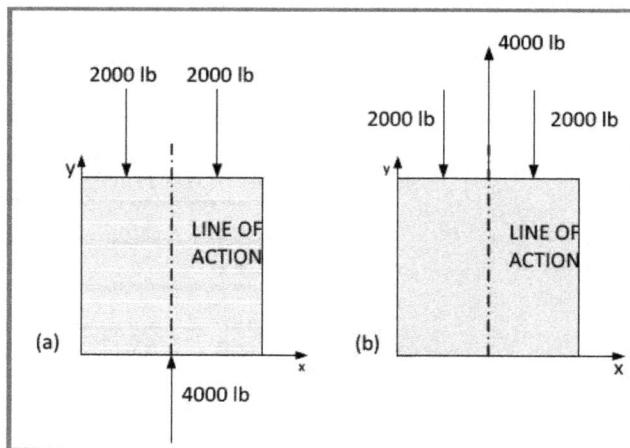

2.5    Draw a disk shaped body and apply a vertical force directed downward at its center. Also show a reactive force at its point of contact with a horizontal support. (a) Describe the condition for equilibrium of the disk. (b) Slide the applied force along its line of action to another location and describe the "new" condition for equilibrium.

2.6    Prepare a drawing showing the reactive forces on the short beam illustrated in the figure to the right.

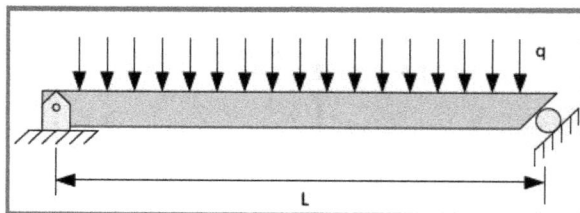

]
2.7    Prepare three drawings showing the reactive forces on each of the three cylinders in the stack shown in the figure to the right. Note that the cylinders are of the same diameter, and they are bonded together at the three contact points.

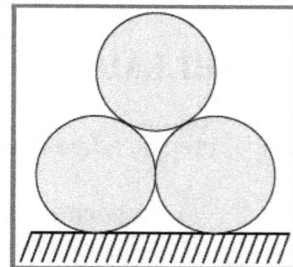

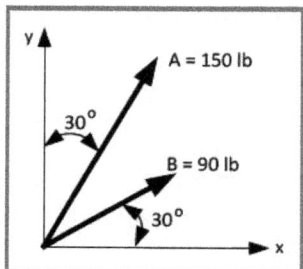

2.8    For the vectors **A** and **B**, illustrated in the figure shown to the left, determine the magnitude and direction of the following resultant vectors:        $S_v = A + B$

2.9   For the vectors **A** and **B,** illustrated in the figure shown to the right, determine the magnitude and direction of the following resultant vectors:

(a)  $S_v = A + B$
(b)  $D_v = A - B$
(c)  $D_v = B - A$.

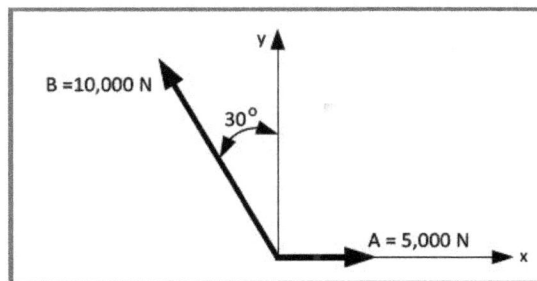

2.10  For the vectors **A** and **B,** illustrated in the figure shown to the left, determine the magnitude and direction of the following resultant vectors:

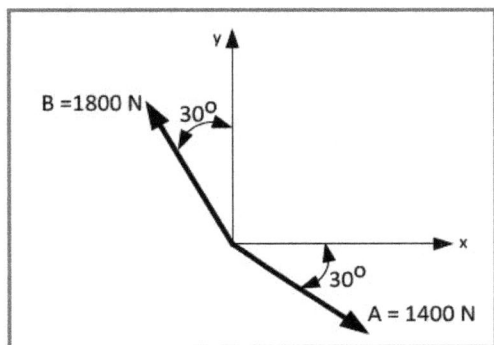

(a)  $S_v = A + B$
(b)  $D_v = A - B$
(c)  $D_v = B - A$.

2.11  Prepare a drawing similar to the one shown below illustrating the vector subtraction $D_v = B - A$.

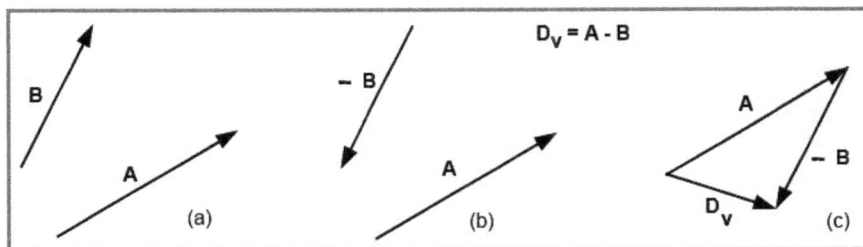

2.12  From the figure shown to the right, determine the Cartesian components of the force vectors: (a) **A** and (b) **B**.

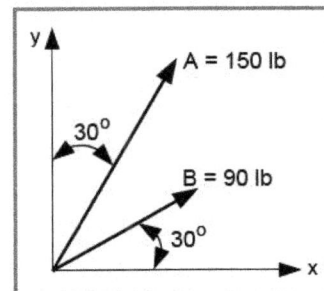

2.13    From the figure shown to the left, determine the Cartesian components of the force vectors: (a) **A** and (b) **B**.

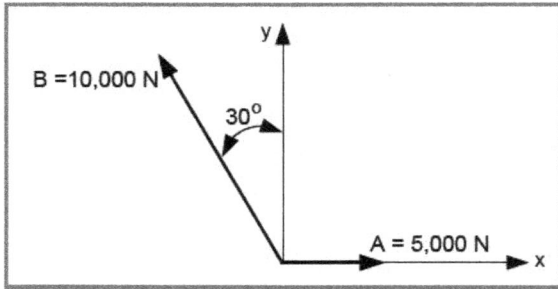

2.14    From the figure shown to the right, determine the Cartesian components of the force vectors: (a) **A** and (b) **B**.

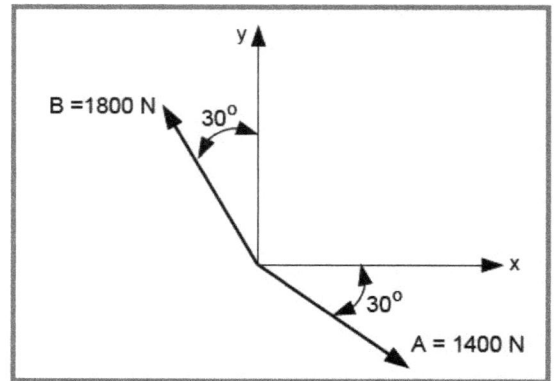

2.15    Using the Cartesian components of vectors **A** and **B**, in the figure shown to the left, determine the magnitude and direction of:

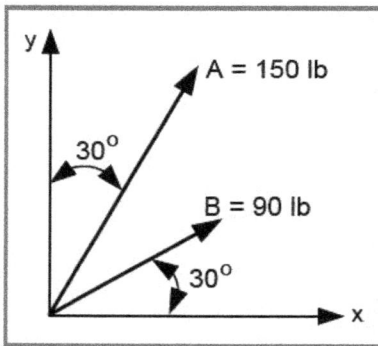

(a)  $S_v = A + B$
(b)  $D_v = A - B$
(c)  $D_v = B - A$.

2.16  Using the Cartesian components of vectors **A** and **B**, in the figure shown to the right, determine the magnitude and direction of:

(a)  $S_v = A + B$
(b)  $D_v = A - B$
(c)  $D_v = B - A$.

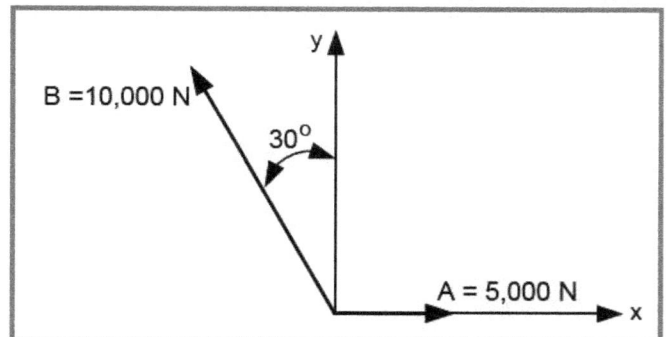

2.17    Using the Cartesian components of vectors **A** and **B**, in the figure shown to the left, determine the magnitude and direction of:

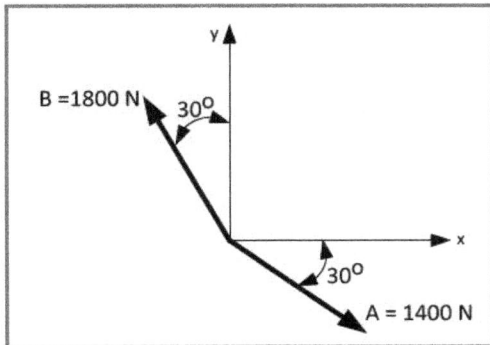

(a)  $S_v = A + B$
(b)  $D_v = A - B$
(c)  $D_v = B - A$.

2.18    Determine the force components along the x and y' axes due to the force of 18,440 lbs that is pulling on the eyebolt shown in the figure to the right.

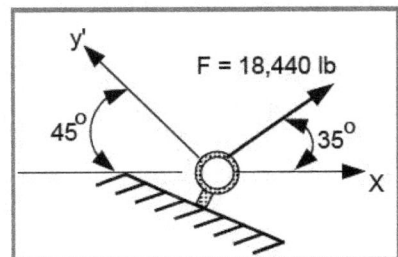

2.19    A linkage, shown in the figure to the left, is subjected to a force of 2,150 N.   Determine the components of the force along the axes of links AB and AC.

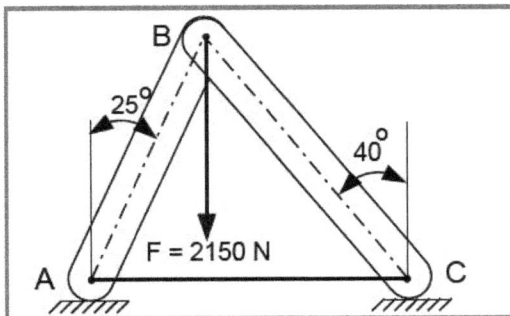

2.20    Determine the angle θ so that the link AB, shown in the figure to the right, has a component of force of 1,250 lb along its axis.  The angle φ is fixed at 6.4 degrees.

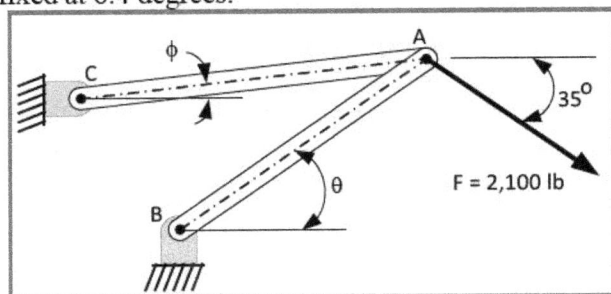

2.21 Determine the vector sum $S_v$ of the three vectors shown in the figure to the right. Specify its direction relative to the x-axis.

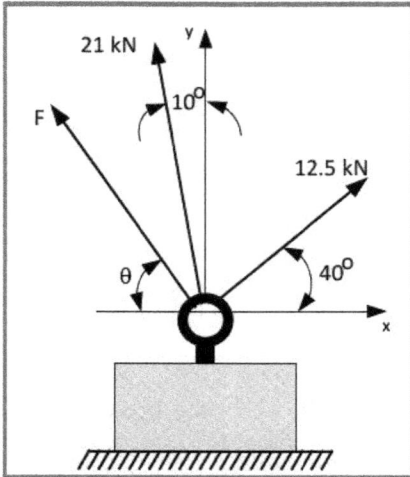

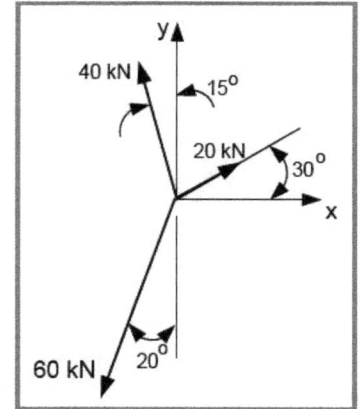

21 kN

F

12.5 kN

$10^o$

θ

$40^o$

2.22 Determine the magnitude and the direction of the force **F**, shown in the figure to the left if the box is to be lifted from the floor without swinging. The weight of the box is 35 kN.

2.23 If θ is fixed at 55°, in the figure above, calculate the maximum weight of the box that can be lifted without it swinging when it clears the floor.

2.24 The gusset plate, illustrated in the figure below, is subjected to four forces due to the attached uniaxial structural members. If the vector sum of the four forces is zero, determine the forces $F_1$ and $F_2$.

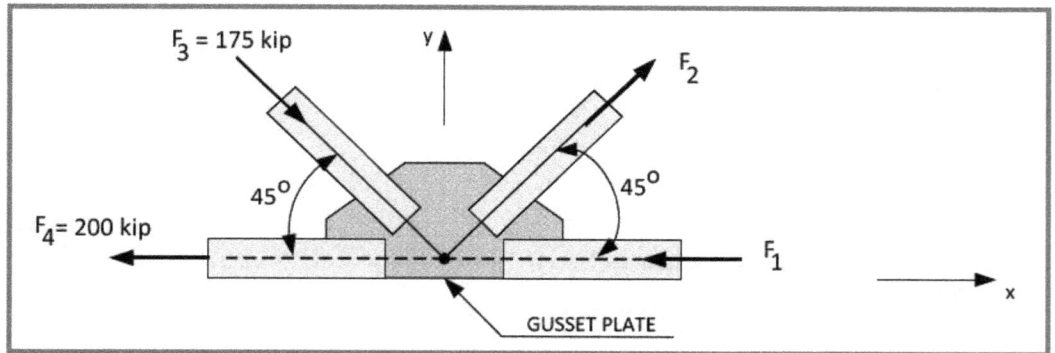

$F_3$ = 175 kip

$F_2$

$45^o$        $45^o$

$F_4$ = 200 kip

$F_1$

GUSSET PLATE

2.25 Three forces act on the bracket, illustrated in the figure to the right. Determine $F_1$ and θ if the magnitude of one of the Cartesian components of the vector sum $S_{vx}$ = 16,200 lb. Note, $S_{vy}$ = 0.

2.26 If the vector sum $S_v$ of the three forces is 17,500 lb, and is oriented at 10° relative to the x-axis, determine $F_1$ and θ. Reference the figure to the right.

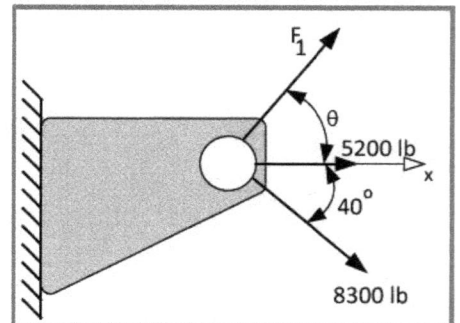

$F_1$

θ

5200 lb

$40^o$

8300 lb

2.27   Define coplanar forces, and show a sketch illustrating four coplanar force vectors.

2.28   Define concurrent forces, and show a sketch illustrating four concurrent force vectors.

2.29   Prepare a sketch of three forces that are:

(a) coplanar, concurrent        (b) non-coplanar, concurrent     (c) coplanar, non-concurrent

2.30   Suppose you exert a force of 125 N on a single ended lug wrench with a 320-mm long handle to loosen a wheel bolt.  What moment (torque) are you applying to the bolt?

2.31   You exert a force of 150 N with each hand on the handles of a double-ended lug wrench to loosen a wheel bolt.  If each handle is 225 mm long, what moment (torque) are you applying to the bolt?

2.32   Determine the moment about point O due to the force shown in the figure to the right.

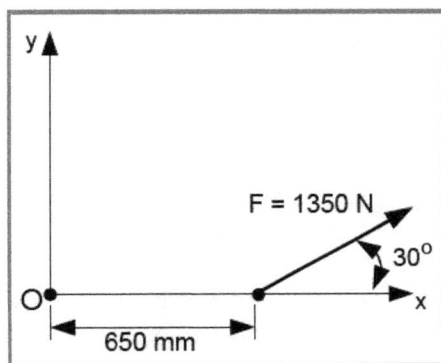

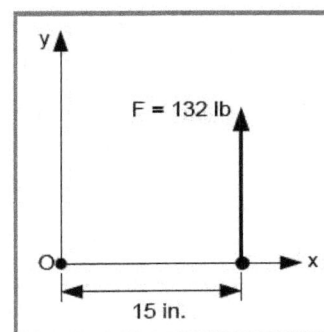

2.33     Determine the moment about point O due to the force shown in the figure to the left.

2.34     Determine the total moment about point O due to the two forces shown in the figure to the right.

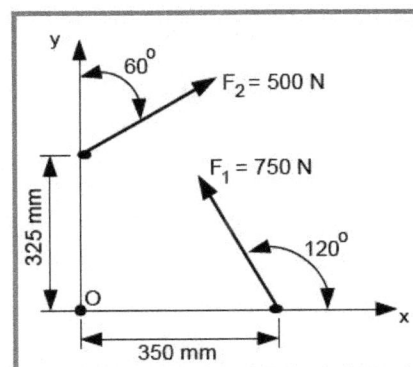

2.35   Express the force **F** or moment **M** in vector format if its Cartesian components are:

    (a)  $F_x = 2,100$ N, $F_y = -2,300$ N, and $F_z = 1,250$ N
    (b)  $F_x = 78$ lb, $F_y = -185$ lb, and $F_z = 124$ lb
    (c)  $M_x = 300$ N-m, $M_y = 400$ N-m, and $M_z = 500$ N-m
    (d)  $M_x = 740$ in-lb, $M_y = 425$ in-lb, and $M_z = -520$ in-lb

2.36   Determine the magnitude and direction, relative to the positive x-axis, for each of the vectors listed:

(a)  $F_x = 1,400$ N, $F_y = 900$ N and $F_z = 0$
(b)  $F_x = 355$ lb, $F_y = 520$ lb and $F_z = 0$
(c)  $M_x = 1,800$ N-m, $M_y = 1,350$ N-m and $M_z = 0$
(d)  $M_x = 655$ ft-lb, $M_y = -335$ ft-lb and $M_z = 0$

2.37   A position vector **r** is constructed in space between the origin (0, 0, 0) and point P.  Write **r** as a Cartesian vector and determine the magnitude (r) and coordinate direction angles ($\alpha$, $\beta$, $\gamma$) if:

(a)  P = (7, 4, 6) in.
(b)  P = (110, 65, 70) mm
(c)  P = (0.120, 0.155, 0.170) m
(d)  P = (−10, 3, −8) ft

2.38   Determine the coordinate direction angles ($\alpha$, $\beta$, $\gamma$) for the following forces:

(a)  **F** = (4 **i** + 5 **j** − 6 **k**) N
(b)  **F** = (−3 **i** + 2 **j** − 6 **k**) lb
(c)  **F** = (−3 **i** − 3 **j** + 7 **k**) kN
(d)  **F** = (5 **i** − 2 **j** − 8 **k**) kip

2.39   A space force **F** has a known magnitude (F) and two known coordinate direction angles ($\alpha$, $\beta$). Write a vector equation for **F**, specify the three components ($F_x$, $F_y$, $F_z$) and prepare a drawing showing the force vector in a three-dimensional Cartesian coordinate system for:

(a)  F = 1,200 N, $\alpha = 60°$ and $\beta = 30°$
(b)  F = 1,500 lb, $\alpha = 75°$ and $\beta = 50°$
(c)  F = 17 kN, $\alpha = 135°$ and $\beta = 60°$
(d)  F = 25 kip, $\alpha = 50°$ and $\beta = 140°$

2.40   Write a vector equation for the summation of two space forces $F_1$ and $F_2$. The force $F_1$ is 22.5 kN in magnitude with coordinate direction angles of $\alpha_1 = 60°$, $\beta_1 = 60°$ and $\gamma_1 = 135°$. The force $F_2$ is 30.0 kN in magnitude with coordinate direction angles of $\alpha_2 = 60°$, $\beta_2 = 30°$ and $\gamma_2 = 90°$. Prepare a drawing showing this resultant force vector in a three-dimensional Cartesian coordinate system.

2.41   Determine the angle between the following pairs of space forces:

(a)  $\mathbf{F_1}$ = (3 **i** + 6 **j** − 3 **k**) N and $\mathbf{F_2}$ = (−2 **i** + 5 **j** + 7 **k**) N
(b)  $\mathbf{F_1}$ = (5 **i** − 1 **j** + 4 **k**) lb and $\mathbf{F_2}$ = (8 **i** − 2 **j** + 3 **k**) lb
(c)  $\mathbf{F_1}$ = (−2 **i** + 6 **j** − 3 **k**) kN and $\mathbf{F_2}$ = (−3 **i** + 4 **j** − 4 **k**) kN

2.42   Determine the magnitude of a single force component produced by two space forces ($F_1$, $F_2$) that is directed along a line of action which lies in the x–y plane and makes an angle of $\theta$ with the positive x-axis:

(a)  $\mathbf{F_1}$ = (3 **i** + 6 **j** − 3 **k**) lb; $\mathbf{F_2}$ = (−2 **i** + 3 **j** + 5 **k**. ) lb; and $\theta = 30°$
(b)  $\mathbf{F_1}$ = (4 **i** + 3 **j** − 2 **k**) kN; $\mathbf{F_2}$ = (3 **i** − 5 **j** + 6 **k**) kN; and $\theta = 40°$
(c)  $\mathbf{F_1}$ = (2 **i** − 4 **j** − 2 **k**) kip; $\mathbf{F_2}$ = (−4 **i** + 3 **j** − 5 **k**) kip; and $\theta = 120°$

2.43   A force vector **F**, with components $F_x$, $F_y$ and $F_z$ is applied to a structure at point Q.  Determine the moment of **F** about the origin O of the coordinate system when:

(a)  $F_x = 4,000$ N, $F_y = 2,500$ N, $F_z = 1,500$ N and Q = (2.3, 3.4, 4.6) m
(b)  $F_x = -2,000$ lb, $F_y = 2,500$ lb, $F_z = 1,200$ lb and Q = (3.4, −2.5, 4.6) ft
(c)  $F_x = 2.3$ kN, $F_y = 5.5$ kN, $F_z = 3.3$ kN and Q = (1.4, 3.9, 2.6) m

2.44 If a structure is loaded at point Q with a force **F**, determine the moment **M$_O$** about the origin O. Prepare a drawing showing the moment vector **M$_O$** in a three-dimensional Cartesian coordinate system. Values for **F** and Q are:

(a) **F** = (125 **i** + 112 **j** + 45 **k**) lb and Q = (2, –3, 15) ft
(b) **F** = (3.8 **i** + 3.3 **j** – 4.5 **k**) kN and Q = (–2.2, 2.6, –1.8) m
(c) **F** = (55 **i** – 89 **j** + 42 **k**) lb and Q = (–3, 4, 8) ft

2.45 The cell phone tower illustrated in the figure to the right is supported by three cables that are maintained with tension forces $F_A = F_B = F_C = 800$ lb. The cables are anchored into the ground plane at locations A, B and C. Because the structural strength and rigidity of the tower is along its axis, we design the cable support system to exhibit a vector sum **S$_v$** that is directed along the axis of the tower from point D to the tower support point O. If the tower height H = 120 ft and the anchor locations are given in the figure, determine the position for the anchor at point B to achieve the design objective.

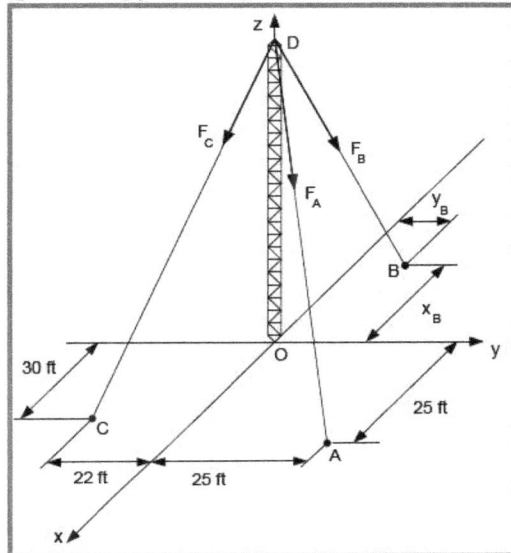

2.46 The cell phone tower, illustrated in the figure to the left, is supported by three cables that are maintained with tension forces $F_A = 800$ lb and $F_C = 1,000$ lb. The cables are anchored into the ground plane at locations A, B and C. Because the structural strength and rigidity of the tower is along its axis, we design the cable support system to exhibit a vector sum **S$_v$** that is directed along the axis of the tower from point D to the tower support point O. If the tower height H = 120 ft and the anchor locations are given in the figure, determine the force $F_B$ that must be maintained in cable BD to achieve the design objective.

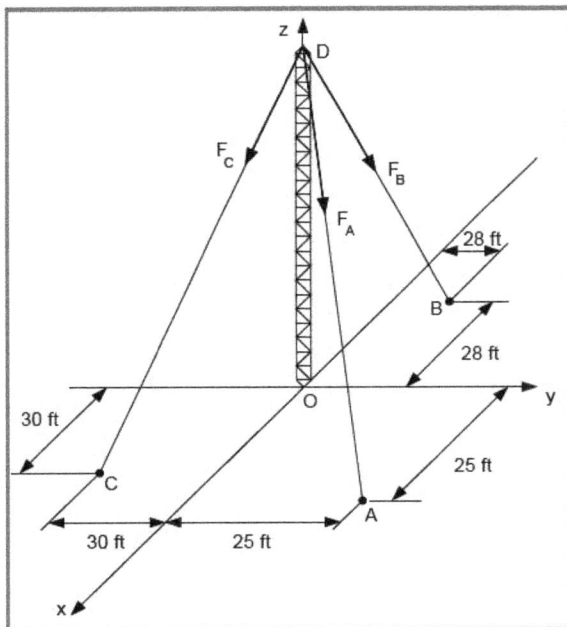

2.47 The boom of a crane extends from its base at point O to its tip at point Q as indicated in the figure below. A three-dimensional coordinate system has been established in this illustration with the point O at the origin and the point Q defined with coordinates (– 1, 3, 8). A force with a magnitude of 21 kN is applied by the boom onto a cable that extends from the tip of the boom to point P. Point P is located on the x–y plane with coordinates shown in the figure below. If the coordinates are expressed in meters, determine the following quantities:

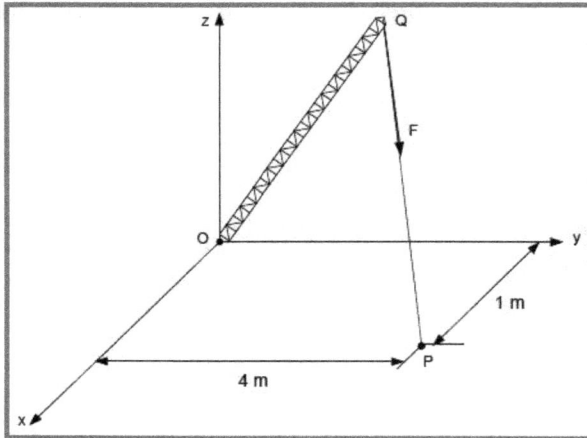

1. The force **F** (in vector form).
2. The force components $F_x$, $F_y$, and $F_z$.
3. The moment $\mathbf{M_O}$ (in vector form).
4. The moment components $M_x$, $M_y$, and $M_z$.
5. The magnitude of the moment ($M_O$).
6. The unit vector giving the direction of $\mathbf{M_O}$.
7. The angle between **F** and boom OQ.
8. The projection of **F** along boom OQ.

2.48 The boom of a crane extends from its base at point O to its tip at point Q as indicated in the figure below. A three-dimensional coordinate system has been established in this illustration with the point O at the origin and the point Q defined with coordinates (3, 2, 18). A force with a magnitude of 7,500 lb is applied by the boom onto a cable that extends from the tip of the boom to point P. Point P is located on the x–y plane with coordinates shown in the figure below. If the coordinates are expressed in feet, determine the following quantities:

1. The force **F** (in vector form).
2. The force components $F_x$, $F_y$, and $F_z$.
3. The moment $\mathbf{M_O}$ (in vector form).
4. The moment components $M_x$, $M_y$, and $M_z$.
5. The magnitude of the moment ($M_O$).
6. The unit vector giving the direction of $\mathbf{M_O}$.
7. The angle between **F** and boom OQ.
8. The projection of **F** along boom OQ.

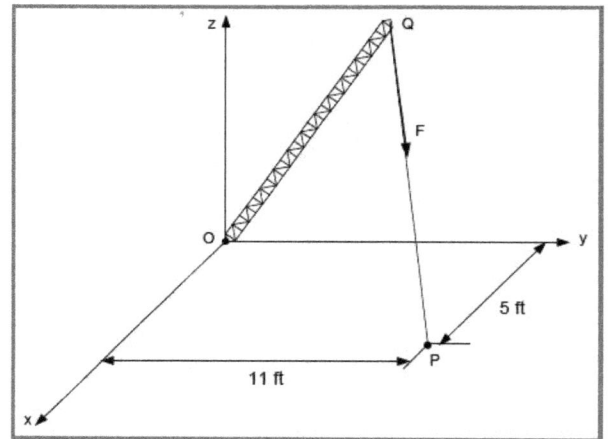

# CHAPTER 3

# FREE BODY DIAGRAMS AND EQUILIBRIUM

## 3.1 INTRODUCTION

Equilibrium is an extremely important concept because it provides us with an approach for determining unknown forces that act on a body. We understand that a body is in equilibrium if it is at rest or traveling at some constant velocity. For example, the beam shown in Fig. 3.1 is at rest because the loads acting downward are resisted by the two reactive forces produced by the supports. The entire assembly is fixed to bedrock.

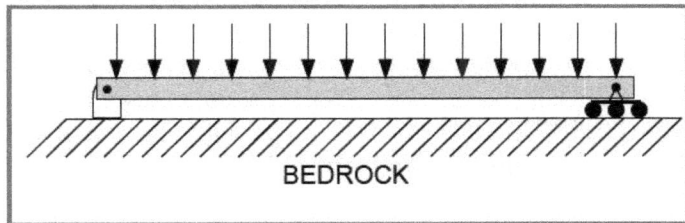

Fig. 3.1 A simply supported beam is at rest and in equilibrium.

Bodies that move may or may not be in equilibrium. If a body is in equilibrium, it moves with a constant velocity and a constant **momentum**[1]. The effect of unbalanced forces ($\Sigma F \neq 0$) acting on a body is to change its momentum by altering its velocity. The rate of that change is proportional to the unbalanced force. Therefore, when a body moves at constant velocity (zero acceleration) it possesses constant momentum and is in equilibrium.

The automobile in Fig. 3.2 is in equilibrium because the forward thrust due to the action of the wheels on the pavement is exactly equal to the rolling friction and drag forces that tend to impede the forward motion of the auto. The sum of the forces in the direction of motion is zero, the momentum is constant, the acceleration is zero, and the auto is in equilibrium although it may be moving at a velocity of 65 MPH.

Fig. 3.2 An automobile traveling at constant velocity is in equilibrium.

## 3.2 EQUATIONS OF EQUILIBRIUM

In Chapters 1 and 2, two equations of equilibrium were described:

$$\Sigma \mathbf{F} = 0 \qquad \text{and} \qquad \Sigma \mathbf{M_O} = 0$$

These relations are represented in a vector format; however, they may also be represented as Cartesian components of the forces and the moments:

---

[1] Momentum, M a vector quantity, is defined as M = mv; where **v** is the velocity of the body.

$$\Sigma F_x = 0; \qquad\qquad \Sigma F_y = 0; \qquad\qquad \Sigma F_z = 0 \qquad\qquad (3.1)$$

$$\Sigma M_x = 0; \qquad\qquad \Sigma M_y = 0; \qquad\qquad \Sigma M_z = 0 \qquad\qquad (3.2)$$

The two vector equations of equilibrium are equivalent to the six scalar equations of equilibrium. We will develop many of the solutions to problems arising in both statics and mechanics of materials by using Eqs. (3.1) and (3.2). The decision regarding which form of the equations to use will be left to your discretion, although in many cases you will find the scalar equations much easier to apply.

Although there are six scalar equations of equilibrium, it is not always necessary to use all six of them to solve an equilibrium problem. By classifying different systems of forces that act on a body, only those equations, which provide relevant information for solving equilibrium problems, may be identified. This classification greatly simplifies the associated equilibrium problems. Force systems are classified as follows:

- Non-coplanar and non-concurrent.
- Non-coplanar and concurrent.
- Coplanar and non-concurrent.
- Coplanar and concurrent.

Let's now examine the relevant equations of equilibrium for each of these force system.

### 3.2.1 Non-coplanar, Non-concurrent Force Systems

The three-dimensional force system, illustrated in Fig. 3.3, is **non-coplanar** because forces with components oriented in the x, y, and z directions act on the body. It is **non-concurrent** because the lines of action of the forces $F_1$, $F_2$ and $F_3$ do not intersect at a common point. Moments are produced by forces $F_1$, $F_2$ and $F_3$, and clearly there are components of the forces in all three directions. When the body in Fig. 3.3 is in equilibrium, the directions and magnitudes of $F_1$, $F_2$ and $F_3$ must satisfy the **six** Cartesian component equations of equilibrium:

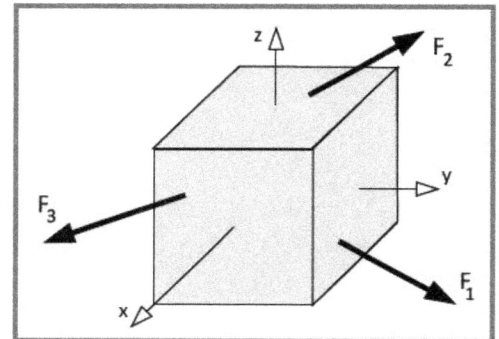

Fig. 3.3 An illustration of a non-coplanar, non-concurrent force system.

The **six** Cartesian component equations of equilibrium are:

$$\Sigma F_x = 0 \qquad\qquad \Sigma F_y = 0 \qquad\qquad \Sigma F_z = 0$$

$$\Sigma M_x = 0 \qquad\qquad \Sigma M_y = 0 \qquad\qquad \Sigma M_z = 0$$

### 3.2.2 Non-coplanar, Concurrent Force Systems

The three-dimensional force system, illustrated in Fig. 3.4, is **non-coplanar** because forces $F_1$, $F_2$ and $F_3$ acting on the body exhibit components in the x, y and z directions.

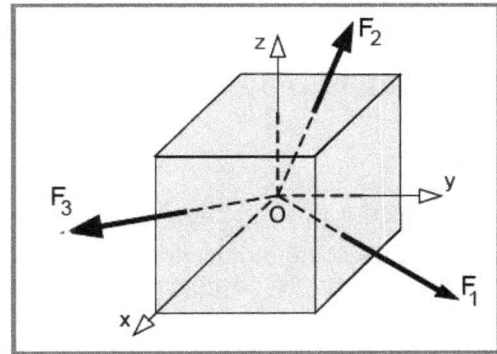

Fig 3.4 Example of a **non-coplanar**, **concurrent** force system.

This force system is **concurrent** because all of the forces pass through a common point (the origin of the Cartesian coordinate system in this case). The force system does not produce moments about the origin; hence, the equilibrium relation $\Sigma\mathbf{M} = 0$ is independent of the magnitude and direction of the forces $\mathbf{F_1}$, $\mathbf{F_2}$ and $\mathbf{F_3}$. However, these forces must satisfy the three remaining Cartesian equations of equilibrium:

$$\Sigma F_x = 0 \qquad\qquad \Sigma F_y = 0 \qquad\qquad \Sigma F_z = 0$$

### 3.2.3 Coplanar, Non-concurrent Force Systems

A coplanar non-concurrent force system is illustrated in Fig 3.5.

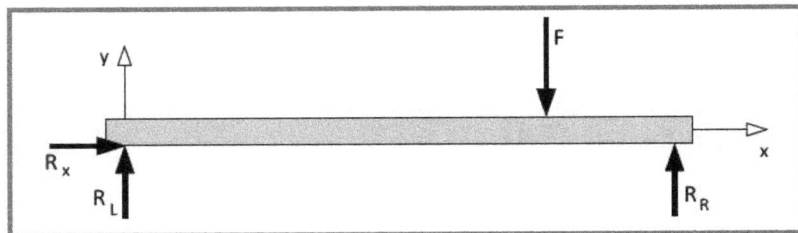

Fig. 3.5 A coplanar, non-concurrent force system.

Clearly, all of the forces acting on the beam lie in the x-y plane making this system **coplanar**. Three of the forces are parallel and their lines of action cannot intersect; hence, the system is **non-concurrent**. In this case, the forces acting on the body, F, $R_x$, $R_L$ and $R_R$, must satisfy three of the six equations of equilibrium, namely:

$$\Sigma F_x = R_x = 0, \qquad \Sigma F_y = 0, \qquad \text{and } \Sigma M_z = 0$$

The remaining three equations of equilibrium are satisfied independently of the solution for F, $R_x$, $R_L$ and $R_R$.

### 3.2.4 Coplanar, Concurrent Force Systems

A **coplanar, concurrent force system** is shown in Fig. 3.6.

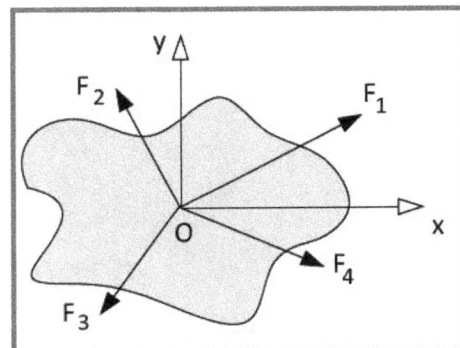

Fig. 3.6 A coplanar and concurrent system of forces.

This force system is **coplanar** because the body and all of the forces acting on it lie in the x-y plane. It is **concurrent** because the lines of action of all the forces pass through point O. This is the simplest of the four force systems because equilibrium is satisfied when:

$$\Sigma F_x = 0 \qquad \text{and} \qquad \Sigma F_y = 0$$

The other four of the six equilibrium relations are satisfied automatically and provide no useful information. Due to concurrency, we recognize that no moments occur about point O and the equations $\Sigma M_x = \Sigma M_y = \Sigma M_y = 0$ are satisfied regardless of the values assigned to the forces. Also, because the forces are coplanar, they all lie in the x-y plane and $\Sigma F_z = 0$.

## 3.3 FREE BODY DIAGRAMS (FBDs)

The equations of equilibrium that were described in the previous section apply to:

- A single body or member.
- A structure made of several members.
- A portion of a multi-member structure formed by a section cut.
- A part of a body or structure that has been formed by section cuts.
- An element removed from a body.

Before utilizing the equations of equilibrium, we must first construct a **free body diagram (FBD)** of the member being analyzed. A FBD is a model representation of the body being analyzed. The purpose of the model is to simplify the physical representation of the structure by omitting fine details that are not necessary for solving the equilibrium problem.

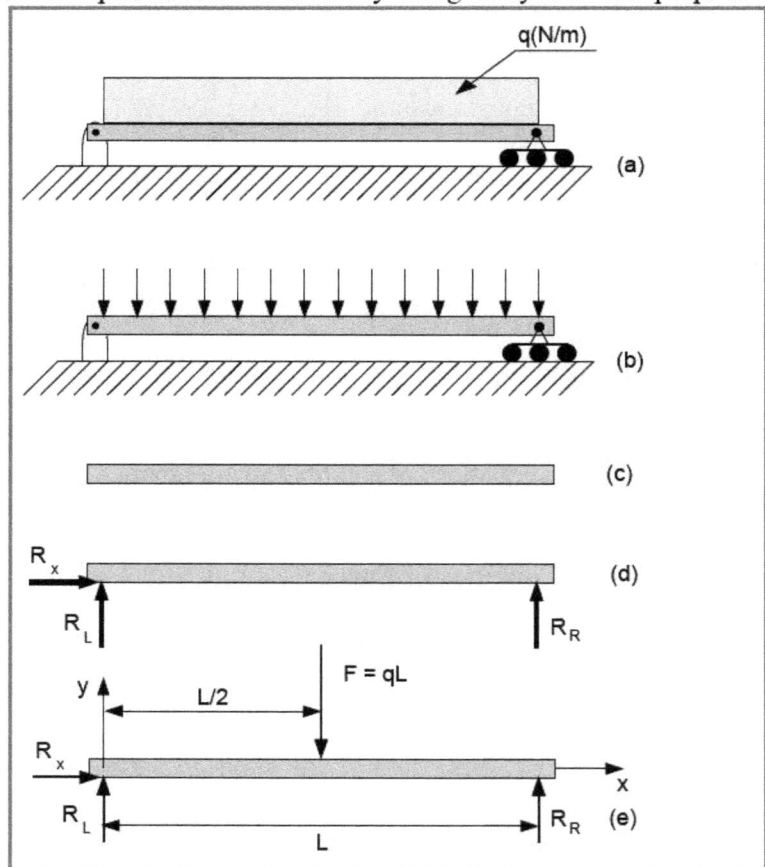

Fig. 3.7 Construction of a FBD for a simply supported uniformly loaded beam.

An example for the construction of a FBD is illustrated in Fig. 3.7. The beam is subjected to a uniformly distributed load of magnitude q (N/m), and supported near each end with simple supports. In Fig. 3.7a, we show the uniformly distributed load with shaded rectangle placed over the span of the beam. The uniformly distributed load may also be represented with a series of arrows as illustrated in Fig. 3.7b. We will use both of these techniques to model uniformly distributed loads. The simple supports are modeled with a pin and clevis on the left side and a roller on the right side. The pin and clevis holds the beam in a fixed position at its left end while the roller permits the beam to expand and/or contract with changes in temperature.

A four-step procedure is employed in drawing the free body of the beam.

1. Isolate the body (a beam in this case) by removing the supports and the uniformly distributed load. We show the isolated beam in Fig. 3.7c.
2. The supports are replaced with the **reaction** loads $R_L$, $R_R$ and $R_x$. The pin and clevis at the left support produces reactions in both the x and y directions as shown in Fig. 3.7d. However, the roller, which is free to rotate, produces a reactive force normal to the surface of the beam (the y direction). At this stage of the analysis, the magnitude of these forces is not known.
3. The uniformly distributed load q applied over the length L of the beam is replaced with a concentrated force F. The magnitude of the force is F = (qL). This force is due to gravity and acts downward as shown in Fig. 3.7e.
4. Finally, we dimension the FBD and establish a coordinate system that will facilitate the equilibrium analysis. In dimensioning the FBD, we place the concentrated force at the location of the centroid (center) of the shaded rectangular area representing the uniformly distributed load (e.g. at L/2 from the left support).

### 3.3.1 Modeling Loads

The FBD is a model of a structure or some part of a structure. To prepare the FBD, we model both the loads that act on the structure and its supports. Let's first consider modeling loads due to gravity, which are the most commonly encountered forces. Suppose we have a block with a mass m as shown in Fig. 3.8a. We modeled the effect of the block with a concentrated force F = mg. This concentrated force is applied at the center of the block (its centroid). Because the force is due to gravity, the direction of the force is downward (in the negative y direction).

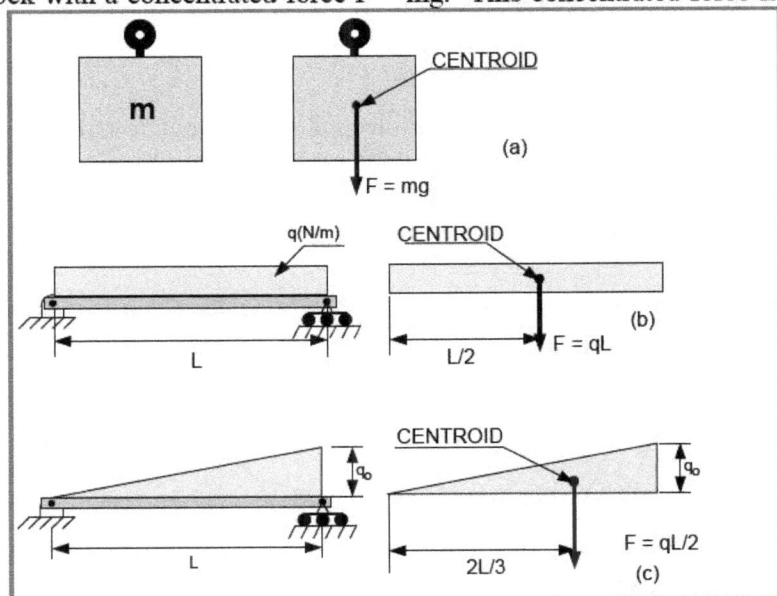

Fig. 3.8 Modeling loads due to gravity.

In Fig. 3.8 b, we encountered loads distributed over the span of a beam. In this illustration, the load is uniformly distributed with a magnitude q expressed in terms of force per unit length (N/m). We represented this uniformly distributed load with a rectangular area of height q and length L. A concentrated force F, the static equivalent of the uniformly distributed load, is given by the area of the rectangular (F = qL). The concentrated force is applied at the centroid of the rectangular area (x = L/2); because the load is due to gravity, the force acts downward.

In Fig. 3.8 c, the load is again distributed over the length of the beam, but it increases as a linear function of x (e.g. $q(x) = q_o x/L$). We represented this load with a triangular area with an altitude $q_o$ and a base of L. A concentrated force, the static equivalent of the distributed load, is given by the area of the triangle as $F = q_o L/2$. It is applied at the centroid of the triangle (x = 2L/3) and acts downward.

When modeling to solve for reaction forces at structural supports, we replace distributed loads with statically equivalent concentrated forces. However, later when we are concerned with internal moments and shear forces, it is not possible to use this simplified modeling technique.

## 3.3.2 Modeling Supports

The pin and clevis and roller are often used to support beams. The pin and clevis is stationary and does not permit the beam to move in either the x or y directions. On the other hand, the roller permits the beam to expand or contract with changes in temperature. When the pin and the roller are removed in the construction of a FBD, they are replaced with reaction forces shown in Fig. 3.9. For the pivot, which restrains motion in the x and y directions, reaction forces $R_L$ (perpendicular to the bottom surface of the beam) and $R_x$ (parallel to the bottom surface of the beam) are required to represent this support. For the roller (on a frictionless surface), a single reaction force perpendicular to the bottom surface of the beam is sufficient for the representation on a FBD.

Fig. 3.9 Modeling the pivot and roller type supports.

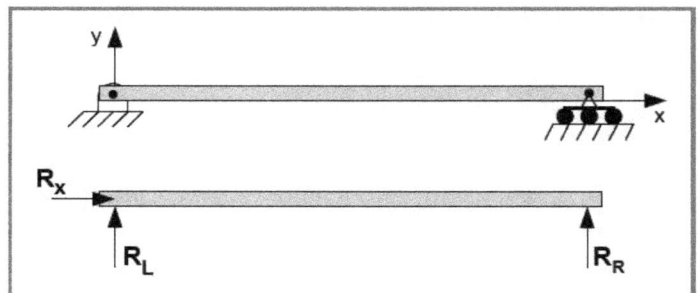

There are many support conditions and connections to structures. In constructing FBDs, we model the structure by removing these supports and replace them with one or more reactive forces or reactive moments. We will list several different types of supports or connections and their reactive forces and moments in the illustrations presented in Fig. 3.10.

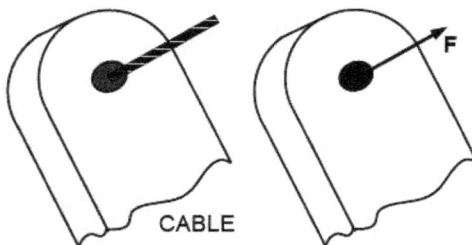

Fig. 3.10a A cable connection is represented with a single tension force acting along the cable in the direction away from the structural element.

CABLE

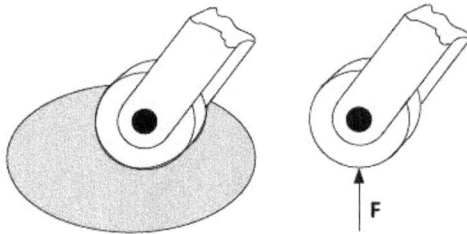

Fig. 3.10b A roller or rocker on a flat surface is represented with a concentrated force that acts perpendicular to the surface at the point of contact.

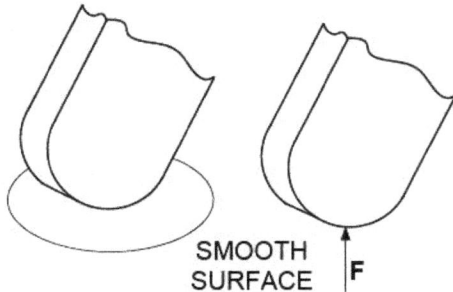

SMOOTH SURFACE

Fig. 3.10c A structural element contacting a smooth (frictionless) surface is modeled with a concentrated force that acts perpendicular to the surface at the point of contact.

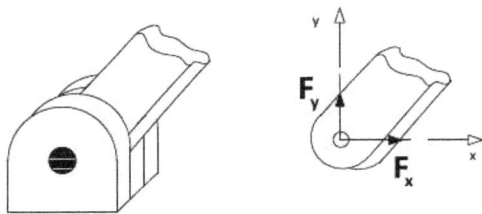

Fig. 3.10d A pin and clevis connection is represented with two Cartesian forces perpendicular to the pin.

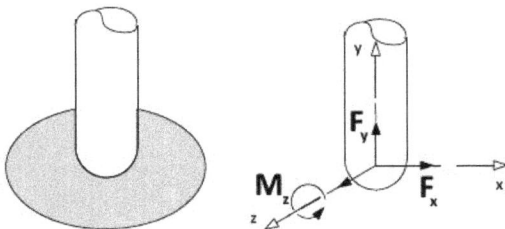

Fig. 3.10e A fixed support is represented with three possible reactions (two Cartesian forces and one Cartesian moment).

## EXAMPLE 3.1

Draw a FBD of the ring that is used to attach the cables to the weight shown in Fig. E3.1.

Fig. E3.1

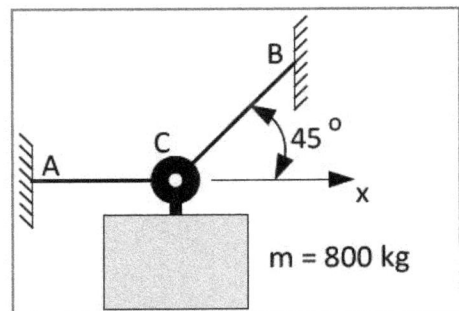

**Solution:**

The tension developed in the cables CA and CB maintains the 800-kg mass in equilibrium. We will determine the tension in the cables later in Section 3.5, but the first step in solving for the cable tensions is to construct the appropriate FBD. In this example, the small ring located at point C is the appropriate free body.

**Step 1.** Isolate the body. The isolated ring is shown in Fig. E3.1a.

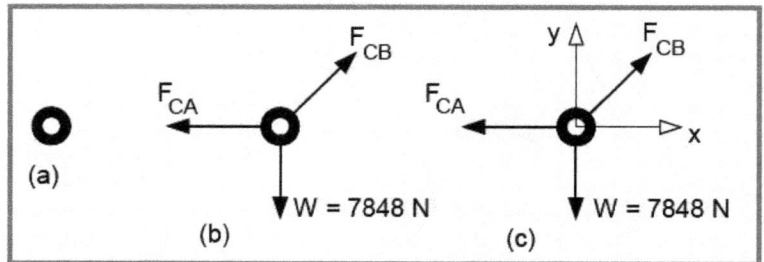

Fig. E3.1a to c

**Step 2.** Replace the cables with force vectors as shown in Fig. E3.1b. The forces $F_{CA}$ and $F_{CB}$ are of unknown magnitude; however, we know they are oriented in the direction of cables CA and CB respectively.

**Step 3.** Replace the mass with a force vector acting vertically downward through the center of gravity of the block. The weight of the mass is given by $W = 800 \text{ kg} \times 9.81 \text{ N/kg} = 7848 \text{ N}$.

**Step 4.** Dimension the FBD and establish a coordinate system. Since the force system is concurrent linear dimensions are not relevant. We show the 45° angle that $F_{CB}$ makes with the x-axis. A x-y coordinate system is placed on the FBD with the origin located at the center of the ring.

Let's consider another example showing the procedure for drawing FBDs. In this instance, we will draw a FBD for a coplanar, non-concurrent force system.

## EXAMPLE 3.2

The beam in Fig. E3.2 is subjected to a distributed load that increases from zero at the left support to $q_o$ at the right support. Draw a FBD of the beam.

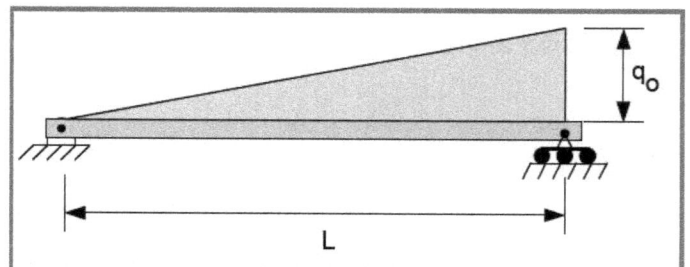

Fig. E3.2

**Solution:**

Again we model the distributed loading with arrows as shown in Fig. E3.2a. Then we utilize the four-step procedure to prepare the FBD.

**Step 1** Isolate the body. The isolated beam is shown below in Fig. E3.2b.

**Step 2** The supports are replaced with the reaction loads $R_L$, $R_R$ and $R_x$. At this stage of the analysis, the magnitude of these forces is not known. However, at the roller, we know that the direction of the reaction forces is normal to the surface of the beam as shown in Fig. E3.2c. At the pivot, reaction forces $R_x$ and $R_L$ are shown in the x and y directions.

**Step 3** Replace the linearly distributed load with a concentrated force F as shown in Fig. E3.2d. The area under the triangle representing the linearly increasing distributed load gives the magnitude of F. Hence;

$$F = (1/2) \, q_o \, L$$

The force F is positioned at $(2/3) \, L$, which corresponds to the location of the centroid of the area of a triangle. Of course, the force is directed downward to coincide with the direction of the gravitational field.

**Step 4.** Dimension the FBD by adding the coordinate system shown in Fig. E3.2d.

Fig. E3.2 a-d

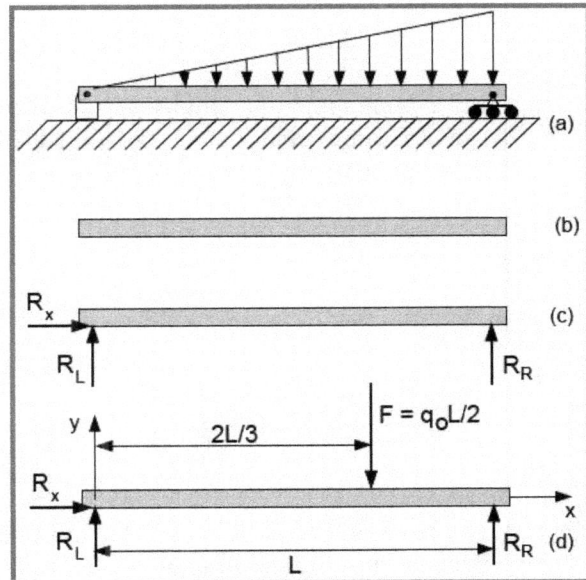

The FBD is complete because:

1. All the forces acting on the beam are shown
2. The forces are located in the proper positions
3. All the required dimensions are given
4. A coordinate system is provided for reference.

You should use this checklist to determine if the free bodies that you prepare in executing the assigned exercise are complete.

## EXAMPLE 3.3

Three smooth (frictionless) cylinders each with a diameter D and a mass of 4 slugs are constrained by a channel with a width W as illustrated in Fig. E3.3. Draw individual FBDs for cylinders A and C if the ratio W/D = 2.5.

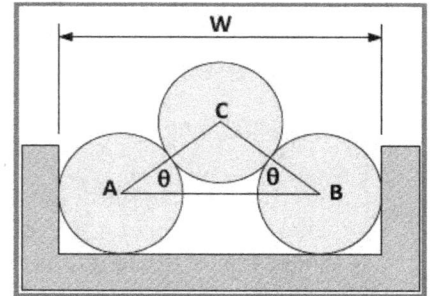

Fig. E3.3

**Solution:**

Before we construct the FBDs it is necessary to determine the stacking angle $\theta$ of the cylinders. From the geometry of the cylinders in the channel, we write:

$$W = D + 2D\cos\theta \qquad \Rightarrow\Rightarrow \qquad \theta = \cos^{-1}[(W/D) - 1]/2 \qquad \text{(a)}$$

$$\theta = \cos^{-1}(2.5 - 1)/2 = 41.41° \qquad \text{(b)}$$

Next let's determine the weight W of one of the cylinders:

$$F = mg = (4 \text{ slug})(32.17 \text{ lb/slug}) = 128.7 \text{ lb} \qquad \text{(c)}$$

An examination of cylinder A indicates that four forces must be placed on the FBD—three contact forces and one gravitational force. We show the FBD for cylinder A in Fig. E3.3a.

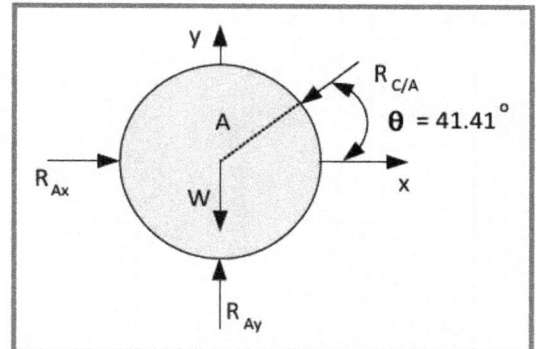

Fig. E3.3a

The reaction forces $R_{Ax}$ and $R_{Ay}$ are due to the surfaces of the channel contacting cylinder A. Because the cylinder is smooth (frictionless), the reaction forces are normal to the contact surface. We have used a double subscript to identify them—the first subscript identifies the cylinder in question and the second gives the direction of the force. Another reaction force $R_{C/A}$ is due to the action of cylinder C on cylinder A. It is normal to the surface of the cylinder at the point of contact. The subscript C/A indicates that cylinder C is acting on cylinder A. The angle $\theta = 41.41°$ is specified on the FBD. Finally the weight W is shown acting downward (in the negative y direction) from the center of the cylinder.

Following similar procedures, we draw the FBD for cylinder C in Fig. E3.3b.

Fig. E3.3b

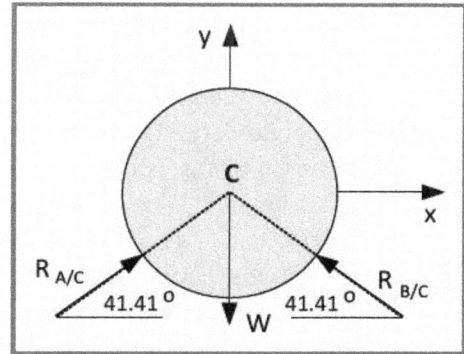

In this instance, we have the two reaction forces $R_{A/C}$ and $R_{B/C}$ both normal to the surfaces in contact. The weight W acts downward from the center of gravity of the cylinder.

Let's use the checklist to ascertain if the FBDs are complete.

- ❑ Are all the forces acting on each cylinder? Yes, we have counted them.
- ❑ Are the forces located in the proper positions? Yes, we have positioned the reaction forces at the points of contact and the gravitational forces at the center of gravity of the cylinders.
- ❑ Are the required dimensions given? In this case the forces are concurrent and dimensions are not necessary. However, the directions of the reaction forces are given by the angle θ.
- ❑ Is a coordinate system provided for reference? Yes, we have placed coordinates on both FBDs with the origins at the center of each cylinder.

## EXAMPLE 3.4

A hex nut is being tightened with the wrench shown in Fig. E3.4. The dimension of the flat on the hex nut is 40 mm. Prepare FBD diagrams for the nut and the wrench.

Fig. E3.4

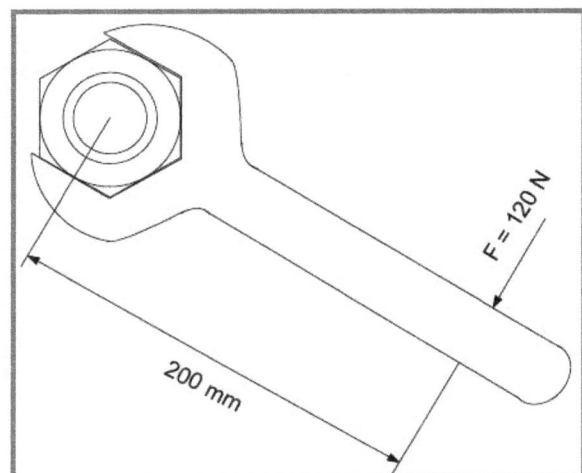

**Solution:**

The free body diagrams of the nut and the wrench are shown in Fig. E3.4a and E3.4b. When the force is applied to the wrench, it contacts the hex nut at two corners producing reaction forces $R_1$ and $R_2$. These forces are perpendicular to the surfaces of the hex nut. The threads on the nut react with those on the bolt to produce a resistive moment (torque) $M_z$ and a reactive force $R_3$. The angle defining the direction of the reaction forces acting on the flats is 60° because of the geometry of the hex configuration. The distance between the forces is the dimension of the flats on the hex nut. The FBD of the wrench is presented in Fig. E3.4b.

   The reaction forces $R_1$ and $R_2$ are applied at locations corresponding to the corners of the hex nut. Of course, they are equal in magnitude and opposite in direction of those shown on the FBD of the hex nut. The force applied to the wrench handle is shown. Dimensions locating the point of application of the force F and the reaction forces are provided. Angles are not shown because they are implied by the configuration of the hex nut.

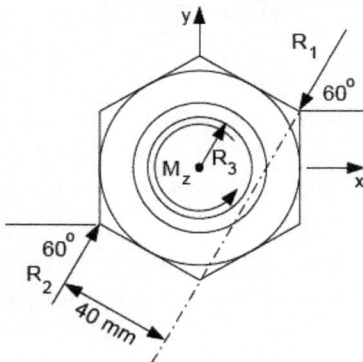

Fig. E3.4a

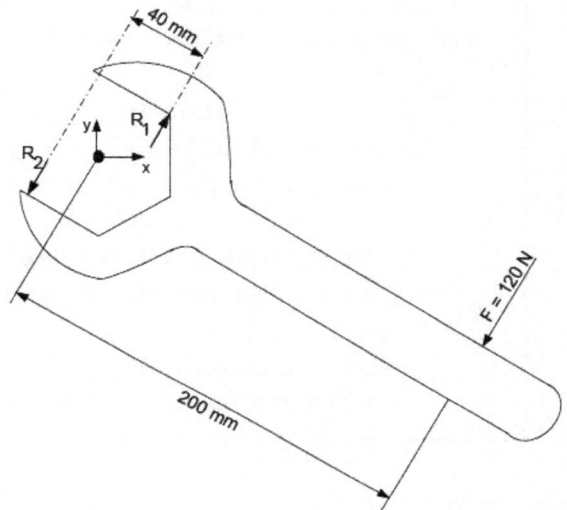

Fig. E3.4b

## EXAMPLE 3.5

Prepare a FBD of the wheeled crane lifting a weight W.

Fig. E3.5

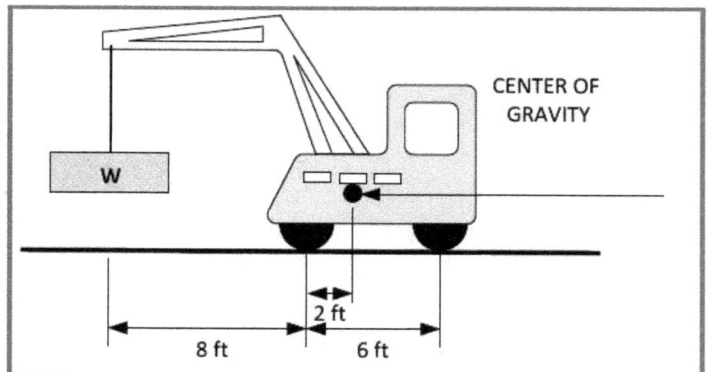

**Solution:**

Let's isolate the body and apply the reaction loads $R_L$ and $R_R$ at the wheels as shown in Fig. E3.5a. Next, the forces $W_L$ and $W_C$ due to gravity are applied at the center of gravity of the weight and the crane, respectively. The dimensions locating the points of application of the forces are specified. Finally, a Cartesian coordinate system with its origin at the center of mass of the crane is established.

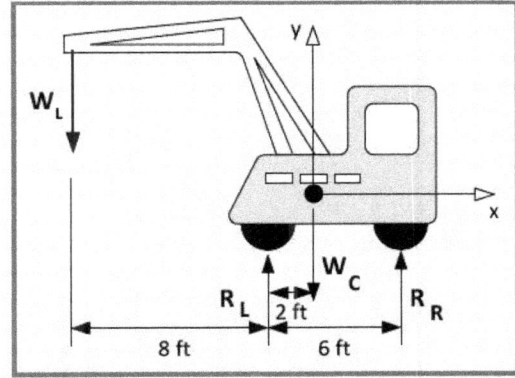

Fig. E3.5a

## EXAMPLE 3.6

A small stadium for football games and other sporting events is constructed with a partial roof covering the seating arrangements. One of the many supports for the roof is illustrated in Fig. E3.6. A uniformly distributed load of q is applied to the inclined member that supports the roof. Prepare a FBD of this structural element. Note that the support is modeled as a two-dimensional structure.

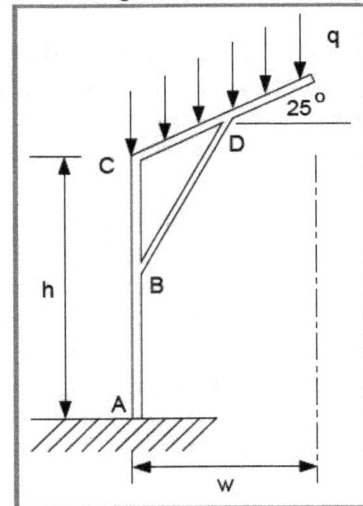

Fig. E3.6

**Solution:**

We first isolate the stadium component by removing the uniformly distributed load and the support. The uniformly distributed load is replaced with a statically equivalent force $F = qw$ applied at a distance $w/2$ from the vertical member. The fixed end support of the vertical member is then replaced with reaction forces $R_x$ and $R_y$ and a reaction moment $M_z$. We show the moment $M_z$ as positive (counterclockwise). Since the structure is coplanar, the other reactions at point A vanish ($R_z = M_x = M_y = 0$). A Cartesian coordinate system is placed on the FBD with its origin at point A. The dimensions of the applied force F are specified. The complete FBD is presented in Fig. E3.6a.

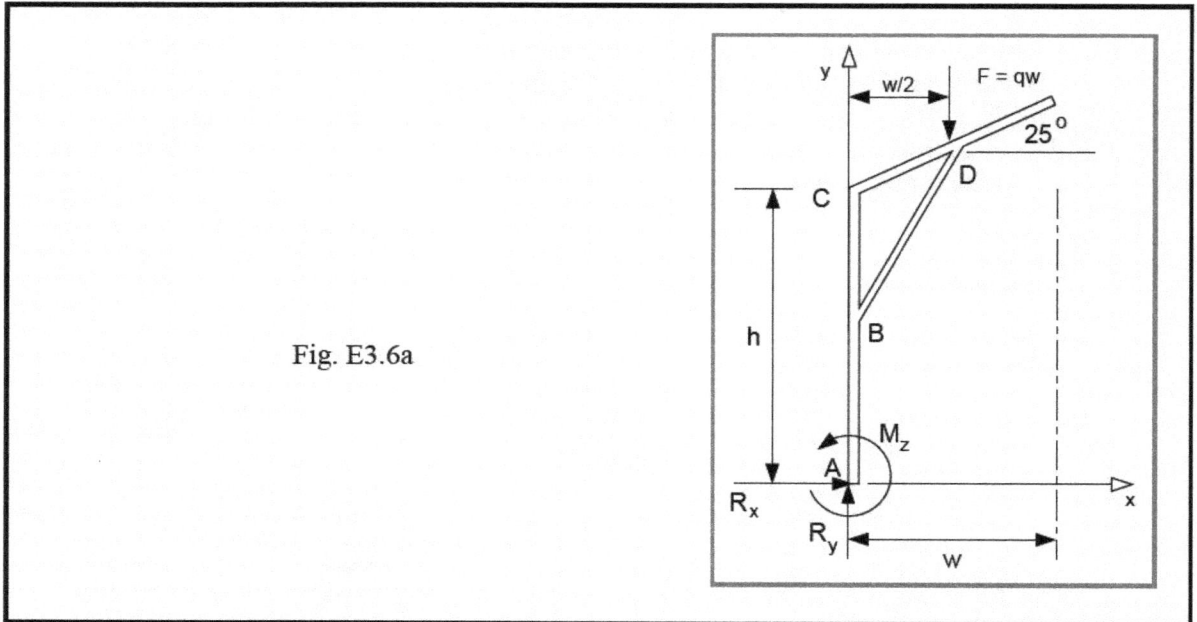

Fig. E3.6a

## 3.4 FBDs OF PARTIAL BODIES

We are very interested in determining internal forces in structural members because these internal forces produce the stresses and deflections that might cause failure of the structure. Consequently, determining these internal forces will be very important to you when designing your structures. The approach followed in studying internal forces is to:

1. Make an imaginary cut through the structural member being studied.
2. Draw a FBD of one part of this member.
3. Account for the effect of the cut away portion of the member by applying internal forces.
4. Solve for the internal forces or moments using the appropriate equations of equilibrium.

This procedure for constructing FBDs of partial bodies is demonstrated in Example 3.7.

### EXAMPLE 3.7

Consider the stepped tension bar loaded with external forces shown in Fig. E3.7. Draw a FBD of the portion of the bar to the left side of section cut A-A.

Fig. E3.7

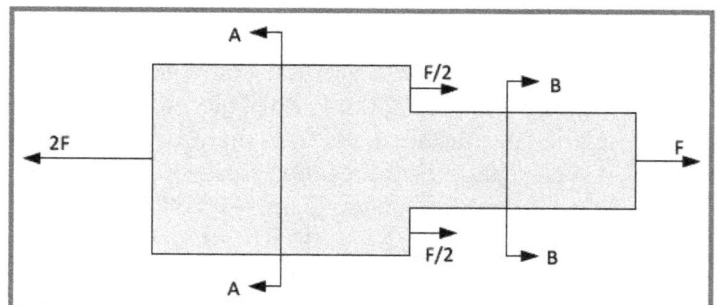

**Solution:**

**Step 1**. Isolate the partial body, which is the part of the tension bar to the left of the section cut A-A as shown in Fig. E3.7a.

**Step 2**. Replace the force 2F on the left part of the partial body as shown in Fig. E3.7b.

**Step 3**. Place an internal force P on the right face of the partial body to account for the right side of the tension bar that has been cut away as shown in Fig. E3.7c.

**Step 4.** Establish a coordinate system. In this case, dimensions are not necessary because the forces are **collinear** and **concurrent**. The complete FBD of the left portion of the tension member is shown below in Fig. E3.7a.

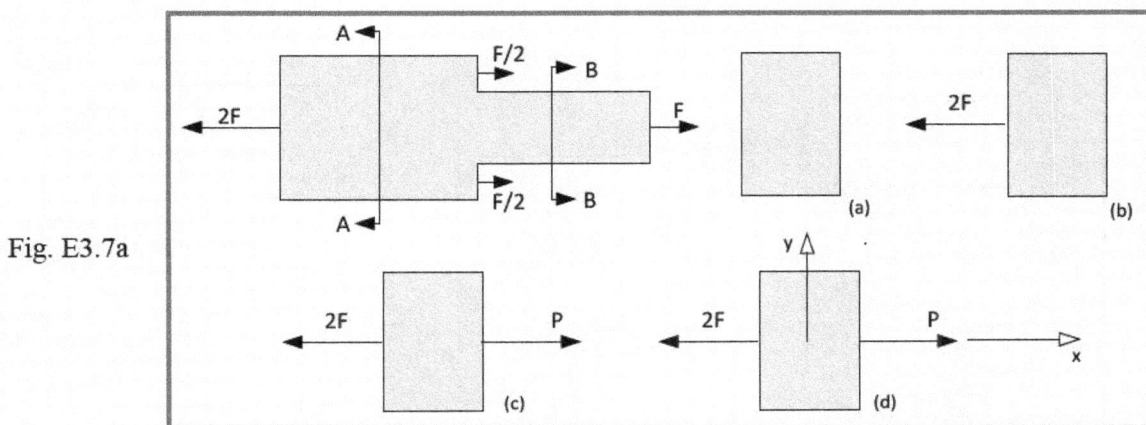

Fig. E3.7a

## EXAMPLE 3.8

Prepare a FBD of the right portion of the tension bar shown in Fig. E3.7 from section A-A to the bar's right hand end.

**Solution:**

**Step 1.** Isolate the right portion of the bar as shown in Fig. E3.8a.

**Step 2**. Replace the force F and the two forces F/2 on the right part of the body as shown in Fig. E3.8b.

**Step 3**. Place an internal force P on the left face of the body to account for the left end of the tension bar that has been cut away as shown in Fig. E3.8c.

**Step 4.** Establish a coordinate system. Because the forces are symmetric about the x-axis, dimensions are not necessary. The complete FBD of the right hand portion of the tension member is shown below in Fig. E3.8c.

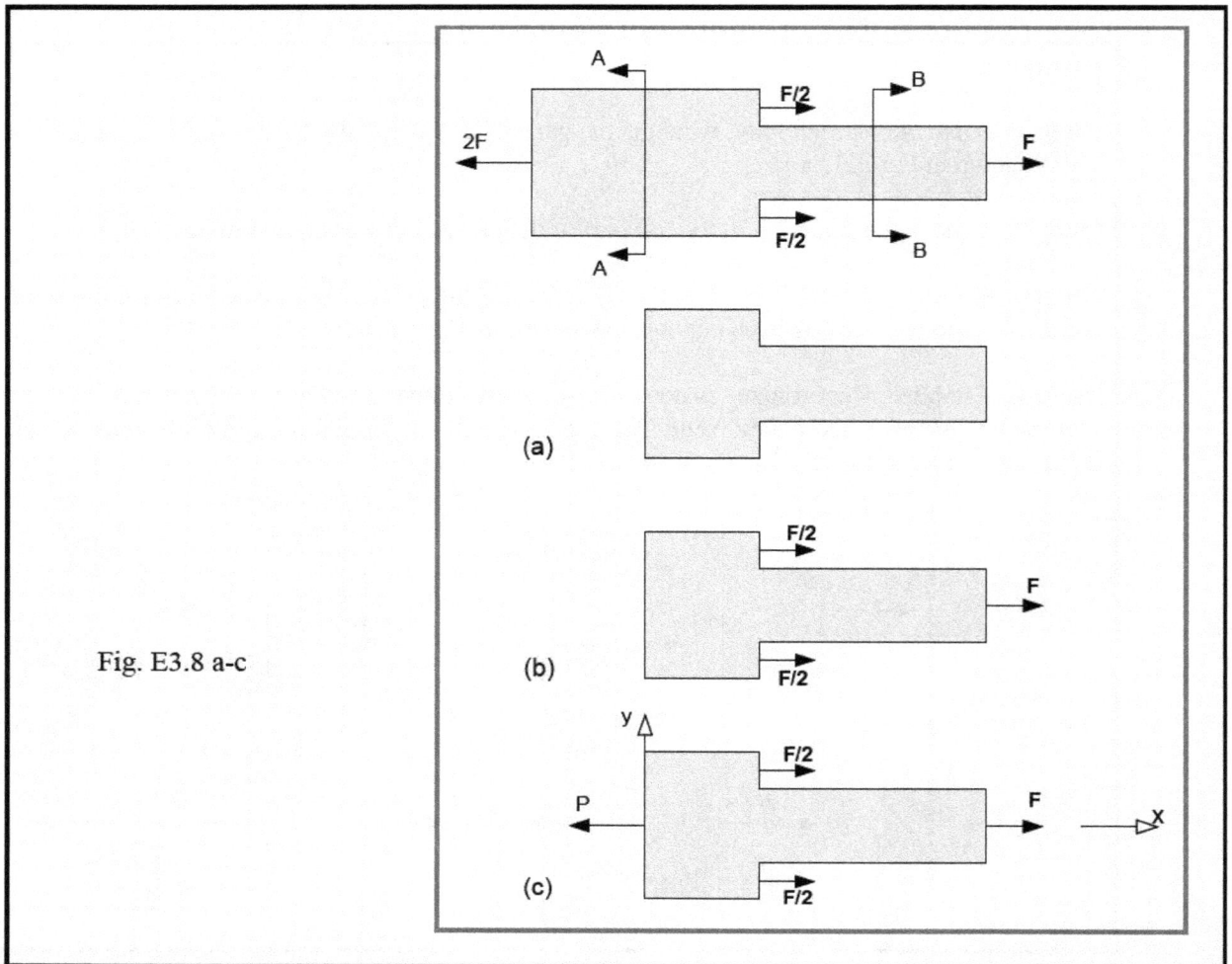

Fig. E3.8 a-c

## EXAMPLE 3.9

Prepare a FBD of the left side of a simply supported beam subjected to a uniformly distributed load as shown in Fig. E3.9.

Fig. E3.9

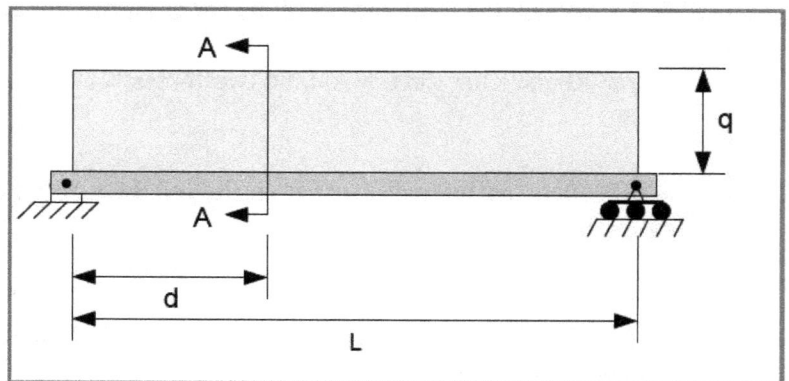

**Solution:**

**Step 1**. Isolate the segment of the beam to the left of section A-A as shown in Fig. E3.9a.

**Step 2**. Replace the uniformly distributed load q with a concentrated force F = qd positioned at d/2 from the left end of the beam.

**Step 3**. Place reaction forces $R_x$ and $R_y$ to account for the effect of the pivot support. Place internal forces P and V and an internal moment $M_z$ on the exposed face to account for the effect of the right end of the beam that has been cut away. We apply a positive force P (in the positive x direction) and a positive moment $M_z$ (counterclockwise). The force V is applied in the negative y direction. The reason for this choice will be explained later in the text.

**Step 4.** Dimension the FBD and establish a coordinate system. In analyzing beams, we usually place the origin of the coordinate system at the left end. The complete FBD of the left hand portion of the beam is shown in Fig. E3.9a.

Fig. E3.9a

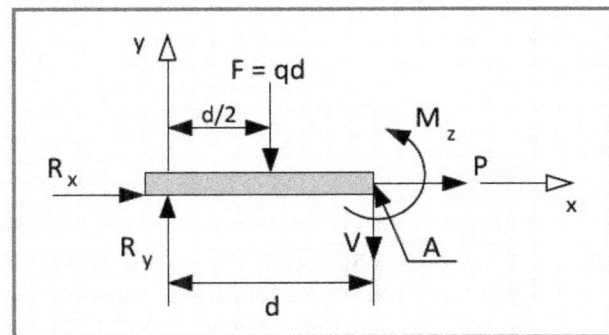

# EXAMPLE 3.10

A new factory for heavy machinery has an assembly line with component parts stored above floor level. The components used in assembly are in elevated bins (#1, #2, and #3) as shown in Fig. E3.10. The elevated bins are suspended from a frame ABCD. Prepare a FBD of the left portion of the frame from Section A-A to the roller support at point A.

Fig. E3.10

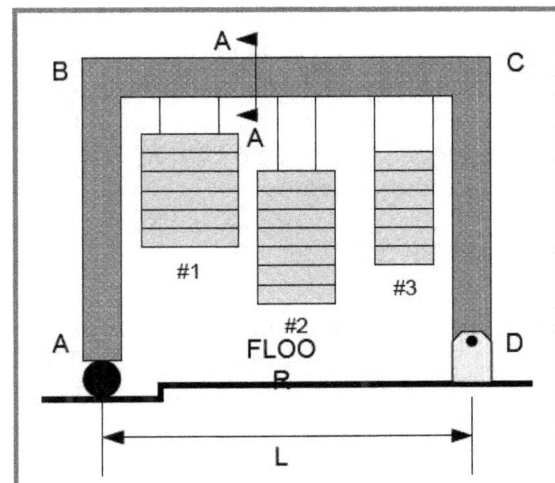

**Solution:**

**Step 1**. Isolate the segment of the frame to the left of section A-A as shown in Fig. E3.10a.

**Step 2**. Replace the loads due to the component parts on elevated bin #1 with concentrated forces $F_1/2$ and $F_1/2$ positioned at the location of the cables supporting this bin.

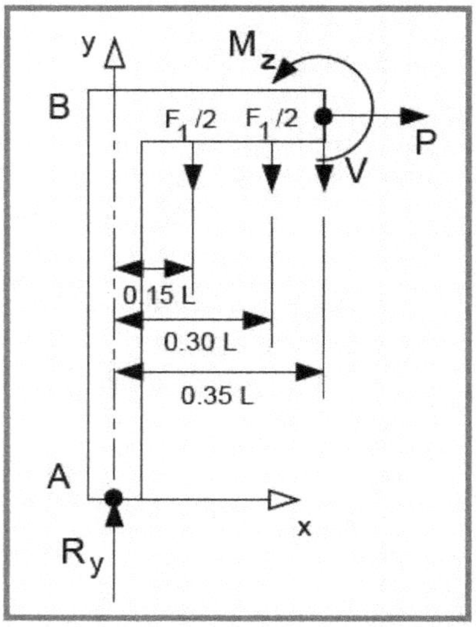

Fig. E3.10a

**Step 3**. Place a reaction force $R_y$ at point A to account for the effect of the roller support. Place an internal forces P and V and an internal moment $M_z$ on the exposed face to account for the right hand portion of the frame that has been cut away. We apply a positive force P (in the positive x direction) and a positive moment $M_z$ (counterclockwise). The force V is applied in the negative y direction. The reason for this choice will be explained later in the text.

**Step 4.** Dimension the FBD and establish a coordinate system. We have placed the origin of the coordinate system at point A and have scaled the dimensions locating the cables for bin 1. The complete FBD of the left hand portion of the frame is shown in Fig. E3.10a.

## 3.5 SOLVING FOR REACTIONS

The FBD provides a model to assist us in writing equilibrium equations that are used in the solution for unknown forces and reactions. The diagram indicates the known and unknown forces and their directions. It also provides the dimensions needed to compute moments. Let's consider the FBDs, which we developed in Examples 3.1 to 3.10, and solve for the unknown forces by using the appropriate equations of equilibrium.

### EXAMPLE 3.11

Recall the 800 kg mass that was supported by cables AC and BC as shown in the illustration in Example 3.1. In Section 3.3, we constructed a FBD for the ring that ties the weight and the cables together. For your convenience it is repeated in Fig. E3.11. Solve for the unknown forces acting on the ring.

Fig. E3.11

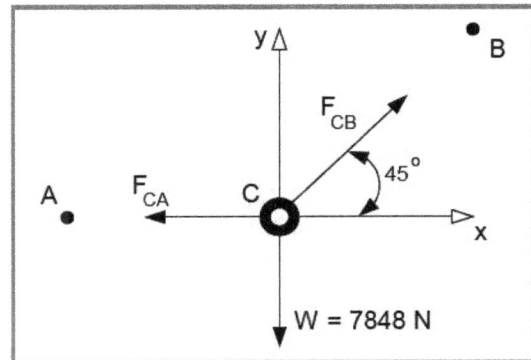

**Solution:**

The solution for the unknown forces and reactions is accomplished using the equations of equilibrium with a simple six-step procedure. We demonstrate this step-by-step approach for the cable-mass arrangement represented by the FBD.

**Step 1**. Prepare a complete FBD of the ring as shown above.

**Step 2**. Classify the force system. In this example, the force system is coplanar and concurrent.

**Step 3**. Write the relevant equations of equilibrium. For coplanar and concurrent force systems, only two of the six equations of equilibrium provide relevant information.

$$\Sigma F_x = 0 \qquad\qquad \Sigma F_y = 0$$

**Step 4.** Substitute the unknowns from the FBD into the relevant equations of equilibrium.

$$\Sigma F_y = 0 \qquad\qquad \Sigma F_x = 0$$

$$\Sigma F_y = F_{CB} \sin (45°) - 7{,}848 \text{ N} = 0 \qquad \Sigma F_x = F_{CB} \cos (45°) - F_{CA} = 0$$

**Step 5**. Execute the solution.

$$\Sigma F_y = F_{CB} \sin (45°) - 7{,}848 \text{ N} = 0 \qquad \Sigma F_x = F_{CB} \cos (45°) - F_{CA} = 0$$

$$F_{CB} = 7{,}848 \text{ N}/0.7071 \qquad\qquad F_{CA} = (11{,}100 \text{ N}) (0.7071)$$

$$F_{CB} = 11{,}099 \text{ N} \Rightarrow 11{,}100 \text{ N} \qquad\qquad F_{CA} = 7{,}848 \text{ N}$$

**Step 6.** Check the solution and interpret the results. In this example, the equations are simple, and we may easily determine if the numerical results are realistic by checking the signs in each of the equilibrium equations. Note that **forces directed in the negative coordinate directions are considered negative quantities** in writing the $\Sigma F$ terms. You should always make certain that the forces are expressed in the appropriate units. In this problem, the mass was given in kg (SI units), and this fact dictates that the force be expressed in newtons (N). Finally, we make a sketch of the force vectors shown in Fig. E3.11a verifying the results.

Fig. E3.11a

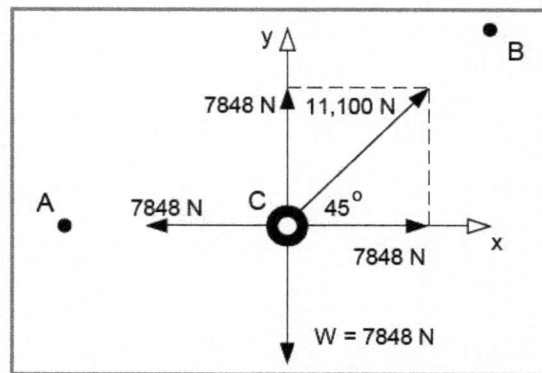

The 11,100 N force produces 7,848 N components in the positive x and y directions, which cancel out the 7,848 N forces, directed in the negative x and y directions. Clearly, the forces in the x and y directions are balanced, and the ring is in equilibrium.

## EXAMPLE 3.12

Next consider the beam that was illustrated in Example 3.2. Recall that we constructed the FBD to model this beam as indicated below:

Fig. E3.12

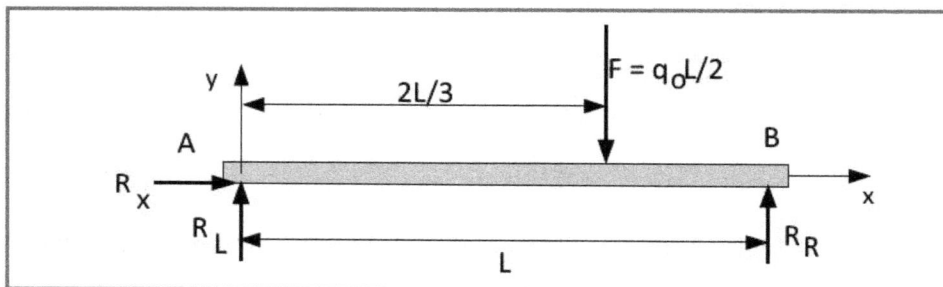

Let's execute the six-step procedure to solve for the unknown magnitudes of the reaction forces $R_x$, $R_L$ and $R_R$.

**Solution:**

**Step 1.** Prepare a complete FBD of the beam as shown above. Note the force F was determined from the area of the triangle representing the linear distributed load as:

$$F = (1/2)\, q_o\, L$$

**Step 2.** Classify the force system. In this problem, the force system is coplanar and non-concurrent.

**Step 3.** Write the relevant equations of equilibrium. For coplanar and non-concurrent force systems, only three of the six equations of equilibrium are relevant.

$$\Sigma F_x = 0 \qquad\qquad \Sigma F_y = 0 \qquad\qquad \Sigma M_A = 0$$

To determine the moments to substitute into the equation $\Sigma M = 0$, it is necessary to select some arbitrary point on the FBD to use as a reference for determining the moments. We have selected point A because it eliminates the moment due to the force $R_L$ and simplifies the resulting equations.

**Step 4.** Substitute the unknowns from the FBD into the relevant equations of equilibrium.

$$\Sigma F_x = 0 \qquad \Rightarrow \qquad R_x = 0$$

$$\Sigma F_y = 0 \qquad \Rightarrow \qquad \Sigma F_y = R_L + R_R - (1/2) \, q_o \, L = 0 \qquad (a)$$

$$\Sigma M_A = 0$$

$$\Sigma M_A = R_R \, L - (1/2) \, q_o \, L \, (2/3)L = 0 \qquad (b)$$

**Step 5.** Execute the solution.

$$\Sigma M_A = R_R \, L - (1/2) \, q_o \, L \, (2/3)L = 0 \qquad (c)$$

$$\Sigma F_y = R_L + R_R - (1/2) \, q_o \, L = 0 \qquad (d)$$

Solving Eq. (c) for $R_R$ yields:

$$R_R = q_o \, L/3 \qquad (e)$$

and Eq. (d) gives:

$$R_L = q_o \, L/6 \qquad (f)$$

**Step 6.** Check the solution and interpret the results. In this example, the solution involves solving two linear algebraic equations. We check the signs in each of the equilibrium equations. Note that **forces directed in the negative coordinate directions are considered negative quantities** in writing the $\Sigma F$ terms. The **moments tending to rotate the beam counterclockwise are treated as positive quantities, and those tending to produce clockwise motion are treated as negative quantities**. You should make certain that the forces and the moments are expressed in the appropriate units. In this problem, no units were specified since the applied loading and the beam's length were given as symbols $q_o$ and L respectively. However, let's suppose that L = 14 ft and $q_o$ = 1200 lb/ft. With L and $q_o$ specified we may determine numerical values and units for $R_L$ and $R_R$ as:

$$R_R = q_o \, L/3 = (1200 \text{ lb/ft})(14 \text{ ft})/3 = 5600 \text{ lb}$$

$$R_L = q_o \, L/6 = (1200 \text{ lb/ft})(14 \text{ ft})/6 = 2800 \text{ lb}$$

$$F = (1/2) \, q_o \, L = (1200 \text{ lb/ft})(14 \text{ ft})/2 = 8400 \text{ lb} = R_R + R_L$$

The result for the force F provides a check on the solution, and the units for all of the results are specified.

Let us continue to develop skills for solving equilibrium problems. For the next example, consider the cantilever beam that is shown in Fig. 3.11. A cantilever beam is a long slender structural member. It is built-in (supported so that the beam cannot rotate) on the left-hand end, and free (not supported in any manner) on the right-hand end. A beam is usually loaded with transverse forces distributed along its length. The cantilever beam in Fig. 3.11 is loaded with a transverse concentrated force F at its free end.

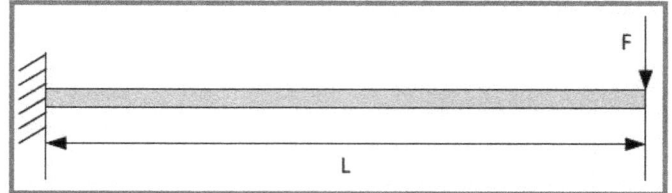

Fig. 3.11 A cantilever beam loaded at its free end with a concentrated force F.

## EXAMPLE 3.13

Determine the reactions (forces and moments) at the built-in end of the cantilever beam illustrated in Fig. 3.11.

**Solution:**

Let's execute the six-step procedure to determine the unknown magnitudes of the reaction forces $R_x$ and $R_L$ and the reaction moment $M_W$ due to the built-in support.

**Step 1.** Prepare a FBD to provide a model of the beam as shown in Fig. E3.13. Note the built-in end support is removed, and replaced with the reaction forces $R_L$ and $R_x$ and the reaction moment $M_W$. The moment $M_W$ is taken as a positive quantity and is shown in the counter clockwise direction. A two-dimensional Cartesian coordinate system was added for reference.

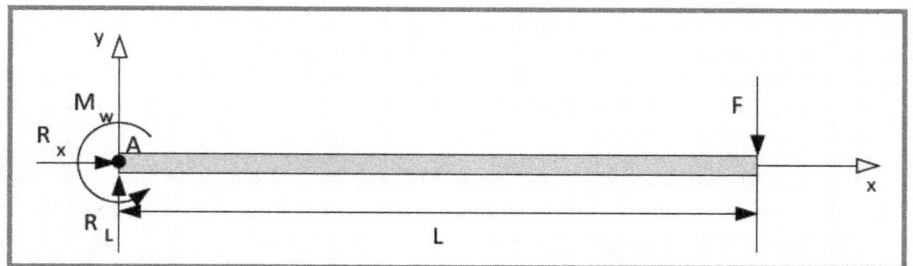

Fig. E 3.13

**Step 2.** Classify the force system. In this problem, the force system is coplanar and non-concurrent.

**Step 3.** Write the relevant equations of equilibrium. For coplanar and non-concurrent force systems, only three of the six equations of equilibrium are relevant.

$$\Sigma F_x = 0 \qquad \Sigma F_y = 0 \qquad \Sigma M_A = 0$$

**Step 4.** Substitute the unknowns from the FBD into the relevant equations of equilibrium.

$$\Sigma F_x = 0$$

$$R_x = 0 \qquad\qquad\qquad \text{(a)}$$

With the applied forces all in the transverse direction, the reaction force $R_x$ at the built-in support will always be zero as shown in Eq. (a).

$$\Sigma F_y = R_L - F = 0 \qquad\qquad\qquad \text{(b)}$$

$$\Sigma M_A = M_W - F\,L = 0 \qquad\qquad\qquad \text{(c)}$$

**Step 5.** Execute the solution. Solving Eq. (a) for $R_L$ yields:

$$R_L = F \qquad\qquad\qquad \text{(d)}$$

and Eq. (c) gives:

$$M_W = FL \qquad\qquad\qquad \text{(e)}$$

**Step 6.** Check the solution and interpret the results. In this example, the mathematics involves solving two very simple linear algebraic equations. We check the signs in each of the equilibrium equations. The forces directed in the negative coordinate directions are considered negative quantities in writing the $\Sigma F$ terms. The moments tending to rotate the beam counterclockwise are treated as positive quantities, and those tending to produce clockwise motion are treated as negative quantities in writing the $\Sigma M$ terms. The reaction moment $M_W$ was assumed to be in the counterclockwise direction. The solution $M_W = FL$ is a positive quantity indicating this assumption was correct. **If we had assumed the incorrect direction, the solution of the equilibrium relation would have resulted in a negative quantity.** You should make certain that the forces and the moments are expressed in the appropriate units. In this problem, no units were specified since the applied loading and the length of the beam were given as symbols F and L, respectively. However, let's suppose that L = 6 m and F = 20 kN. Determining numerical values and units for $R_L$ and $M_W$, we find:

$$R_L = F = 20 \text{ kN}$$

$$M_W = F\,L = (20 \text{ kN})(6 \text{ m}) = 120 \text{ kN-m}$$

For another example, let's determine the reaction forces at the supports for the truss bridge structure shown in Fig. 3.12. The truss bridge is supported at one end with a pin and clevis and at the other with a roller. **The roller is often used to support one end of a long structure because it rotates to accommodate the changes in the structure's length due to temperature changes.** With the roller support, the truss in Fig. 3.12 is free to expand and contract with temperature, and thermal stresses due to constraint do not develop.

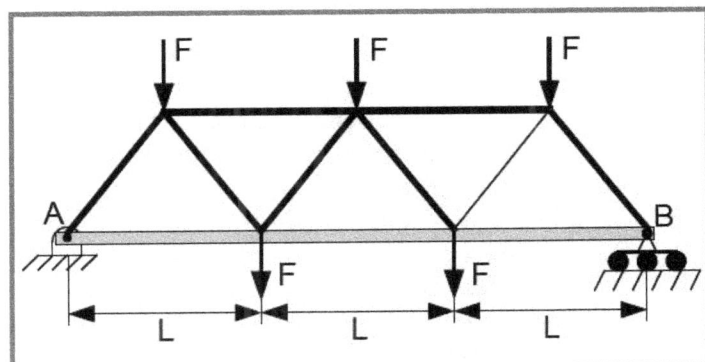

Fig. 3.12 The truss is loaded at the intersection of the members with a load of F.

## EXAMPLE 3.14

Determine the reaction forces at the supports for the truss presented in Fig. 3.12.

**Solution:**

**Step 1.** Prepare a FBD to provide a model of the truss as shown in Fig. E3.14. Note the supports were removed and replaced with the reaction forces $R_L$ and $R_R$. A two-dimensional Cartesian coordinate system was added for reference.

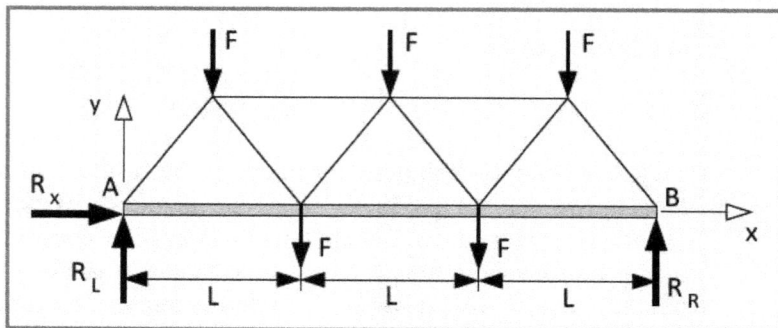

Fig. E3.14

**Step 2.** Classify the force system. In this instance, the force system is coplanar and non-concurrent.

**Step 3.** Write the relevant equations of equilibrium. For a coplanar and non-concurrent force system, only three of the six equations of equilibrium are relevant.

$$\Sigma F_x = 0 \qquad\qquad \Sigma F_y = 0 \qquad\qquad \Sigma M_A = 0$$

**Step 4.** Substitute the unknowns from the FBD into the relevant equations of equilibrium.

$$\Sigma F_x = 0 \qquad \Rightarrow \qquad R_x = 0 \tag{a}$$

$$\Sigma F_y = R_L + R_R - 5F = 0 \tag{b}$$

$$\Sigma M_A = R_R (3L) - F(L/2) - FL - F(3L/2) - F(2L) - F(5L/2)\ 0 \tag{c}$$

**Step 5.** Execute the solution. Solving Eq. (c) for $R_R$ yields:

$$R_R = (2.5)\ F \tag{d}$$

and Eq. (b) gives:

$$R_L = (2.5)\ F \tag{e}$$

**Step 6.** Check the solution and interpret the results. Again, the mathematics involves solving two simple linear algebraic equations. We check the signs in each of the equilibrium equations. The moments tending to rotate the truss counterclockwise are treated as positive quantities, and those tending to produce clockwise motion are treated as negative quantities in writing the $\Sigma M$ terms. Once again, if we had assumed the incorrect direction for the reaction forces, the solution of the equilibrium relation would have been a negative quantity. You should make certain that the forces and the moments are expressed in the appropriate units. In this problem, no units were specified since the applied loading and the lengths of the truss segments are given as symbols F and L, respectively. However, let's suppose that L = 10 m and F = 12 kN. Determining numerical values and units for $R_L$ and $R_R$, we find:

$$R_R = R_L = (2.5) \, F = 30 \text{ kN}$$

Notice that the reaction forces are equal. We expect this result because the geometry of the truss and its loading are symmetric. Also recognize that the reaction forces are independent of the length of the truss. Can you explain the reason for the reaction forces to be independent of the length of the truss?

## EXAMPLE 3.15

Using the FBDs presented in Fig. E3.15a and Fig. E3.15b, determine the contact forces between the cylinders described in Example 3.3.

**Solution:**

**Step 1**. Prepare a complete FBD of the cylinders A and B.

~~**Step 2**. Classify the force system. The force system acting on each cylinder is coplanar and concurrent.~~

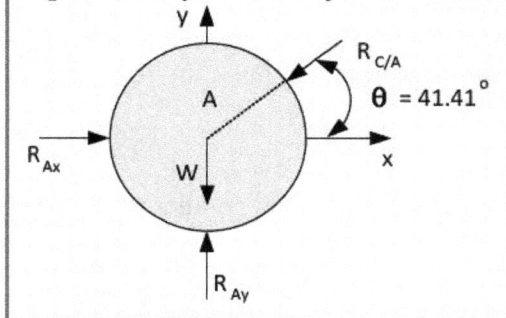

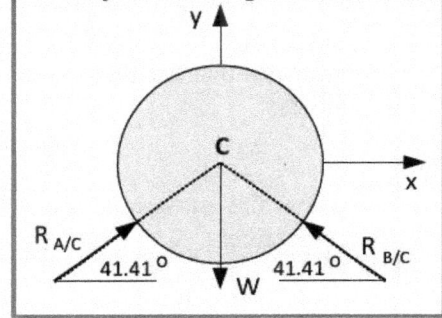

**Step 3**. Write the relevant equations of equilibrium. For coplanar and concurrent force systems, only two of the six equations of equilibrium provide relevant information.

$$\Sigma F_x = 0 \qquad\qquad \Sigma F_y = 0$$

**Step 4.** Consider cylinder C first, substitute the unknowns from the FBD, presented in Fig. E3.3b, into the relevant equations of equilibrium, and execute the solution.

$$\Sigma F_x = R_{A/C} \cos (41.41°) - R_{B/C} \cos (41.41°) = 0$$

$$R_{A/C} = R_{B/C} \qquad\qquad (a)$$

$$\Sigma F_y = R_{A/C} \sin (41.41°) + R_{B/C} \sin (41.41°) - 128.7 = 0$$

$$R_{A/C} = R_{B/C} = 128.7/[2 \sin (41.41°)] = 97.29 \text{ lb} \qquad (b)$$

**Step 5.** Consider cylinder A, substitute the unknowns from the FBD, shown in Fig. E3.3a, into the relevant equilibrium equations and execute the solution.

$$\Sigma F_x = R_{Ax} - R_{C/A} \cos (41.41°) = 0$$

$$R_{Ax} = (97.29)(0.7500) = 72.97 \text{ lb} \qquad (c)$$

$$\Sigma F_y = R_{Ay} - W - R_{C/A} \sin (41.41°) = 0$$

$$R_{Ay} = 128.7 + (97.29)(0.6614) = 193.1 \text{ lb} \qquad\qquad \text{(d)}$$

**Step 6.** Check the solution and interpret the results. In this example, the equations are simple, and we may easily determine if the numerical results are realistic by checking the signs in each of the equilibrium equations and by repeating the calculations. Note that **forces directed in the negative coordinate directions are considered negative quantities** in writing the $\Sigma F$ terms. You should always make certain that the forces are expressed in the appropriate units.

## EXAMPLE 3.16

For the hex nut and wrench described in Example 3.4, determine the torque acting at the threads of the nut and the reaction forces $R_1$ and $R_2$.

**Solution:**

Let's refer to the FBD for the wrench presented in Fig. E3.4b. An examination of the FBD indicates that the force system is coplanar and non-concurrent. Writing the equilibrium relations yields:

$$\Sigma F_\perp = R_1 - R_2 - 120 = 0 \qquad\qquad \Sigma F_\| = 0 \qquad\qquad \text{(a)}$$

where the subscripts $\perp$ and $\|$ refer to directions perpendicular and parallel to the wrench axis, respectively.

$$R_1 = R_2 + 120 \qquad\qquad \text{(b)}$$

$$\Sigma M_O = 20(R_1 + R_2) - (200)(120) = 0 \qquad\qquad \text{(c)}$$

From Eq. (a) and Eq. (b), we solve for $R_1$ and $R_2$ to obtain:

$$R_1 = 660 \text{ N} \qquad \Rightarrow \qquad R_2 = 540 \text{ N} \qquad\qquad \text{(d)}$$

To determine the torque T applied to the nut, refer to the FBD in Fig. E3.4a.

$$\Sigma M_O = M_z - 20 R_1 - 20 R_2 = 0$$

$$T = M_z = 20(660 + 540) = 24,000 \text{ N-mm} = 24 \text{ N-m} \qquad\qquad \text{(e)}$$

The positive sign for the moment $M_z$, which acts on the threads of the nut, indicates that the direction assigned to $M_z$ in Fig. E3.4a was correct.

## EXAMPLE 3.17

Determine the reactions at the wheels of the crane using the FBD presented in Fig. E3.17. The weight lifted is 1,200 lb and the wheeled crane weighs 6,400 lb.

Fig. E3.17

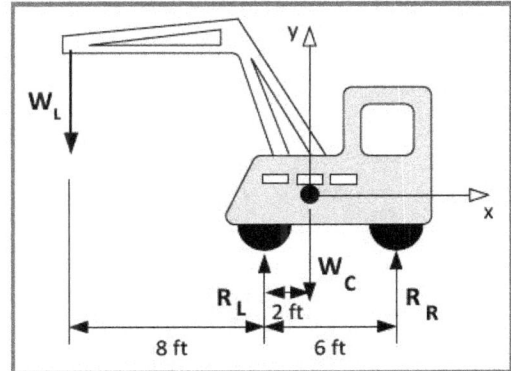

**Solution:**

Let's refer to the FBD for the crane presented in Fig. E3.17. An examination of the FBD indicates that the force system is coplanar and non-concurrent. Writing the equilibrium relations yields:

$\Sigma F_x = 0$ is satisfied regardless of the values determined for the reaction forces.

$$\Sigma F_y = R_L + R_R - W_C - W_L = 0 \tag{a}$$

$$R_L = W_C + W_L - R_R \tag{b}$$

$$\Sigma M_O = (6 - 2)R_R - 2R_L + 10W_L = 0 \tag{c}$$

Substituting Eq. (b) into Eq. (c) and solving for $R_R$ yields:

$$R_R = (1/3)(W_C - 4W_L) \tag{d}$$

$$R_L = (1/3)(2W_C + 7W_L) \tag{e}$$

Substituting $W_C = 6,400$ lb and $W_L = 1,200$ lb into Eqs. (d) and (e) gives:

$$R_R = 533.3 \text{ lb} \qquad \Rightarrow\Rightarrow \qquad R_L = 7,067 \text{ lb} \tag{f}$$

The results obtained may be checked by substituting the numerical values for $R_L$, $R_R$ $W_C$ and $W_L$ into Eq. (a) and verifying the equality.

## EXAMPLE 3.18

Using the FBD in Fig. E3.18, determine the reactions at the base of the vertical member that supports the roof of the football stadium.

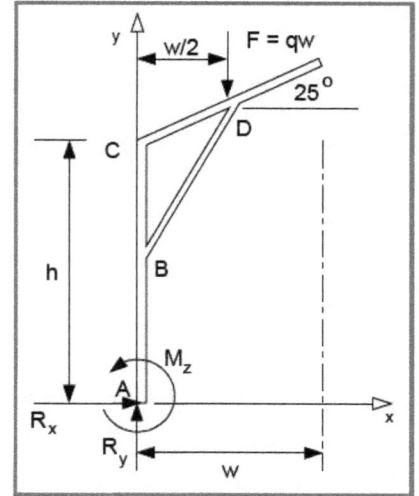

Fig. E3.18

**Solution:**

The force system is classified as coplanar and non-concurrent. Accordingly we may write three equilibrium relations.

$$\Sigma F_x = 0 \qquad \Rightarrow \qquad R_x = 0 \qquad\qquad (a)$$

$$\Sigma F_y = R_y - qw = 0 \qquad \Rightarrow \qquad R_y = qw \qquad\qquad (b)$$

$$\Sigma M_A = M_z - (qw)(w/2) \Rightarrow \qquad M_z = qw^2/2 \qquad\qquad (c)$$

A numerical value for the uniformly distributed load q has not been provided, nor has the value for the dimension w. How are we to check the solution? Consideration of the units of the quantities in Eq. (b) and Eq. (c) provides useful information. In Eq. (b), we know the reaction R must have a force unit F, the distributed load q a unit of F/L, and the distance w a unit of L. We then write the equation to check the units as:

$$F = (F/L)L \qquad \Rightarrow \qquad F = F \qquad\qquad (d)$$

Similarly for the moment M in Eq. (c), it must have a unit FL. We check the homogeneity of the units in Eq. (c) by:

$$FL = (F/L)(L)^2 \Rightarrow \qquad FL = FL \qquad\qquad (e)$$

The results from Eqs. (d) and (e) confirm that the units in our solution are homogenous as required.

## 3.6 FORCES IN CABLE AND PULLEY ARRANGEMENTS

Pulleys and cables[2] are often used in lifting devices such as cranes, derricks, and elevators. The pulley is like a wheel because it rotates about a shaft with very low friction. In fact, we will assume that the pulley is frictionless. At the perimeter of the pulley, a groove is cut to contain the cable and to prevent the cable from slipping off. The essential details of a pulley are illustrated in Fig. 3.13.

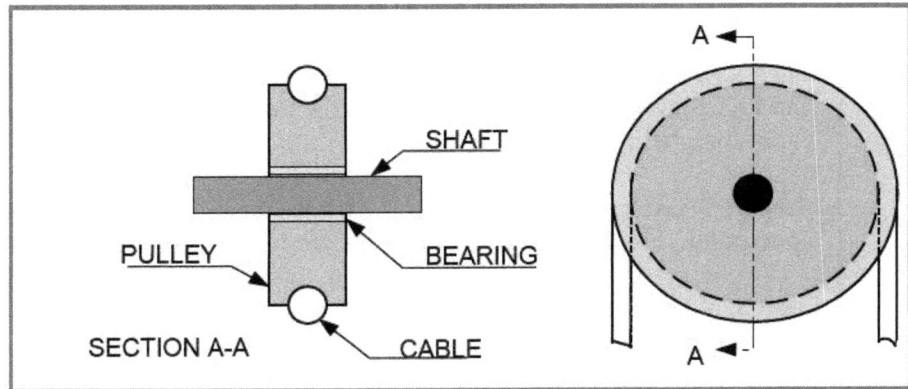

Fig. 3.13 Details of a cable and pulley assembly.

The pulley/cable system has two primary advantages. First, it can change the direction of a force. For instance, as we pull down on the cable at the left side of the pulley in Fig. 3.13, the cable on the right side will move up. Using this property, we may transport masses to significant heights. Second, pulleys and a cable properly arranged act as a machine and amplify the force capability of a person or a mechanical device. Let's consider two different examples that illustrate the advantages of a cable/pulley system either to change the direction or to amplify a force.

### EXAMPLE 3.19

The cable pulley system presented in Fig. E3.19 is being used to lift a mass of 15 slugs from the floor. Determine the force required to lift this mass and the reaction force acting on the pin in the pulley.

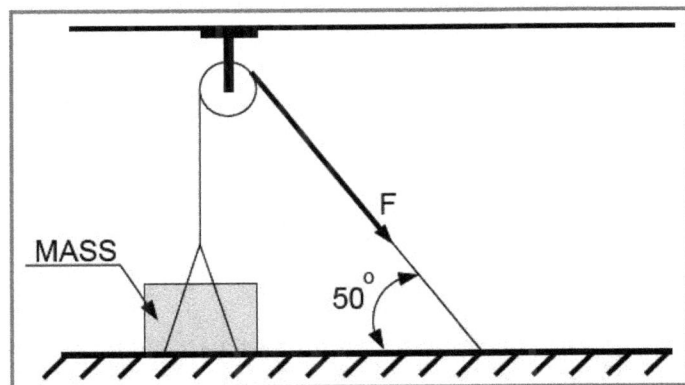

Fig. E3.19

---

[2] A cable is like a rope except that wire is used for the cable's fibers instead of hemp or some other organic material.

**Solution:**

**Step 1.** Prepare two FBDs of the pulley/cable system (one of the pulley and the other of the mass) as shown in Fig. E3.19a. After lifting the mass from the floor, we cut the cable from the mass and replace it with the force $F_1$. A force W, acting downward, equal to the weight of the mass is applied. The weight of the mass is determined from Eq. (1.8) as:

$$W = mg = (15)(32.17) = 482.6 \text{ lb}$$

Then we remove the support attachment from the pulley and replace it with force components $F_{2x}$ and $F_{2y}$. Since we have cut the cable to the mass, it is necessary to apply a force $F_1$ to replace the cable completing the FBD of the pulley.

**Step 2.** Classify the force systems. For the mass, the force system is collinear, coplanar and concurrent. For the pulley, the force system is coplanar and non-concurrent.

**Step 3.** Write the relevant equations of equilibrium. For the mass, only one of the six equations of equilibrium is relevant.

$$\Sigma F_y = 0$$

For the pulley, only three of the six equations of equilibrium are relevant.

$$\Sigma F_x = 0 \qquad\qquad \Sigma F_y = 0 \qquad\qquad \Sigma M_O = 0$$

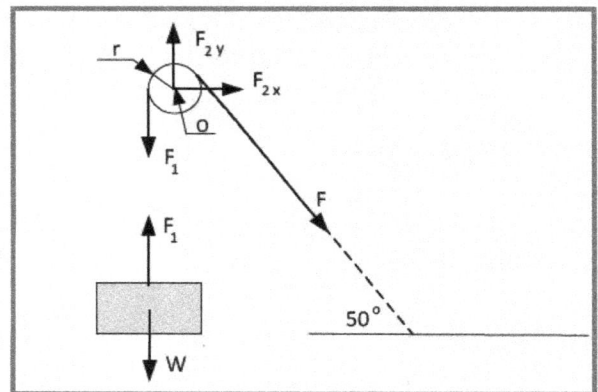

**Step 4 and 5.** Substitute the unknowns from the FBD into the relevant equations of equilibrium for both the mass and the pulley and then execute the solutions. For the mass, we write:

$$\Sigma F_y = F_1 - W = 0$$

$$F_1 = W = 482.6 \text{ lb} \tag{a}$$

The pulley, with a radius r, is assumed to be frictionless. We write moments about the center O of the pulley to obtain:

$$\Sigma M_O = (F_1)(r) - (F)(r) = 0$$

$$F = F_1 = 482.6 \text{ lb} \tag{b}$$

It is clear from Eq. (b) that **the tension in the cable remains constant across the pulley** even though the direction of the forces F and $F_1$ are different.

Next, consider the sum of the forces in the x and y directions for the pulley.

$$\Sigma F_x = F_{2x} + F \cos 50° = 0$$

$$F_{2x} = - (482.6)\cos 50° = - 310.2 \text{ lb}$$

(c)

The minus sign indicates that the direction for $F_{2x}$ in Fig. E3.19a is not correct. The force component is actually in the negative x direction.

Next, consider the sum of the forces in the y direction.

$$\Sigma F_y = F_{2y} - F \sin 50° - F_1 = 0$$

$$F_{2y} = 482.6 (1 + 0.7660) = 852.3 \text{ lb}$$

(d)

$$F_2 = [F_{2x}^2 + F_{2y}^2]^{1/2} = [(-310.2)^2 + (852.3)^2]^{1/2} = 907.0 \text{ lb}$$

**Step 6.** Check the solution and interpret the results. Again, the mathematics involves solving simple linear algebraic equations. We check the signs in each of the equilibrium equations and repeat the calculation. We check to ascertain that the correct units for each of the unknown forces have been assigned. We compare the forces F, $F_1$ and $F_2$ and find that their magnitudes are reasonable.

## EXAMPLE 3.20

For the pulley and cable arrangement shown in Fig. E3.20, determine the force F required to maintain a weight W of 10 kN in equilibrium.

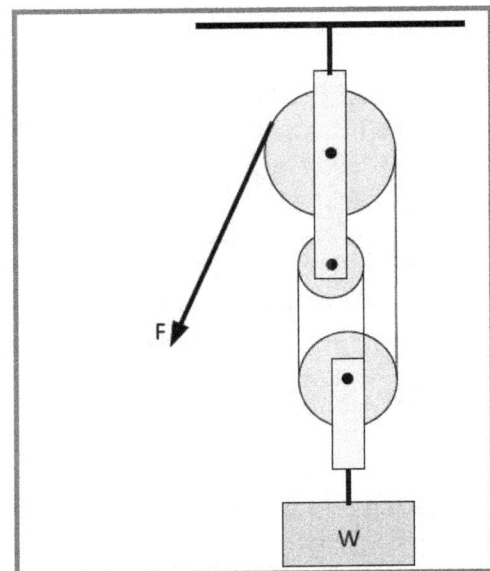

Fig. E3.20

**Solution:**

**Step 1.** Prepare a FBD of the pulley/cable system as shown in Fig. E3.20a. We cut the three cables above the lower pulley and apply a force F to each cable. Recall that the tension in the cable is a constant across a pulley when the friction at the shaft is zero. A force W = 10 kN, acting downward replaces the weight.

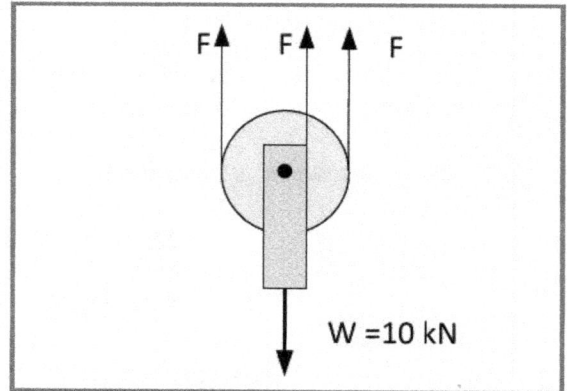

**Step 2.** Classify the force systems. For the lower pulley, the force system is coplanar and non-concurrent.

**Step 3.** Write the relevant equations of equilibrium. For the pulley, only one of the six equations of equilibrium is relevant.

$$\Sigma F_y = 0$$

**Step 4 and 5.** Substitute the unknowns from the FBD into the relevant equations of equilibrium for the pulley and then execute the solutions.

$$\Sigma F_y = 3\,F - W = 0$$

$$F = W/3 = 3.333 \text{ kN} \qquad\qquad (a)$$

**Step 6.** Check the solution and interpret the results. Again, the mathematics involved is trivial. What is important is the realization that this pulley/cable arrangement permits one to lift a weight of 10 kN with a force of only 3.33 kN. The pulley/cable arrangement of Fig. E3.20 is a machine enabling one to amplify applied forces. While the forces are amplified, the work to lift the weight remains constant. The forces are decreased by a factor of three but the distance through which the force is moved to lift the weight is increased by that same factor. No work is gained by a pulley/cable arrangement.

# EXAMPLE 3.21

Consider the pulley arrangement that supports blocks A, B and C as shown in Fig. E3.21. Blocks A and B each have a mass of 80 kg. The angle $\beta = 30°$. Determine the angle $\theta$ and the mass of block C if the system is in equilibrium.

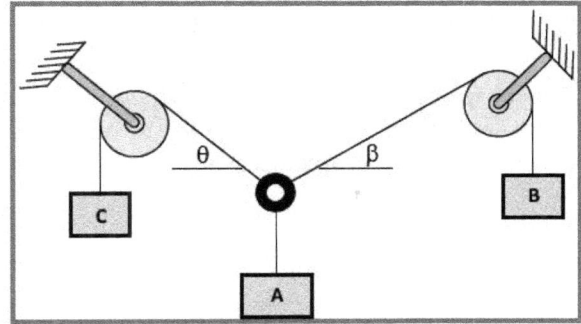

Fig. E3.21

**Solution:**

Prepare a FBD of the ring as shown in Fig. E3.21a.

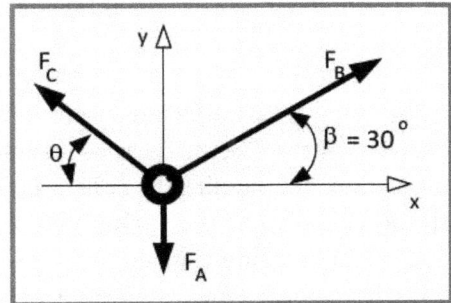

Fig. E3.21a

An examination of the FBD indicates that the force system is coplanar and concurrent. Accordingly, we may write:

$$\Sigma F_x = 0 \Rightarrow \quad F_B \cos(30°) - F_C \cos\theta = 0 \tag{a}$$

$$\Sigma F_y = 0 \Rightarrow \quad F_B \sin(30°) + F_C \sin\theta - F_A = 0 \tag{b}$$

Since we know the mass $m_A = m_B = 80$ kg for blocks A and B, we may write:

$$F_A = F_B = m_A \, g \tag{c}$$

Substitute Eq. (c) into Eq. (b) and solve for $F_C$:

$$F_C = [1/(\sin\theta)][F_A - F_B/2] = m_A g/2\sin\theta \tag{d}$$

Substitute Eq. (d) into Eq. (a) and solve for the angle $\theta$:

$$m_A g \cos(30°) - [m_A g/(2\sin\theta)]\cos\theta = 0 \quad \Rightarrow \quad \tan\theta = 1/[(2)(0.8660)] = 0.5774$$

$$\theta = 30° \tag{e}$$

Finally, to determine the mass of block C, substitute Eq. (e) into Eq(d):

$$F_C = m_A \, g/(2\sin\theta) = m_C \, g \quad \Rightarrow \quad m_C = 80/[(2)(0.500)] = 80 \text{ kg} \tag{f}$$

## 3.7 FORCES IN SPRINGS

Helical springs are often used to develop forces and to store energy in mechanical systems. These springs are usually wound from wire in the form of a helix as shown in Fig. 3.14. The spring shown in Fig. 3.14 develops a tension force because the hooks at both ends enable the spring to be stretched.

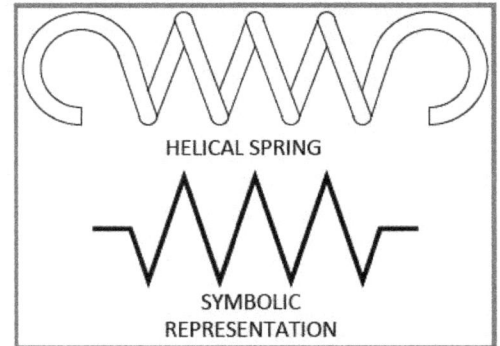

HELICAL SPRING

SYMBOLIC
REPRESENTATION

Fig. 3.14 A tension type helical spring.

If the spring is manufactured with flat ends, it is employed to support compression forces that squeeze the coils together. The stiffness or rigidity of the spring is controlled by the size of the wire used in winding the spring, the number of coils and the inside and outside diameters of the helix. The stiffness of the spring is known as the **spring rate** k. The force F required to elongate a tension spring or to squeeze a compression spring is given by:

$$F = k\delta \qquad (3.3)$$

where $\delta$ is the amount of extension or compaction of the helical spring.

The extension (or compression) $\delta$ of the spring is established by:

$$\delta = L_d - L_o \qquad (3.4)$$

where $L_d$ and $L_o$ are deformed and original length of the spring, respectively.

Because helical springs are difficult to draw, we represent them with a zigzag line drawing as illustrated in Fig. 3.14.

### EXAMPLE 3.22

Suppose a tension spring with an original length of 9 inches is extended to a deformed length of 10.6 inches. Determine the tension force required to stretch the spring to its deformed length if the spring rate is 28 lb/in.

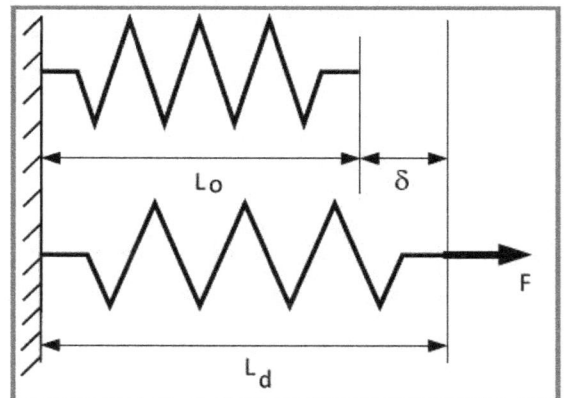

**Solution:**

From Eq. (3.4), we may write:

$$\delta = L_d - L_o = 10.6 - 9 = 1.6 \text{ in.} \qquad (a)$$

Substitute the results from Eq. (a) into Eq. (3.3) to obtain:

$$F = k\delta = (28)(1.6) = 44.8 \text{ lb} \qquad (b)$$

## EXAMPLE 3.23

Suppose a compression spring with an original length of 350 mm is compacted to a deformed length of 273 mm. Determine the compression force required to compress the spring to its deformed length if the spring rate is 85 N/mm.

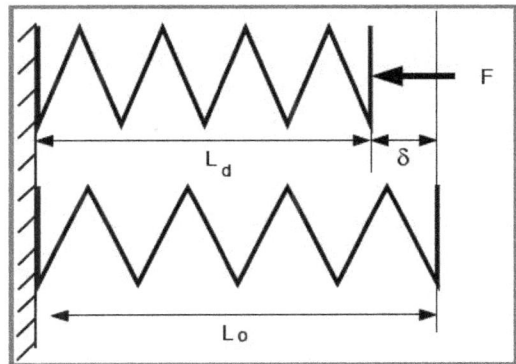

**Solution:**

From Eq. (3.4), we may write:

$$\delta = L_d - L_o = 273 - 350 = -77 \text{ mm} \qquad (a)$$

Substitute the results from Eq. (a) into Eq. (3.3) to obtain:

$$F = k\delta = (85)(-77) = -6.545 \text{ kN} \qquad (b)$$

Note the negative sign indicates that the force acting on the spring is compressive.

## 3.8 SOLVING FOR INTERNAL FORCES

We now have learned to use FBDs and the equations of equilibrium to solve for cable tension and reactions at the supports of beams and other structures. Let's now consider using the equilibrium relations to solve for the internal forces that develop within a structural member due to external loading.

## EXAMPLE 3.24

Consider the stepped tension bar loaded with external forces illustrated in Fig. E3.24.

    (a) Determine the internal force P that occurs in the thick portion of the stepped bar.

    (b) Determine the internal force P that occurs in the thin portion of the stepped bar.

Fig. E3.24

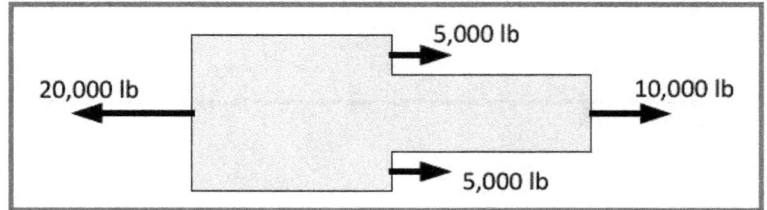

**Solution**:

First, let's execute the six-step procedure to solve for the unknown magnitude of the internal force P in the thicker portion of the stepped tension bar.

**Step 1**. Prepare an appropriate FBD of the bar as shown in Fig. E3.24a. We begin this FBD by making a section cut A-A, which frees the left end of the bar. We will use the left end of the bar with the equilibrium relations to determine the internal force P. The FBD of the left end of the bar is also presented in Fig. E3.24a.

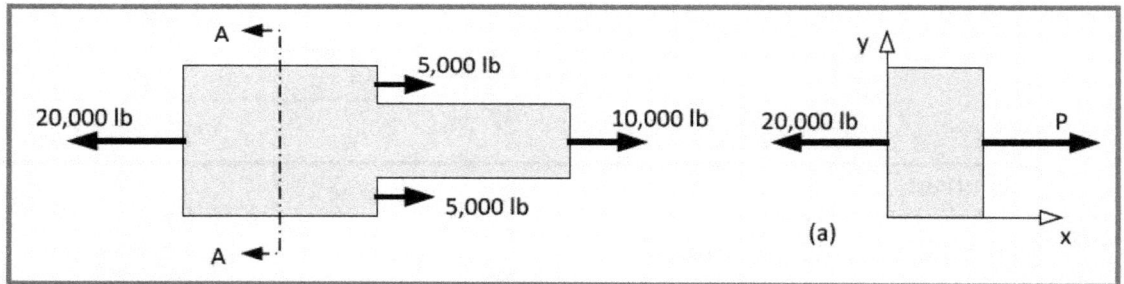

Fig. 3.24a

**Step 2.** Classify the force system. In this problem, the force system is coplanar and parallel.

**Step 3.** Write the relevant equations of equilibrium. In this case, only one of the six equations of equilibrium is relevant.

$$\Sigma F_x = 0$$

It is evident from inspection that the relations $\Sigma F_y = 0$ and $\Sigma M = 0$ are satisfied regardless of the value determined for the unknown force P.

**Step 3.** Substitute the unknowns from the FBD into the relevant equations of equilibrium.

$$\Sigma F_x = 0$$

$$\Sigma F_x = P - 20,000 \text{ lb} = 0$$

**Step 5.** Execute the solution, which is obvious in this example.

$$P = 20,000 \text{ lb}$$

**Step 6.** Check the solution. In this problem the equations were so simple that they were solved by inspection. We have included the units in the result. Since the bar is in tension, we understand that the internal force must be positive as indicated in our solution.

Making a section cut across the thinner end of the bar and repeating the six-step process described above solves part (b) of this example. The result obtained is $P = 10,000$ lb over the length of the thinner portion of the stepped tension bar. A force diagram showing the value of P over the length of the stepped tension bar is shown in Fig. E3.24b.

We have now explored the **solution space** for this problem, and have plotted all of the possible values of P occurring in the tension bar in the force-position diagram shown above. By preparing a graph of the solution for each position along the length of the bar, it is possible to determine the locations where the critical loads will occur. It will be these locations that are of concern when designing structures to resist failure.

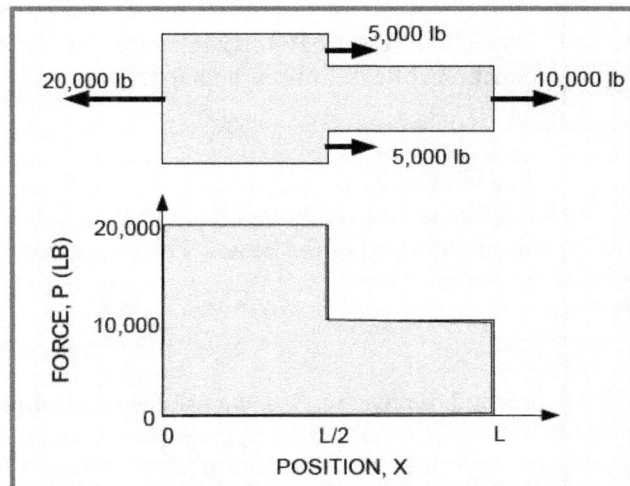

Fig. E3.24b

This example has shown that it is very easy to determine the internal forces in a tension bar. Let's consider a more complex internal force problem—a simply supported beam with a transverse concentrated force that is shown in Example 3.25.

## EXAMPLE 3.25

Determine the internal force and internal moment developed in a simply supported beam with a transverse concentrated load at the locations defined by x = 2 m, x = 3.5 m and x = 5.0 m, where x is defined from the left end of the simply supported beam shown in Fig. E3.25.

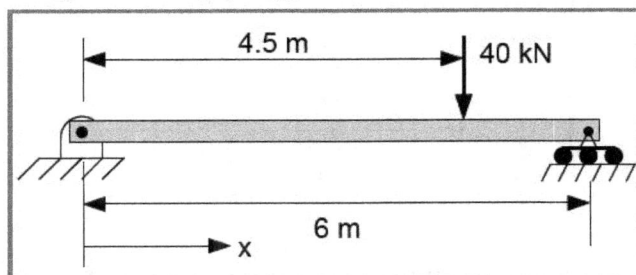

Fig. E3.25

**Solution:**

Let us begin the solution by solving for the reactions $R_x$, $R_L$ and $R_R$ due to the simple supports. We construct the FBD for the entire beam following the procedure described in previous examples.

- Remove the supports.
- Add the reaction forces $R_x$, $R_L$ and $R_R$.
- Define a coordinate system.

The FBD for the beam is shown in Fig. E3.25a.

Fig. E3.25a

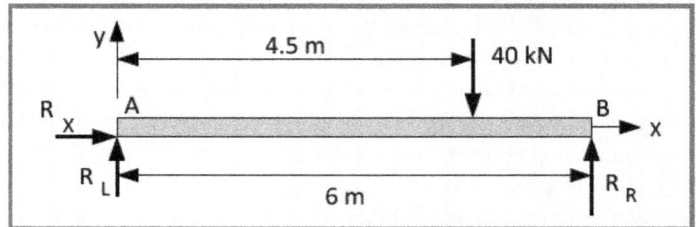

Inspection of this FBD indicates that the force system is coplanar and non-concurrent. Hence the equations of equilibrium that apply are:

$$\Sigma F_x = 0 \qquad \Sigma F_y = 0 \qquad \Sigma M = 0$$

The relation $\Sigma F_x = 0$ enables us to show that $R_x = 0$. The relation $\Sigma M = 0$ is very useful since we are free to choose the point about which the moments are to be determined. Let's select point B at the right end of the beam. Then we write:

$$\Sigma M_B = (40 \text{ kN})(1.5 \text{ m}) - (6 \text{ m})R_L = 0$$

$$R_L = (40 \text{ kN/6 m})(1.5 \text{ m}) = 10 \text{ kN}$$

To find the reaction $R_R$, we use the remaining equation of equilibrium.

$$\Sigma F_y = R_L + R_R - 40 \text{ kN} = 0$$

$$R_R = 40 \text{ kN} - 10 \text{ kN} = 30 \text{ kN}$$

Now that the external reaction forces acting at the beam supports have been determined, we proceed to solve for the internal forces at x = 2 m by making a section cut A-A through the beam at this location (see figure below). We then remove the left portion of the beam, freed by the section cut, as a free body. The FBD, which is the model for the equilibrium analysis, is shown in inset (b) of this figure.

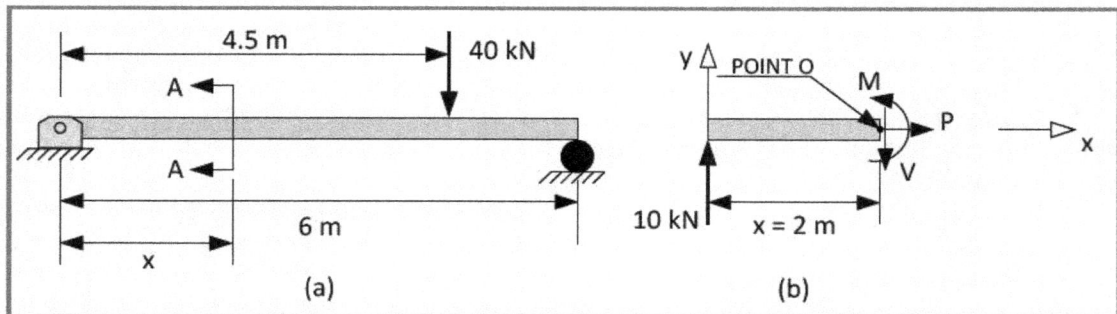

The FBD shows an internal force P (tensile) in the positive x direction, and another internal shear force V, which we will consider positive when it acts in the negative y direction. An internal moment $M_z$ about the point O is shown indicating rotation in the counterclockwise direction. We have drawn the arrows representing P, V and M assuming they are all positive quantities.

An examination indicates the force system is coplanar and non-concurrent. The appropriate equations of equilibrium are:

$$\Sigma F_x = 0 \qquad \Sigma F_y = 0 \qquad \Sigma M_o = 0$$

From the relation $\Sigma F_x = 0$, we find:

$$\Sigma F_x = P = 0$$

It is clear that the internal axial force P in a transversely loaded beam vanishes.

From the relation $\Sigma F_y = 0$, we find:

$$\Sigma F_y = 10 \text{ kN} - V = 0$$

$$V = + 10 \text{ kN}$$

The positive sign for V indicates our assumption was correct—this vertical force is in the negative y direction. The internal force V is a shear force, which we will discuss in much more detail in a later course titled Mechanics of Materials.

The relation $\Sigma M_O = 0$ permits us to solve for the internal moment $(M_z)_A$.

$$\Sigma M_O = (M_z)_A - (10 \text{ kN})(2 \text{ m}) = 0$$

$$(M_z)_A = 20 \text{ kN-m}$$

In reality, this internal moment is not produced by a rotating force vector as symbolized in the FBD. Instead, it is produced by a linear distribution of bending stresses that act normal to the surface exposed in section A-A. The rotating force vector is used to simplify the representation of the moment in the FBD while capturing the rotational effect it has on the body.

Let's continue our analysis of the internal forces in the beam by considering the location x = 4.5 m. Note that we freed the left portion of the beam with the section cut B-B. The cut was at x = 4.5 m, but incrementally to the left side of the 40 kN concentrated force as shown in insert (a) of Fig. E3.25c. Accordingly, the 40 kN force is not included in the FBD shown in inset (b) of Fig. E3.25c.

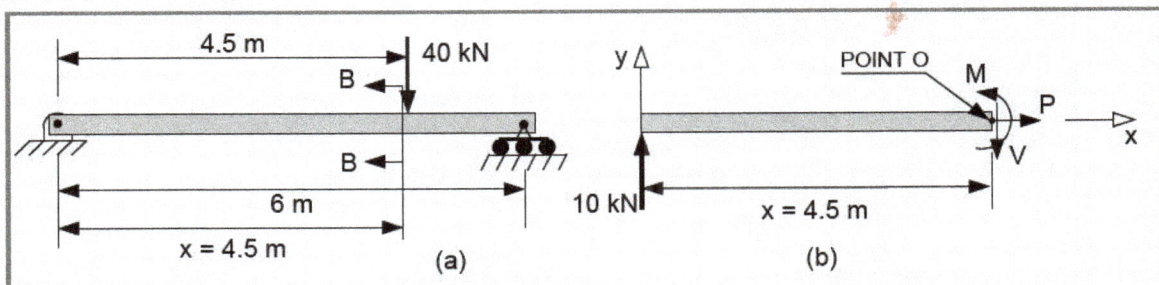

Fig. E3.25c

An examination indicates the force system is coplanar and non-concurrent. The appropriate equations of equilibrium are:

$$\Sigma F_x = 0 \qquad \Sigma F_y = 0 \qquad \Sigma M_o = 0$$

From the relation $\Sigma F_x = 0$ we again find:

$$\Sigma F_x = P = 0$$

From the relation $\Sigma F_y = 0$ we find:

$$\Sigma F_y = 10 \text{ kN} - V = 0$$

$$V = +10 \text{ kN}$$

This result is identical to that determined for V at x = 2 m. Indeed, the shear force V is a constant for the region of the beam 0 < x < 4.5 m.

The relation $\Sigma M_O = 0$ permits us to solve for the internal moment $(M_z)_B$.

$$\Sigma M_O = (M_z)_B - (10 \text{ kN})(4.5 \text{ m}) = 0$$

$$(M_z)_B = 45 \text{ kN-m}$$

If we examine the results for $M_z$ at x = 2 and 4.5 m, it is clear that the moment increases linearly with x. Indeed, we can write an equation showing the variation of the moment with location x as:

$$M_z = 10 \text{ x} \qquad \text{for } 0 < x < 4.5$$

The moment $M_z$ is in units of kN-m when x is measured in meters. We must constrain this relation for $M_z$ to the range of x covered by the FBDs that have been considered in the analysis. At this stage in the analysis, it is clear that the moment increases from zero at the left end of the beam to a maximum at x = 4.5 m. To determine the internal forces and moments in the beam for values of x > 4.5 m, it is necessary to consider a third FBD of the beam with a section cut C-C located at x = 5 m as presented in Fig. 3.25d.

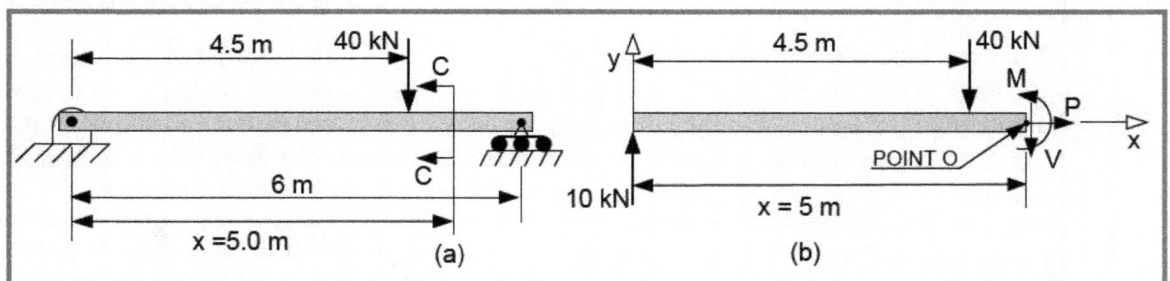

Fig. E32.5d

Following the procedure established previously, we again find:

$$\Sigma F_x = P = 0$$

From the relation $\Sigma F_y = 0$, we determine V as:

$$\Sigma F_y = 10 \text{ kN} - V - 40 \text{ kN} = 0$$

$$V = -30 \text{ kN}$$

Observe that the sign for the shear force is negative. This fact indicates that our assumption for V oriented in the negative y direction was not correct. For $x > 4.5$ m, the shear force is directed upwards in the positive y direction. The shear force V undergoes a step change when x locates a point incrementally to the right of the concentrated load of 40 kN.

The relation $\Sigma M_O = 0$ permits us to solve for the internal moment $(M_z)_C$.

$$\Sigma M_O = (M_z)_C - (10 \text{ kN})(5 \text{ m}) + (40 \text{ kN})(0.5 \text{ m}) = 0$$

$$(M_z)_C = 30 \text{ kN-m}$$

It is interesting to observe that the value of $M_z$ at $x = 5$ m is less than its value at $x = 4.5$ m. If we let $x = 6$ m, it is easy to show that $M_z = 0$. Indeed, at both supports $M = 0$ because the beam is free to rotate at its ends. The simple supports cannot provide constraint against rotation and the moments must vanish at their locations.

In Example 3.25, we examined the internal forces and moments in the transversely loaded beam at three different positions along its length. We also observed that the internal moments vanish at the ends of the beam. Let's use this information to draw the diagrams presented in Fig. 3.15 showing the shear force V and the internal moment M as a function of position along the length of the beam.

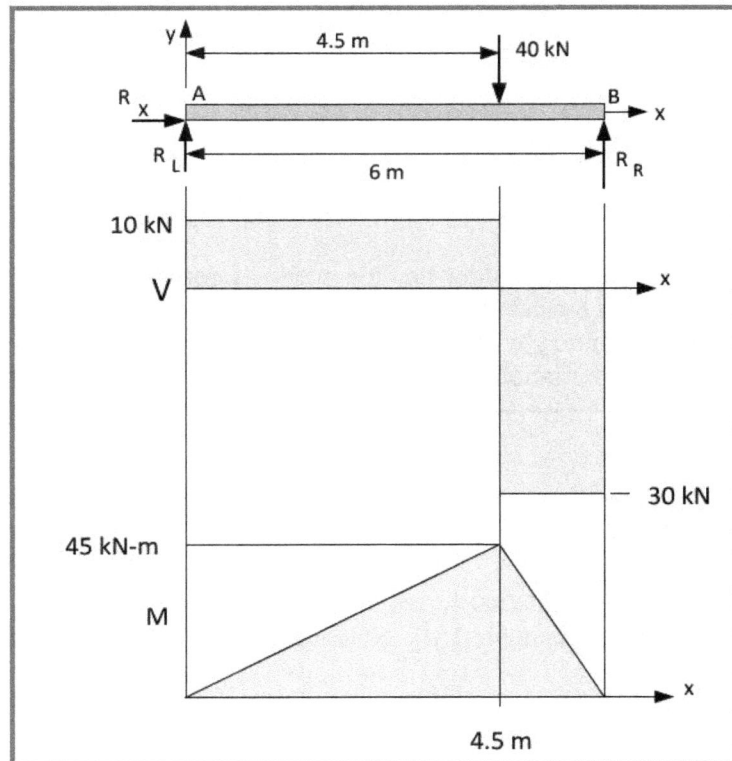

Fig 3.15 Shear and bending moment diagrams for a simply supported beam with a single transverse concentrated force at $x = 4.5$ m.

We will describe shear and bending moment diagrams in much more detail in the next Mechanics course Mechanics of Materials.

## EXAMPLE 3.26

Determine the internal forces and moment in the uniformly loaded beam presented in Fig. E3.26. Solving for the reactions at the pivot support at the left end of the beam gives $R_x = 0$ and $R_y = qL/2$.

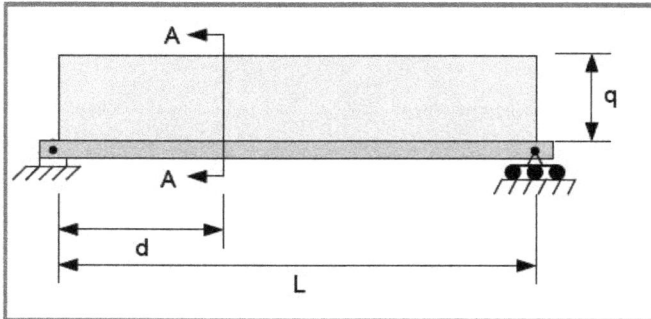
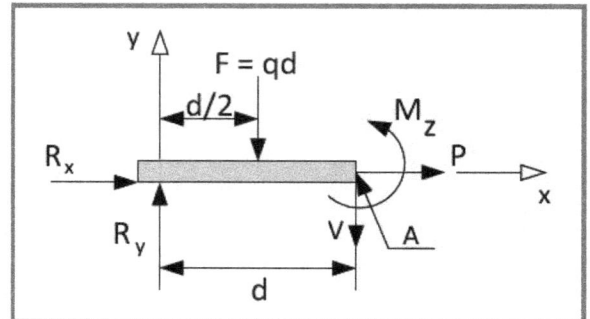

Fig. E3.26                    Fig. E3.26a

**Solution:**

Examination of Fig. E3.26a, shows that the force system is coplanar and non-concurrent. Hence, we write three equilibrium equations to solve for the three unknowns P, V and $M_z$.

$$\Sigma F_x = R_x + P = 0 \qquad\Rightarrow\Rightarrow\qquad P = -R_x = 0 \qquad\qquad (a)$$

$$\Sigma F_y = R_y - qd - V = 0 \qquad\Rightarrow\Rightarrow\qquad V = qL/2 - qd = q(L/2 - d) \qquad (b)$$

$$\Sigma M_A = M_z - R_y\, d + (Fd)/2 = 0 \qquad\Rightarrow\Rightarrow\qquad M_z = (qL/2)d - (qd)(d/2)$$

$$M_z = (qL/2)d - (qd)(d/2) = (qd/2)(L - d) \qquad\qquad (c)$$

The results show that the internal force V and the internal moment $M_z$ depend on the dimension d that locates the section cut. V is a maximum when d = 0, and $M_z$ is a maximum when d = L/2. You may wish to construct a graph of V or M as a function of d as it ranges from 0 to L, to show the variation of the internal forces and moment over the beam's length.

## EXAMPLE 3.27

Determine the internal forces and moment at section A-A in the frame of the factory building described previously in Example 3.10. Note that L = 120 ft, $R_y$ = 60 tons, and $F_1$ = 25 tons.

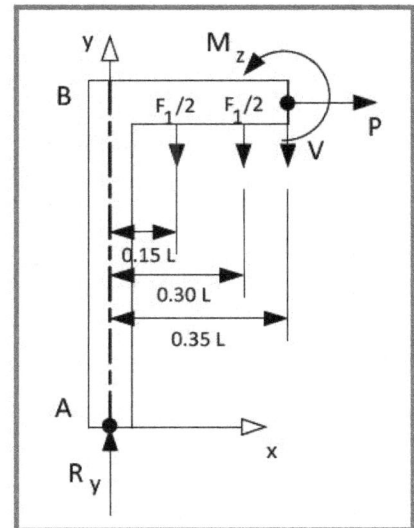

Fig. E3.27

---

**Solution:**

Examination of the FBD presented in Fig. E3.27, shows that the force system is coplanar and non-concurrent. Accordingly, we may write three equilibrium equations to solve for the three unknowns P, V and $M_z$.

$$\Sigma F_x = P = 0 \tag{a}$$

$$\Sigma F_y = R_y - F_1/2 - F_1/2 - V = 0$$

$$V = (60)(2000) - (25)(2000) = 70,000 \text{ lb} \tag{b}$$

$$\Sigma M_O = M_z - R_y (0.35L) + (F_1/2)(0.05L + 0.20L) = 0$$

$$M_z = (2000)(120)[(60)(0.35) - (25/2)(0.25)] = 4.290 \times 10^6 \text{ ft-lb} \tag{c}$$

The results show that the internal force V and the internal moment $M_z$ depend on the position of the section cut.

---

## EXAMPLE 3.28

The C-clamp shown in Fig. E3.28 is tightened onto a block. If the screw applies a force F = 600 lb to the block, determine the internal forces at the section cut A-A along the straight portion of its back.

Fig. E3.28

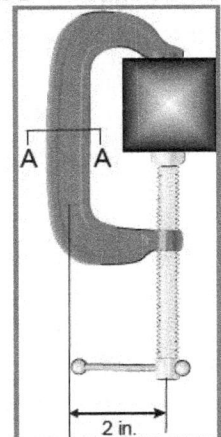

**Solution:**

Next classify the force system as coplanar, non-concurrent. Accordingly, we have three relevant equations of equilibrium. Writing these relations yields:

$$\Sigma F_x = V = 0 \tag{a}$$

$$\Sigma F_y = F - P = 0 \Rightarrow \Rightarrow \qquad P = F = 600 \text{ lb} \tag{b}$$

$$\Sigma M_O = 2F + M_z = 0 \qquad \Rightarrow \Rightarrow \quad M_z = -2F = -1,200 \text{ in.-lb} \tag{c}$$

The negative sign for $M_z$ indicates that the direction of the rotating vector shown in Fig. E3.28a is not correct. This rotating vector should be in the clockwise direction on the FBD to maintain equilibrium of the top portion of the C-clamp.

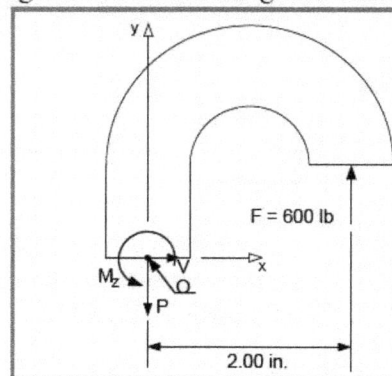

Fig. E3.28a

## 3.9 SUMMARY

The two equilibrium equations in vector form, $\Sigma \mathbf{F} = 0$ and $\Sigma \mathbf{M} = 0$, may be represented by the six equations involving Cartesian components of the forces and moments as:

$$\Sigma F_x = 0, \qquad \Sigma F_y = 0, \qquad \text{and } \Sigma F_z = 0 \tag{3.1}$$

$$\Sigma M_x = 0, \qquad \Sigma M_y = 0, \qquad \text{and } \Sigma M_z = 0 \tag{3.2}$$

It is useful in selecting the appropriate equilibrium relations to classify the force system acting on a structure into one of the following categories.

- Coplanar and concurrent.
- Coplanar and non-concurrent.
- Non-coplanar and concurrent.
- Non-coplanar and non-concurrent.

When the force system is classified it is apparent which of the six Cartesian equilibrium relations provide meaningful results.

It is difficult to overemphasize the importance of preparing complete and correct FBDs. The FBD represents a model of the structure under consideration. It provides a guide in writing the equilibrium equations, information on pertinent dimensions, and a record of the assumptions regarding the direction of the forces and moments. A systematic method for constructing FBDs includes the following four-step procedure:

1. Isolating the body.
2. Replacing supports with reaction forces and/or moments.
3. Replacing distributed loads with concentrated loads positioned at the centroid of the area representing the distributed load.
4. Dimensioning the diagram and defining a coordinate system.

Methods for solving equilibrium problems in statics are covered in detail. We suggest a systematic six step technique that includes:

1. Prepare a complete FBD of the component under consideration.
2. Classify the force system.
3. Write the relevant equations of equilibrium.
4. Substitute the known and unknown quantities shown in the FBD into the relevant equations of equilibrium.
5. Execute the solution.
6. Check the solution and interpret the results.

Many examples are provided to demonstrate techniques for solving for reaction forces in structures with coplanar force systems. The following relevant equilibrium relations were employed depending on the classification of the force systems.

**Coplanar and concurrent:**    $\Sigma F_x = 0$    $\Rightarrow$    $\Sigma F_y = 0$

**Coplanar and non-concurrent:**    $\Sigma F_x = 0$    $\Rightarrow$    $\Sigma F_y = 0$    $\Rightarrow$    $\Sigma M_z = 0$

The methods employed in the analysis of non-coplanar force systems will be described in Chapter 7.

An analysis of cable/pulley arrangements was made to illustrate the ability of pulleys to change the direction of an applied force and with certain arrangements to amplify an applied force. Also the helical spring was introduced as a device that develops a force in proportion to its deformation $\delta$. The relation governing this force is given by:

$$F = k\delta \qquad\qquad (3.3)$$

A method of sectioning structural members provides a means to isolate portions of the structure. FBDs of these partial members yield models that are analyzed to solve for the internal forces and moments. Examples are provided to show solution techniques for determining internal forces in tension members and internal shear forces and bending moments in beams, as well as in other components and structures. An extended example was covered to introduce the concept of shear force and bending moment diagrams that will be used extensively in the course Mechanics of Materials.

## PROBLEMS

3.1 An auto weighing 17,500 N is traveling on a straight highway down a hill with a grade of 3% at a constant velocity of 85 km/h. Construct a FBD of the auto showing all of the forces acting on it. A highway with a grade of 3% elevates 3 m for every 100 m in the horizontal direction.

3.2 A sports utility vehicle weighing 21,500 N is traveling on a straight highway up a hill with a grade of 2% at a constant velocity of 95 km/h. Construct a FBD of the vehicle showing all of the forces acting on it. A highway with a grade of 2% elevates 2 m for every 100 m in the horizontal direction.

3.3 An auto weighing 3,400 lb is traveling on a straight highway up a hill with a 2% grade at a constant velocity of 65 MPH. Construct a FBD of the auto showing all of the forces acting on it. A highway with a grade of 2% elevates 2 ft for every 100 ft in the horizontal direction.

3.4 A sports utility vehicle weighing 4,500 lb is traveling on a straight highway down a hill with a 3.5% grade at a constant velocity of 70 MPH. Construct a FBD of the auto showing all of the forces acting on it.

3.5 Prepare a sketch illustrating the force system described below and write its relevant equilibrium relations.

(a) Coplanar and concurrent          (c) Coplanar and non-concurrent

(b) Non-coplanar and concurrent      (d) Non-coplanar and non-concurrent

3.6 Construct a FBD for the mass and two-cable assembly shown in the figure to the right. Also determine the tension forces in cables AC and BC.

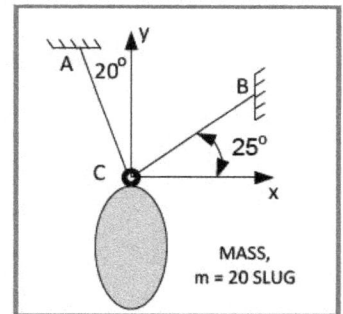

3.7 Construct a FBD for the mass and two-cable assembly shown in the figure to the left. Also determine the tension forces in cables AC and BC.

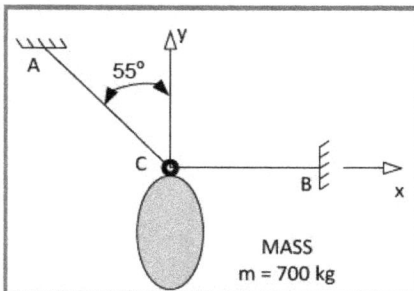

3.8 Construct a FBD for the mass and three-cable assembly shown in the figure to the right. Determine the tension forces in cables AC, BC and CD. Assume cable CD supports 30% of the mass.

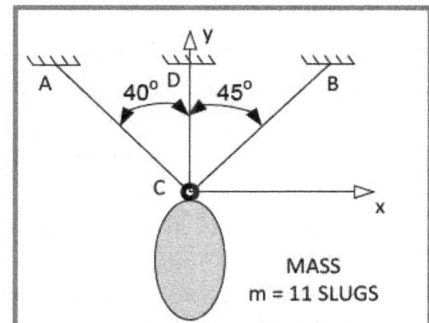

3.9 Determine the forces in cables AC and BC that support the signal lights at a traffic intersection as illustrated in the figure to the left. The signal lights have a mass m = 40 kg. The dimensions for the cable arrangement are given in meters.

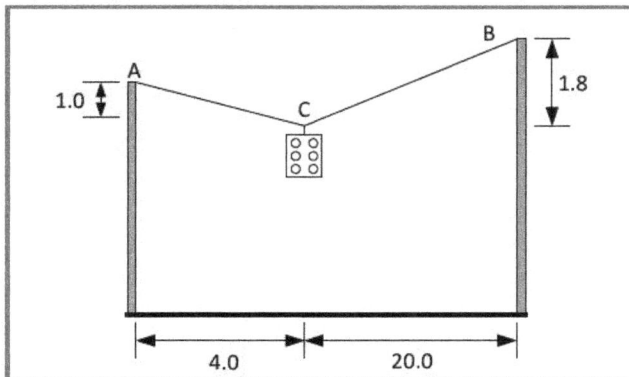

3.10 A container A that holds a 1.0 ton weight ton is lifted from the floor with a cable and pulley system shown in the figure to the right. Determine the equation for the force F required to lift the container as a function of $\alpha$. Evaluate this equation to determine the force F if $\alpha = 20°$. Comment on the effectiveness of this cable pulley system.

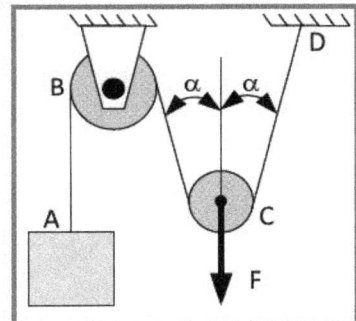

3.11 A container holding a mass m = 850 kg is lifted from the floor with a cable and pulley system shown in the figure to the right. Determine the force F required to lift the container as a function of $\alpha$. Evaluate this equation to determine the force F if $\alpha = 17°$. Comment on the effectiveness of this cable pulley system.

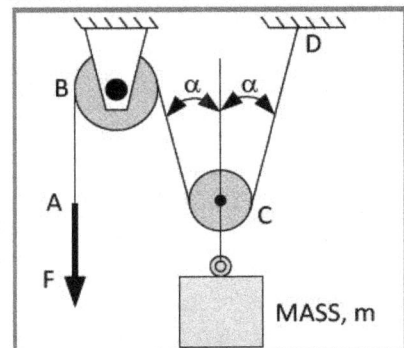

3.12 Using the pulley and cable arrangement shown in the figure to the right, design a lifting arrangement that requires only a force of 1,000 lb to lift a container weighing 1,800 lb. Assume that the pulleys are frictionless.

3.13 Using pulleys and cables, design a lifting arrangement that requires only a force of 500 lb to lift a one-ton container. Assume that the pulleys are frictionless.

3.14 For the pulleys and cable arrangement shown in Problem 3.11, determine the force required to lift a 4.5 kN weight as a function of the angle $\alpha$ the cables make with pulley C. Let the angle $\alpha$ vary from zero to 90°. It is suggested that you use a spreadsheet for these calculations and prepare a graph showing the results.

3.15 A differential hoist is illustrated in the figure below and to the right. At the top, two pulleys with radii $r_1$ and $r_2$ are fastened together so they turn as a single unit. A continuous cable passes around the smaller pulley, with a radius $r_2$, and then around the lower pulley with a radius $r_3$, and finally around the larger pulley with a radius $r_1$. We assume that the pulleys are frictionless relative to the shafts passing through their hubs. We also assume that the cables do not slip in the grooves of the pulleys. If the diameter of the lower pulley is D $= 2r_3 = r_1 + r_2$, determine the equation for the applied force F in terms of the load W, which is maintained in equilibrium.

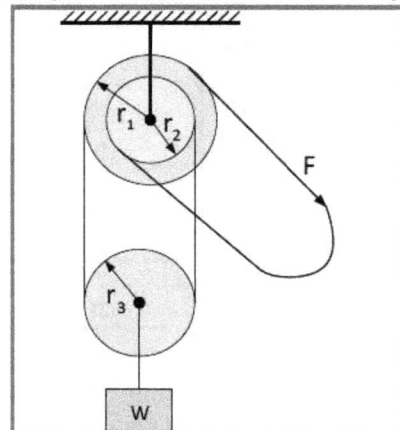

3.16 An artist has designed a bell using a long piece of pipe as the vibrating sound source. The pipe is to be supported using the wire ring arrangement shown in the figure to the left. Determine the force in each of the wires. The pipe has a mass of 43 kg.

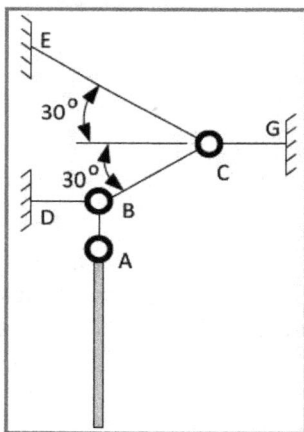

3.17 The pulley and cable, shown in the figure to the right, support a mass m = 12 slug suspended at point A. The cable B-A-C is 65 ft long and sags prior to the addition of the pulley and the container.

(a) If the pulley diameter is small relative to the span of 50 ft between the walls at B and C, determine the height h associated with the equilibrium position of the pulley and container.

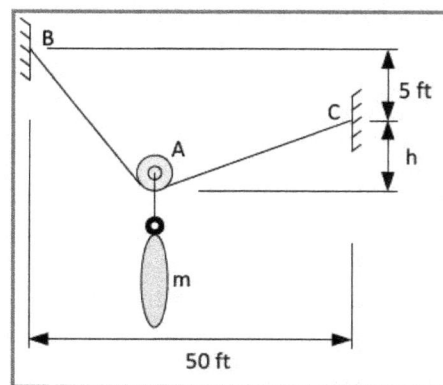

(b) If the pulley and container are released from point C, describe the motion, which you would observe as the system reaches equilibrium.

(c) If the pulley and container are released from point B, describe the motion, which you would observe as the system reaches equilibrium.

3.18 Two masses $m_B$ and $m_C$ are supported by the cable arrangement shown in the figure to the right. Determine the mass $m_C$ and the forces in the cables AB, BC and CD if the mass $m_B$ = 16.0 kg.

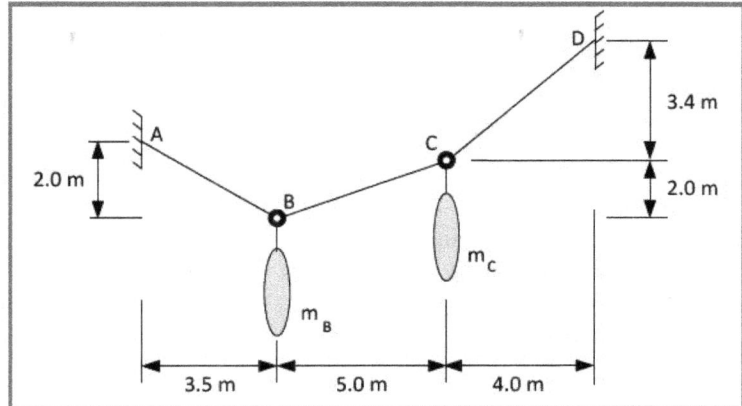

3.19 A bungee cord is stretched across a river as shown in the figure below. A stuntman weighing 140 lb is to cross the river by tightrope walking on the bungee cord. If the cord has a linear spring constant k = 60 lb/in., determine if the stuntman will remain dry.

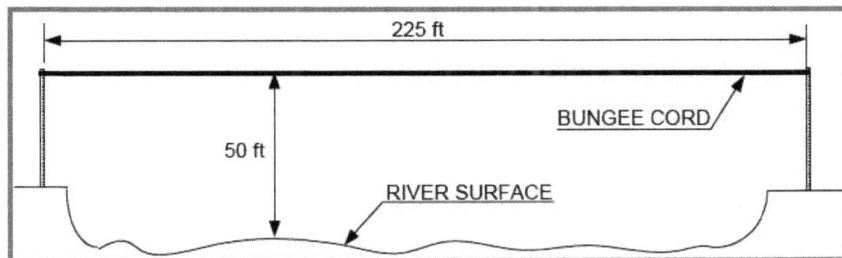

3.20 For the conditions described in Problem 3.19, determine the spring constant of the bungee cord if the stuntman is to barely wet the soles of his shoes.

3.21 Three springs are connected in series as shown in the figure to the right. Determine the extension of the entire arrangement if a force of 23.4 lb is applied to the last spring. Also determine the effective spring rate for the three springs in this series arrangement.

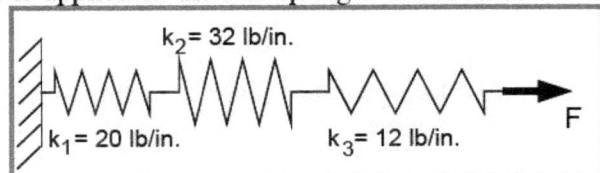

3.22 Three springs are connected in series as shown in the figure to the left. Determine the extension of the entire arrangement if a force of 70 N is applied to the last spring. Also determine the effective spring rate for the three springs in the series arrangement.

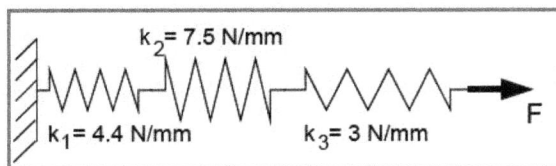

3.23 For the parallel spring arrangement determine the deflection of the lower bar shown in the figure to the right. The spring rates $k_A$ = 12 N/mm. and $k_B$ = 23 N/mm. The force F = 1145 N. Also determine the effective spring rate for the three springs in parallel.

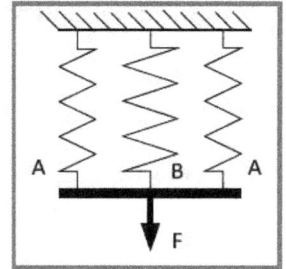

3.24 Three springs are used to support the rigid bar with a mass m as illustrated in the figure to the right. If $k_1$ = 20 lb/in., $k_2$ = 28 lb/in., $k_3$ = 36 lb/in. and m = 1.3 slug, determine the amount each spring elongates. The bar is constrained and remains parallel to the top surface to which the springs are attached.

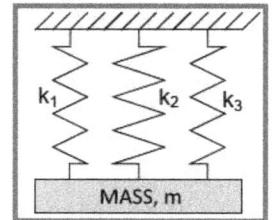

3.25 A cable and a spring are connected together at point C as shown in the figure to the left. Determine the equilibrium position of point C if a force F = 400 N is applied. The undeformed length L of both the cable and the spring before the application of the load is shown in the figure together with the spring rate k.

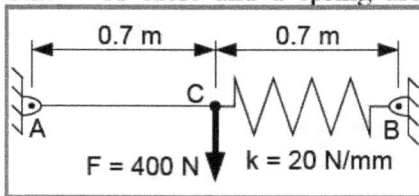

3.26 For the beam, shown in the figure to the right, solve for the reactions at the pin and clevis and the roller.

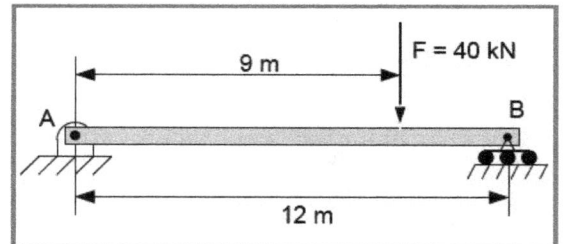

3.27 For the beam, shown in the figure to the left, solve for the reactions at the pin and clevis and the roller.

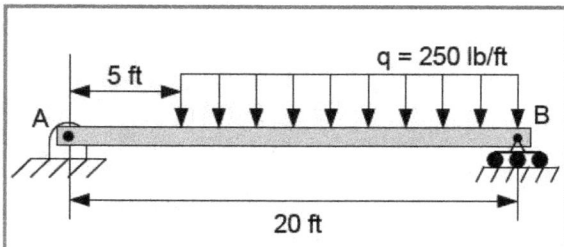

3.28 For the beam, shown in the figure to the right, solve for the reactions at the pin and clevis and the roller.

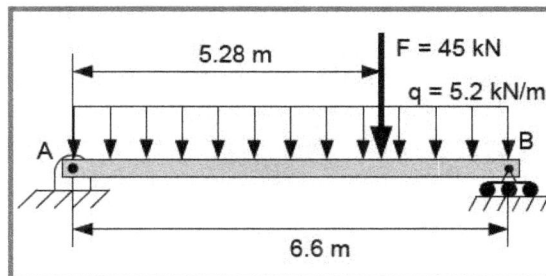

3.29 For the beam, shown in the figure to the right, solve for the reactions at the built-in end of the cantilever beam.

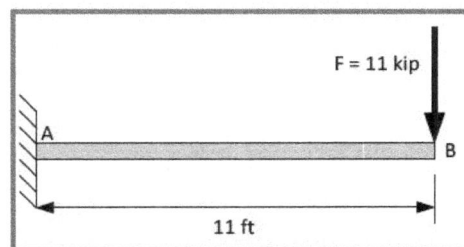

3.30 For the beam, shown in the figure to the left, solve for the reactions at the built-in end of the cantilever beam.

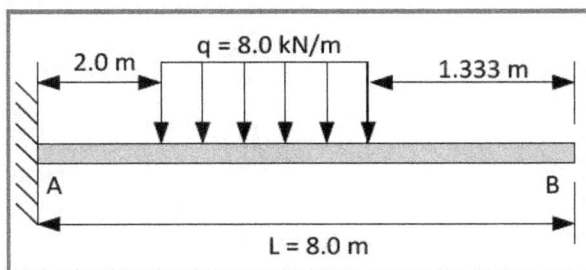

3.31 For the beam, shown in the figure to the right, solve for the reactions at the two supports.

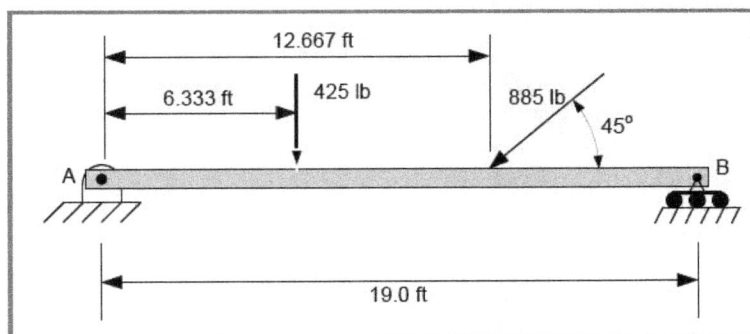

3.32 For the beam, shown in the figure below, determine the reactions at the two supports. The dimensions of the beam and the mass are given in the figure. Note that the arm CD is rigidly attached to the beam. Also, the radii of the two pulleys are the same and equal to 0.25 m.

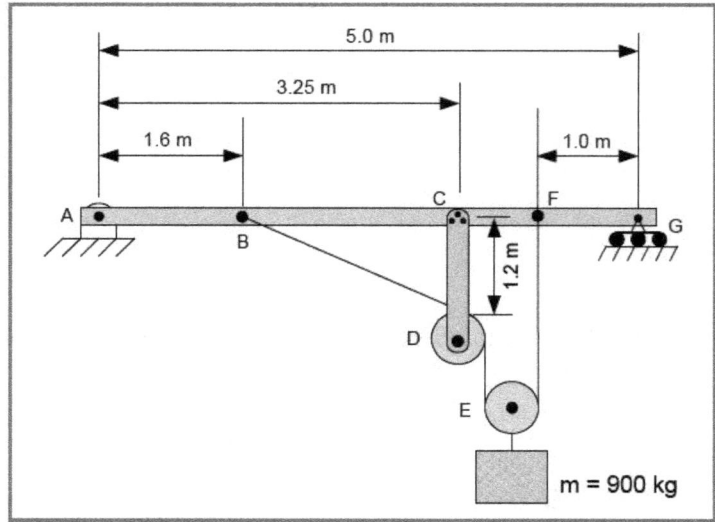

3.33 For the truss, shown in the figure to the right, determine the reactions at the two supports.

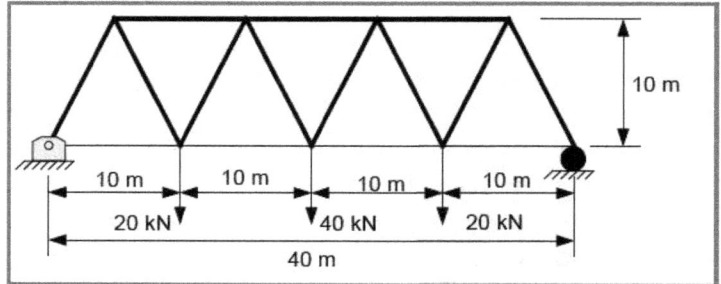

3.34 A long slender pole is to be erected by a winch and cable arrangement that is mounted on a truck, as illustrated in the figure to the left. If the pole weighs 2,000 lb and is 40 ft long, determine the reactions at point O and the force in the cable. Consider angles θ varying from 20° to 75°. A pivot is attached to the base of the pole so that it may be rotated without a resisting moment (i.e. acts like a pin).

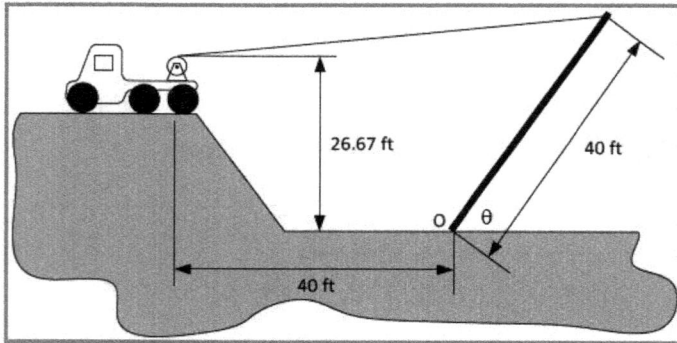

3.35 For the crane, described in the figure to the right, determine the maximum load $W_L$ that can be lifted before the crane tips. Note, the crane weights 30 kN.

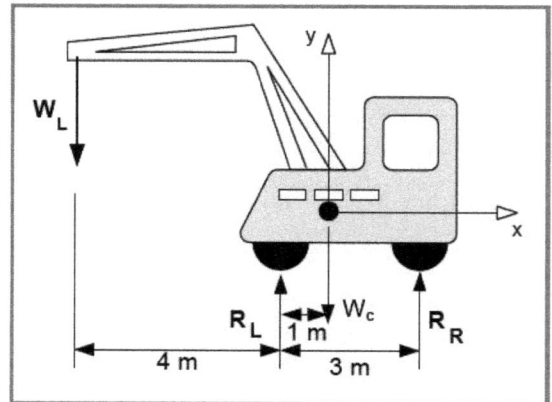

3.36    For the crane, illustrated in the figure to the left, determine the maximum load that can be lifted before the crane tips.  Note, the crane weights 3.5 ton.

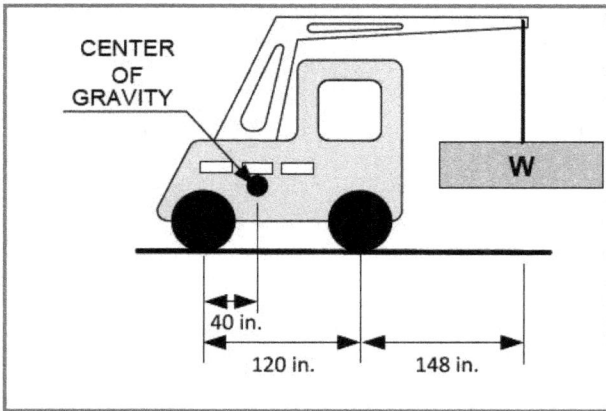

3.37    Determine the reaction forces and moments at the support for the stadium structure shown in the figure to the right.

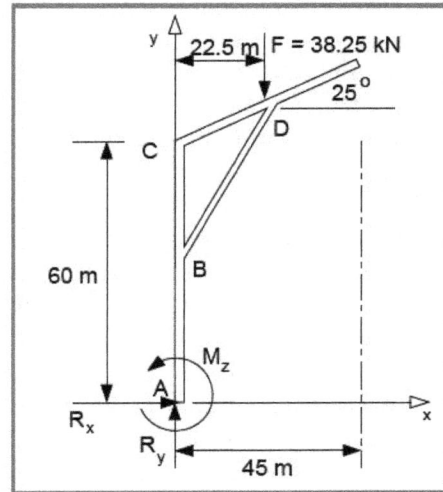

3.38    Determine the reaction forces at pins A and B that support the gusset plate shown in the figure to the left.  Assume that the vertical components of the reaction forces at the pins are equal (i.e. $R_{Ay} = R_{By}$)

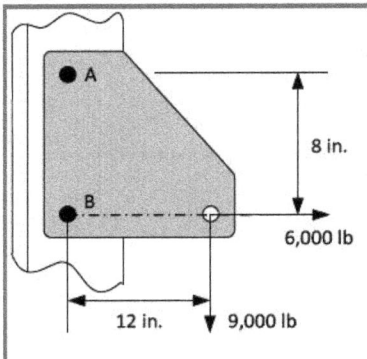

3.39    For the tension bar, shown in the figure to the right, determine the internal tension force P at sections A-A and B-B.

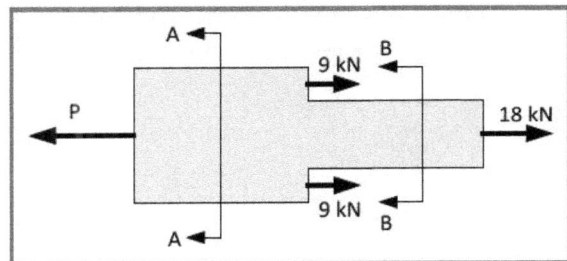

3.40    For the beam, shown in the figure to the left, determine the internal shear force V and the internal moment M at the position $x = L/2$.

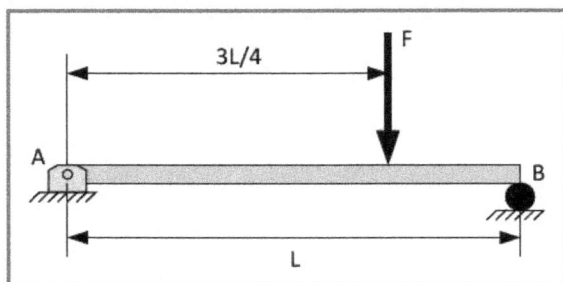

3.41    For the beam, shown in the figure above and to the left, determine the internal shear force V and the internal moment M as a function of position for $0 < x < L$. Prepare a graph of M and V as a function of x.

3.42    For the beam, shown in the figure to the right, determine the internal shear force V and the internal moment M at position $x = L/3$. Present your results in terms of q and L, which are known quantities.

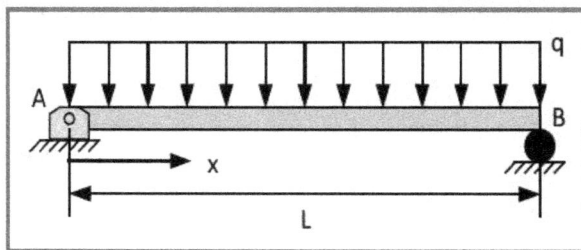

3.43    For the beam, shown, in the figure to the right, determine the internal shear force V and the internal moment M as a function of position x. Prepare a graph of M and V as a function of x.

3.44    For the beam, shown in the figure to the right, determine the internal shear force V and the internal moment M at position $x = 3L/4$. Present your results in terms of F and L, which are known quantities.

3.45    For the beam, shown in the figure to the right, determine the internal shear force V and the internal moment M as a function of position x. Prepare a graph of M and V as a function of x.

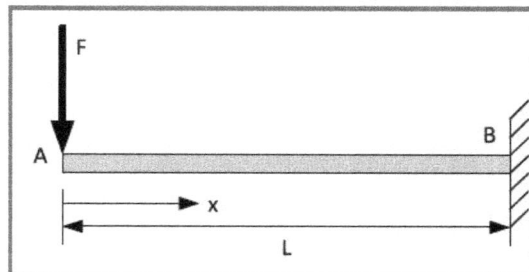

3.46    For the beam, shown in the figure to the lower right, determine the internal shear force V and the internal moment M at the position $x = 12$ ft.

3.47    For the beam, shown in the figure to the right, determine the internal shear force V and the internal moment M as a function of position x. Prepare a graph of M and V as a function of x.

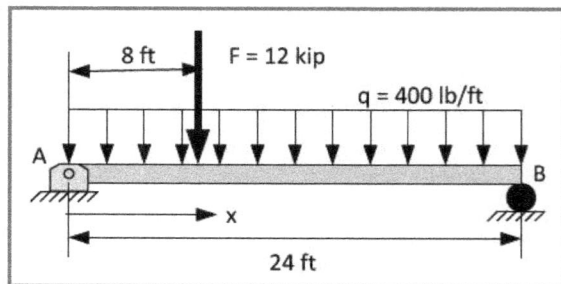

3.49 A new factory for heavy machinery has an assembly line with component parts stored above floor level. The components used in assembly are in elevated bins (#1, #2, and #3) as shown in the figure to the right. The elevated bins are suspended from a frame ABCD. Prepare a FBD of the left portion of the frame from Section A-A to the roller support at point A. Determine the internal forces and moment at section A-A as a function of the loads $W_1$, $W_2$ and $W_3$. Locate Section A-A at position $x = 3L/8$ and consider the loads $W_1$, $W_2$ and $W_3$ as known quantities. Assume that the centers of each bin are located at the following points: $x_1 = L/4$, $x_2 = L/2$ and $x_3 = 3L/4$.

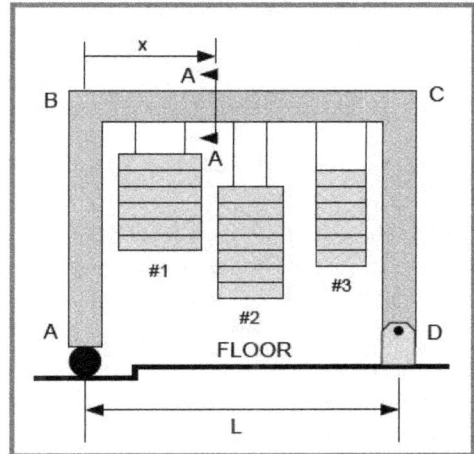

# CHAPTER 4

# AXIALLY LOADED STRUCTURAL MEMBERS

## 4.1 INTRODUCTION

Large structures like bridges, sports stadiums, skyscrapers, etc. are fabricated from many small structural elements. There are several reasons for constructing big structures from much smaller components. First, manufacturing facilities limit the size (width and thickness) of structural elements that are produced. The rolls used in steel and aluminum mills to produce bars, plates and sheets usually are only six to eight feet long. Even more limiting is the requirement for transport from the manufacturing facility to the construction site. If highways are used for transport, state law limits both the width and length of the structural members.

In this chapter we will consider two types of long thin structural members—flexible and stiff. Both types of these structural members will be subjected to axial loading. Wire rope or cables, which are flexible, are capable of supporting only tensile loading that tends to stretch the cable. However, rods or bars, which are stiff[1], are capable of supporting both tensile and compressive loading.

As structural elements, wire rope, cable, rods and bars subjected to axial loading have a significant advantage. They are subjected to a uniform state of stress over their cross sectional area. Also, if they are two-force members loaded at their two ends, the stresses are uniform over their length. Thus, the entire volume of the rod or bar is subjected to the same stress, which is the optimum condition for minimum weight design.

Examples of structural applications of long thin members in bridge construction are numerous. For instance, very high-strength wire rope is used in the construction of suspension bridges and in cable stayed bridges. Also, many bridges are designed with trusses deployed on the two sides of the structure. These trusses are fabricated from many bars or rods connected together at bolted or welded joints. We analyze these rods and bars using the same equations established for wire rope and cables. The difference is that the rods and bars are sufficiently stiff to support compressive stresses.

## 4.2 CHARACTERISTICS OF CABLES, RODS AND BARS

What characteristics do cables, wire ropes, rods and bars have in common and how do they differ? Cables, rods and bars exhibit lengths that are very long when compared to the dimensions of the cross section. A cable differs from a rod or a bar because it is flexible whereas the rod and bar are stiff. **Cables** have a circular cross section that is uniform along its length. A cable is made from many strands of wire that are twisted together to form a larger diameter very high-strength structural member. Even though a typical cable is larger in diameter than a wire, it is treated as a flexible member capable of supporting only tensile forces. We will use wire, wire rope and cable interchangeably in this text. The cross sectional area of **bars and rods**[2] may be of arbitrary shape.

---

[1]Rods or bars actually stretch or compress under the action of axial loads, but this deformation is small and is neglected in determining the loads and stresses. However, this very small deformation is considered when determining the strain and deflection of the long thin structural members.

[2]In most applications, the cross section of a rod is circular and that of a bar rectangular; however, this distinction is not always maintained and one can expect to find cross sections of arbitrary shape for both rods and bars.

Because cables are flexible, they can be wrapped about pulleys, formed into loops, and sometimes knotted. Indeed, they are so flexible that we cannot push on them because they buckle. It is possible to pull, but not push. This fact means that internal tensile forces develop within a thin flexible member but not compression forces. However, rods and bars are sufficiently stiff to support compressive forces and stresses. They are used as tie rods when loaded in tension, and as columns when loaded in compression. We will assume in this chapter that the cross section of the column is adequate and that failure by buckling is not an issue.

We are considering the simplest structural application—the use of long thin members under axial loading. The geometry of a structural member (long length with small cross sectional dimensions) leads us to constrain the direction of the loading and to make an important assumption about the deformation of the member that significantly simplifies the analysis.

1. The internal and external forces supported by the member act along its length.
2. Plane sections before loading remain plane after loading and subsequent deformation of the member.

Constraining the loads in the uniaxial direction is consistent with the title of the chapter. We are considering only the effects of axial loading on cables, rods and bars. Later in a course titled Mechanics of Materials, we will consider bending of beams induced by transverse loading.

The assumption that plane sections before loading remain plane after loading has also been confirmed by experiment. Suppose we draw a straight scribe line on a long thin wire or bar as shown in Fig. 4.1 a. This line represents the edge of a plane through the cross section of the member. Now apply an axial load F to the wire as indicated in Fig. 4.1b. The wire will stretch a small amount $\delta$, but the line remains straight and the plane through the cross section remains plane (i.e. flat). This is a very important observation because it implies that the normal stresses $\sigma$ are uniformly distributed across this plane.

Fig. 4.1 A section line drawn across the member remains straight after loading.

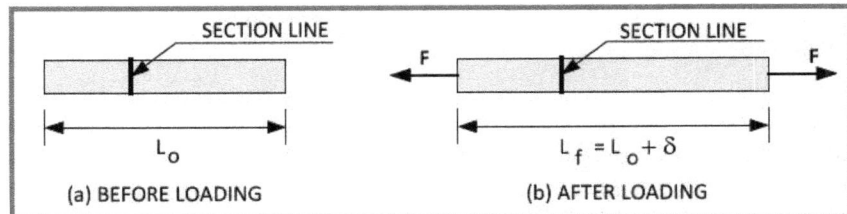

SECTION LINE

SECTION LINE

F

F

$L_o$

$L_f = L_o + \delta$

(a) BEFORE LOADING

(b) AFTER LOADING

## 4.3 DESIGN ANALYSIS METHODS

In the design analysis, we compute the axial forces acting on the uniaxial member, the amount it stretches under load, and the stresses developed. Next, we compare the stresses acting on the structural member with its strength to determine a safety factor. The size of the safety factor is evaluated to determine if this member is safe to be employed in a structure used by the public. This chapter describes methods that will enable you to size cables, rods and bars to provide structures with adequate safety margins.

We will divide this presentation into two parts—the first dealing with cables and the second involving rods and bars. The reason for the separation is the difference in the structural applications. The flexible cables are often used with pulleys for lifting and for high-strength tension members. Rods and bars are employed as tie rods or columns in structural applications. While the applications differ, the equations governing the behavior of the flexible and the stiff members are the same.

### 4.3.1 Design Analysis of Cables and Wire Rope

What happens if we pull on a wire and continue to pull with increasing force? The wire stretches, and stretches still more until it fails by breaking. Let's explore the stretch of the wire, and the consequences of different types of failure of the wire in subsequent sections.

*Stretch of Wire under Load*

When we pull on a wire it stretches as shown in Fig. 4.1b. We define $\delta$, the **stretch** or **extension** of the wire, as:

$$\delta = L_f - L_o \tag{4.1}$$

where $L_f$ is the length of the wire under load and $L_o$ is the original length.

A FBD of a segment of wire is shown in Fig. 4.2a. The equilibrium relation ($\Sigma F_x = 0$) indicates that the internal force P is equal to the external force F for these uniaxial members.

Fig. 4.2 FBD of a long thin member subject to axial load.

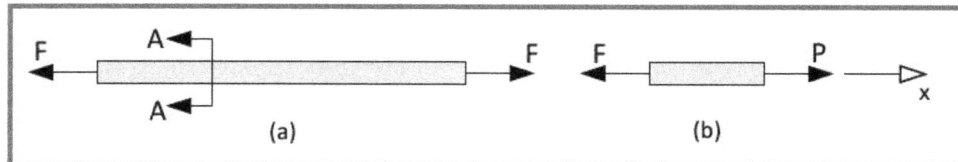

We determine the stretch $\delta$ from:

$$\delta = (PL)/(AE) \tag{4.2}$$

where P is the internal force in the wire, in N or lb, L is the length of the wire, in m or in, $A = \pi r^2$ is the cross sectional area of the wire, in $m^2$ or $in^2$, r is the radius of the wire, in m or in, and E is the **modulus of elasticity**, in GPa or psi.

We will derive Eq. (4.2) later in this section. It is important at this time to understand that the extension of the wire $\delta$ is a very small quantity if the wire is fabricated from a metal such as steel or aluminum. To demonstrate the magnitude of the extension, consider the following example.

### EXAMPLE 4.1

A No. 30 gage[3] steel music wire (0.080 in. in diameter) is employed to lift a weight of 500 lb. If the wire is 6 ft. in length, determine the amount that the wire stretches under load.

> **Solution:** In Appendix B, we note that the modulus of elasticity for steel is listed as $E = 30 \times 10^6$ psi. Next, let's substitute the values for P, L, $A = \pi r^2$ and E into Eq. (4.2) to obtain:
>
> $$\delta = (PL)/(AE) = (500 \text{ lb})(6 \text{ ft})(12 \text{ in/ft})/[\pi (0.040 \text{ in.})^2 (30 \times 10^6 \text{ lb/in}^2)]$$
>
> $$\delta = 0.2387 \text{ in.}$$

---

[3]There are several standards for gage sizes that define the diameter of wire or the thickness of sheet metal. Four of the commonly referenced standards are presented in Appendix A.

Is 0.2387 in. a small extension of the wire? Small is a relative term and must be compared to another value to be judged. Let's compare the amount of this extension to the original length of the wire by computing the ratio:
$$\delta/L_o = 0.2387/(6)(12) = 0.003316 \text{ or } 0.3316\%$$

The 500 lb load stretches the wire by less than a one third of one percent. That is a very small extension when compared to the original length of the wire.

Let's conduct a simple experiment and observe the behavior of a wire under the action of a monotonically increasing tensile load. If we measure the load F applied to the wire, the extension $\delta$ of the wire, and equate $F_{EXT}$ to $P_{INT}$, we can construct the graph shown in Fig. 4.3 a.

Fig. 4.3 Graph of load versus stretch in (a) and stress versus strain in (b).

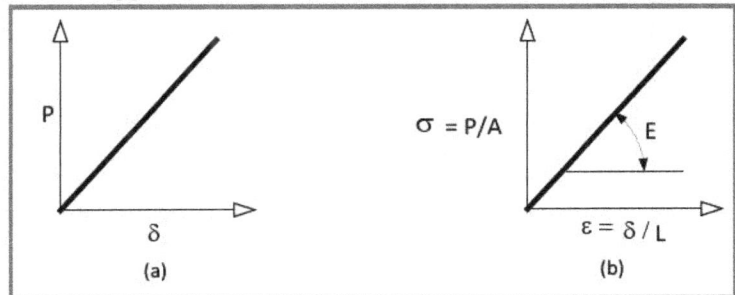

Note a linear relationship exists between P and $\delta$ as indicated by Eq. (4.2). This P-$\delta$ curve is a straight line until the wire begins to fail by **yielding**. The linear portion of the P-$\delta$ relation is the **elastic region** of the load-deflection response. Equation (4.2) is valid only in this elastic region.

Let's modify the axes in Fig. 4.3a by dividing the force P by the area A and the extension $\delta$ by the length L as indicated below:

$$\varepsilon = \delta/L \tag{4.3}$$

$$\sigma = P/A \tag{4.4}$$

where $\varepsilon$ is the strain and $\sigma$ is the stress in the wire, respectively.

These relations show that the **normal strain**, $\varepsilon$, is the change in length divided by the original length; strain is a dimensionless quantity. The normal stress $\sigma$ is the internal force P divided by the area A, over which it acts. The stress is expressed in terms of Pa (Pascal) in SI units or psi (lb/in$^2$) in U. S. Customary units. We have shown the graph of stress versus strain in Fig. 4.3b. As expected there is a linear relation between stress and strain until the stress is sufficient to cause the wire to yield. The **slope** of the $\sigma$ - $\varepsilon$ line is the **modulus of elasticity** E of the material from which the wire is fabricated. The **modulus of elasticity** is a **material property**, which is independent of the shape of the body.

It is evident from the linear response in the stress-strain diagram that:

$$\sigma = E\varepsilon \tag{4.5}$$

This stress-strain relation is known as **Hooke's law**. It is named after Robert Hooke who is credited with the discovery of elasticity in the 17$^{th}$ century. A word of caution—Hooke's law is valid only for uniaxial states of stress that arise in long thin structural members. We will introduce another more complex form of the stress-strain relation to accommodate multi-axial stress fields in a later chapter.

Let's combine Eqs. 4.3, 4.4 and 4.5 in the manner shown below:

$$\sigma = P/A = E\varepsilon = (E\delta/L)$$

and solve this expression for $\delta$ to derive Eq. (4.2):

$$\delta = (PL)/(AE) \qquad\qquad (4.2 \text{ bis})$$

## EXAMPLE 4.2

Determine the strain that develops in a 7 m long wire, which is stretched by 12 mm.

> **Solution:**
>
> Using Eq. (4.3), we write:
>
> $$\varepsilon = \delta/L = 12 \times 10^{-3} \text{ m} / 7 \text{ m} = 0.001714 \text{ dimensionless}$$
>
> Note that strain is dimensionless because we divided a change in length by the original length. Also, observe that we have determined the strain without knowledge of the material used in fabricating the wire. Strain is a geometric concept. The strain imposed on a wire or rod may be determined without knowledge of the load or the modulus of elasticity if the amount of deformation is known.

## EXAMPLE 4.3

Determine the strain that develops in a 10 ft long wire that is subjected to a force of 160 lb. The wire size is Gage 10 (0.10189 in. in diameter) and it is fabricated from an aluminum alloy 2024-T4.

> **Solution:**
>
> Using Eq. (4.3), we write:
>
> $$\varepsilon = \delta/L$$
>
> Recall Eq. (4.2):
>
> $$\delta = (PL)/(AE)$$
>
> Combining Eqs. (4.2) and (4.3) yields:
>
> $$\varepsilon = P/(AE) \qquad\qquad (4.6)$$
>
> From Appendix B, we find the modulus of elasticity E for aluminum alloy 2024-T4 is $10.4 \times 10^{6}$ psi. Next, we compute the cross sectional area of the wire as:
>
> $$A = \pi d^2 /4 = \pi(0.1019)^2 /4 = 0.008155 \text{ in.}$$
>
> Substituting values for P, A and E into Eq. (4.6) gives:
>
> $$\varepsilon = 160 \text{ lb}/(0.008155 \text{ in.}^2 \times 10.4 \times 10^{6} \text{ lb/in.}^2) = 0.001886$$

In this example, it was necessary to use the value of the modulus of elasticity in determining the strain. Why? The modulus of elasticity was needed because the deformation (stretch) of the wire was not specified in the problem statement. It was necessary to determine the deformation of the wire from the load using Eq. (4.2). This step in the analysis required us to introduce the modulus of elasticity.

## *Internal Forces and Stresses*

Internal forces that develop within a structural member when it is subjected to external loads generate stresses. In Chapter 2, we introduced the concept of internal forces and showed that they are determined from the appropriate equations of equilibrium in Chapter 4. We also indicated the necessity for making an imaginary section cut to expose the cross sectional area of the member over which the stresses act. In this section, we emphasize the connection between the external forces, internal forces, and the stresses that develop at some point in the member.

To illustrate these connections, examine Fig 4.4, which shows an axial member subjected to external axial forces F.

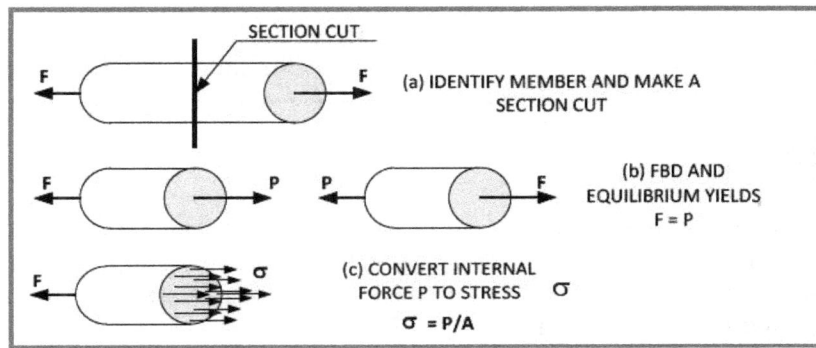

Fig. 4.4 An illustration showing the transition from external forces to internal forces to stresses.

1. We identify the member of interest and the forces acting on the body in Fig. 4.4a.
2. A section cut is made at a location of interest along the length of the member. We then remove one end or the other to construct the FBD depicted in Fig. 4.4b.
3. Next, the equilibrium relation $\Sigma F_x = 0$ is written to prove that $F_{EXT} = P_{INT}$.
4. Finally, from Eq. (4.4) ($\sigma = P/A$), we convert the internal force P into the normal stress $\sigma$ as indicated in Fig. 4.4c.

In this illustration, the internal force P is a constant over the length of the uniaxial member and the location of the section cut is not important. However, you should be aware that in many applications the internal forces often vary from one location to another. In these situations the location of the section cut is vitally important. Also observe in Fig. 4.4c that the distribution of the stress $\sigma$ over the cross sectional area exposed by the section cut is uniform. The uniform distribution is a result of two conditions:

- Plane sections remain plane for long thin members subjected to axial loading.
- Internal moments do not develop under the applied external loading.

In the most general case of loading, we relate the internal force and the normal stress over an area A by:

$$P_{INT} = \int \sigma \, dA \qquad (4.7)$$

However, when the stress is uniformly distributed as it is in this illustration Eq. (4.7) reduces to:

$$P_{INT} = \sigma \int dA = \sigma A \qquad (4.8)$$

## EXAMPLE 4.4

Determine the stress in a 1.5 mm diameter wire subjected to an axial load of 250 N.

**Solution:** From Eq. (4.4), we write

$$\sigma = P/A$$

From a FBD identical to the one shown in Fig. 4.4b and the equilibrium relation $\Sigma F_x = 0$, we understand that:

$$F = P = 250 \text{ N}$$

The area $A = \pi d^2/4 = \pi(1.5 \text{ mm})^2/4 = 1.767 \text{ mm}^2$. Substituting these values into Eq. (4.4) gives:

$$\sigma = 250\text{N}/ 1.767 \text{ mm}^2 = 141.5 \text{ N/mm}^2 = 141.5 \text{ MPa}$$

Note that one $\text{N/mm}^2$ is equivalent to one MPa. Later in this Chapter, we will present an interpretation of this result after we describe in detail the strength of structural materials.

## EXAMPLE 4.5

A large diameter wire rope fabricated from steel with a cross sectional area of 6.4 $\text{in}^2$ is to support a portion of the roadway on a suspension bridge. The highway engineers have specified that the maximum load imposed on the cable is not to exceed 150 ton. Determine the stress and strain in the wire rope when subjected to the specified load.

**Solution:**

Let's compute the stress from Eq. (4.4) as:

$$\sigma = P/A = (150 \text{ ton})(2000 \text{ lb/ton})/6.4 \text{ in.}^2 = 46,880 \text{ psi} = 46.88 \text{ ksi}$$

Next, calculate the strain $\varepsilon$ from Eq. (4.6) to obtain:

$$\varepsilon = P/(AE) = 300,000 \text{ lb}/[(6.4 \text{ in}^2)(30 \times 10^6 \text{ lb/in}^2)] = 1.563 \times 10^{-3}$$

In determining the strain, we obtained the value of $E = 30 \times 10^6$ psi for steel from Appendix B. We also introduced ksi, a unit in the U. S. Customary system. The conversion factor between psi and ksi is given by $\Rightarrow$ 1 ksi = 1,000 psi.

## *Failure*

Suppose we take a length of wire, grip it in a universal-testing machine, and pull until it fails. The behavior of the wire under increasing load depends on the material from which the wire was drawn[4]. Nearly all materials exhibit a **linear elastic** response like that shown in Fig. 4.3 at lower stress levels; however, at higher stresses significantly different behavior is observed for different materials. We have classified these behaviors as:

---

[4]Drawing a circular rod through a smaller diameter die produces wire. The drawing process improves the surface finish and enhances the strength of the wire.

- **Brittle** with abrupt failure and small, elastic deformations.
- Yielding with **strain hardening** and plastic deformation prior to rupture.
- Yielding with **strain softening** and plastic deformation prior to rupture.

These three behaviors are illustrated graphically in Figs. 4.5 and 4.6.

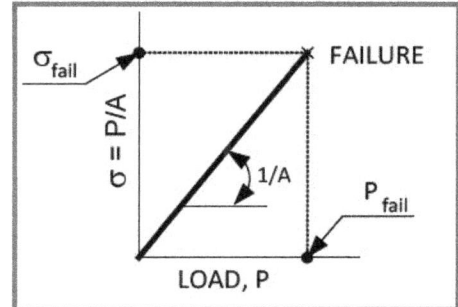

Fig. 4.5 Linear-elastic response until brittle failure.

The graph in Fig. 4.5 indicates the linear elastic response of a brittle material. The stress increases linearly with both axial load and strain. The slope of the stress-load line is the reciprocal of the area (1/A), and the slope of the stress-strain line is the modulus of elasticity E as indicated in Fig. 4.3. When we increase the load to a critical value $P_{fail}$, the uniaxial member breaks. The failure occurs at a specific value of stress, $\sigma_{fail}$. We define this failure stress as the **ultimate tensile strength**, $S_u$ of the material of the wire.

$$S_u = \sigma_{fail} = P_{fail}/A \qquad (4.9)$$

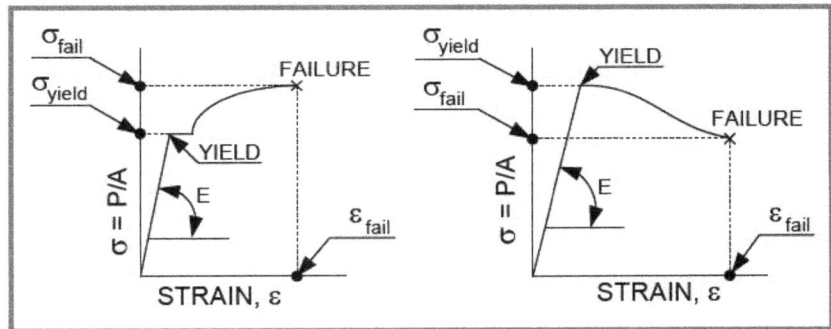

Fig. 4.6 Materials yield with strain hardening (left) and strain softening (right).

The two graphs in Fig. 4.6 illustrate material behavior when yielding occurs. The graph to the left in Fig. 4.6 is typical of a material that yields and then strain hardens. As the axial load on the wire increases, the stress increases linearly until the material begins to yield at $\sigma_{yield}$. At that point, the uniaxial member continues to stretch, but the load and stress remain essentially constant. After some degree of post-yield stretch, the material stiffens and the load and stress begin to increase. The stress increases until the wire ruptures at $\sigma_{fail}$.

We establish two strengths for the wire from this graph. The **yield strength** $S_y$ given by:

$$S_y = \sigma_{yield} = P_{yield}/A \qquad (4.10)$$

and the **ultimate tensile strength,** $S_u$ given by Eq. (4.9).

The graph to the right side in Fig. 4.6 is typical of a material that yields and then strain softens. As the load increases, the stress increases linearly until the material begins to yield at $\sigma_{yield}$. At that point, the uniaxial member continues to stretch, but the load and stress remain essentially constant. After some degree of post-yield stretch, the material softens and the load and stress begin to decrease with a continued increase in the deformation. The strain increases until the wire ruptures at $\sigma_{fail}$.

## Strength and Safety Factor

We have defined two different strengths (i.e. yield and ultimate tensile) in the previous section. Since a structural member may fail by excessive deformation, we may choose to limit the stress applied to the structure so that it is less than the yield strength $S_y$. On the other hand in some structures, we can tolerate plastic deformation in one or more members without compromising the structure's function. In these cases, we can tolerate stresses exceeding the yield strength, $S_y$ but they must be less than the ultimate strength, $S_u$. For example, when designing a bridge, it is important that its shape remain fixed under normal service loads. In this case, the yield strength is specified as the maximum limit for the design stress in the bridge to prevent post-yield deformations. However, in designing a bridge to remain standing during a collision with a freighter or during an earthquake, the ultimate strength of the structural members is the important criterion.

In designing a structural member to carry a specified load, we always size the member so that the **design stress**, $\sigma_{design}$, is less than the strength based on either the yield or failure criteria. It would not be prudent to permit the design stress to equal the strength of the member. To size our structural members so they are safe, we usually employ a **factor of safety, SF** in the analysis. We define the safety factor as the ratio of a strength divided by the design stress. This definition leads to the two relations given below:

$$SF_u = S_u / \sigma_{design} \tag{4.11}$$

$$SF_y = S_y / \sigma_{design} \tag{4.12}$$

where $S_u$ the ultimate strength of the material, in units of MPa or psi, $S_y$ the yield strength of the material, in units of MPa or psi, $\sigma_{design}$ the design stress given by $P_{design}/A$, in units of MPa or psi ,$P_{design}$ the load that is specified in designing the structural member, in N or lb.

Once again, the value for the factor of safety depends upon the application. Ideally, you would like the factor of safety to be as large as possible. However, this desire for excessive safety factors must be balanced by practical considerations such as economics, aesthetics, functionality, and ease of assembly. In designing, you may be concerned with the cost and weight of its components. The factor of safety you decide to specify should reflect these concerns.

## EXAMPLE 4.6

A hoisting cable with a diameter of 7/16 in. is fabricated from many strands of an improved plow steel wire. The manufacturer of the wire certifies its breaking load as 16,500 lb. Determine the strength of the wire used in the manufacture of the cable. Also discuss the assumption pertaining to the wire rope made in the analysis and its implication on the strength of the strands of wire.

> **Solution:** Recall Eq. (4.9), which gives:
>
> $$S_u = \sigma_{fail} = P_{fail} / A$$
>
> The cross sectional area of the cable will be less than $A = \pi d^2/4 = 0.1503$ in$^2$. Substituting this value for the area into Eq. (4.9) yields:

$$S_u = 16{,}500 \text{ lb}/ 0.1503 \text{ in.}^2 = 109.8 \text{ ksi}$$

We have assumed the cross sectional area of the wire rope to be equivalent to that of a solid wire 7/16 in. in diameter. Wire rope is made of many very small diameter wires that are twisted together to form strands. The strands in turn are formed in a helix about a fiber core. The wire rope in this example has a designation of 6 × 19 (6 strands with 19 small wires in each strand). The cross section of this 6 by 19-wire rope is illustrated below.

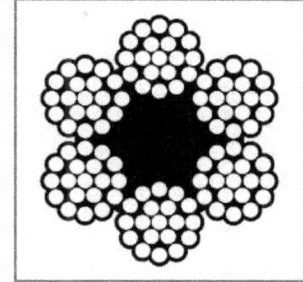

Fig. 4.7 A wire rope is fabricated from many strands of small-diameter high-strength wire.

The assumption of a solid cross sectional area overestimates the cross sectional area, A by a factor of more than two. The ultimate tensile strength of the small diameter wire used to form the strands of wire for a typical cable is in excess of 200 ksi.

## EXAMPLE 4.7

A solid hard drawn copper wire exhibits an ultimate tensile strength of 390 MPa. If the diameter of the wire is 1.6 mm, determine the load required to break the wire.

**Solution:** Recall Eq. (4.9), which gives:

$$S_u = \sigma_{fail} = P_{fail}/A$$

Rearrange this relation and substitute known quantities to give:

$$P_{fail} = S_u A = (390 \times 10^6 \text{ N/m}^2)(\pi/4)(1.6 \text{ mm})^2 \times 10^{-6} \text{ m}^2/\text{mm}^2 = 784.1 \text{ N}$$

An axial load of approximately 800 N is required to rupture the hard drawn copper wire.

## EXAMPLE 4.8

A monofilament, nylon fishing line is rated at 10-lb test. If the line is 0.012 inch in diameter, determine the strength of the nylon in the form of a small diameter line. Comment on the effect that the filament geometry has on the strength of a polymer like nylon-6/6.

**Solution:** Recall Eq. (4.9), which gives:

$$S_u = \sigma_{fail} = P_{fail}/A$$

$$S_u = (4)(10)/[\pi (0.012)^2] = 88{,}420 \text{ psi}$$

Monofilaments of polymers like nylon-6/6 are drawn from a melt into thin fibers or lines. In this process, the long molecules of the polymer are aligned with the axis of the filament thus enhancing its strength.

## EXAMPLE 4.9

A No. 14 gage (0.080 in. diameter) black annealed steel wire is listed in a material handbook with a yield strength of 220 MPa and an ultimate tensile strength of 340 MPa. Determine the axial load that will cause the wire to yield.

> **Solution:** Recall Eq. (4.10), and rearrange it to give:
>
> $$S_y = \sigma_{yield} = P_{yield} / A$$
>
> $$P_{yield} = S_y A \qquad (a)$$
>
> Since we have mixed units in the problem statement, let's convert the units for the diameter of the wire to the SI system and determine the cross sectional area A in $mm^2$.
>
> $$A = \pi d^2/4 = \pi\,(0.080\text{ in.})^2\,(25.4\text{ mm/in.})^2\,/4 = 3.243\text{ mm}^2$$
>
> Next, substitute into Eq. (a) to obtain:
>
> $$P_{yield} = S_y A = (220\text{ MPa})[(N/(mm^2)/MPa)](3.243\text{ mm}^2) = 713.4\text{ N}$$

## EXAMPLE 4.10

A 5/8 in. diameter stainless steel (type 304) wire rope with a $6 \times 19$ configuration is listed in a catalog with a breaking load of 35,000 lb. If a safety factor of 2.4 is to be employed in the design of a structure using this type of cable, specify the design load.

> **Solution:** Recall Eqs. (4.9) and (4.11) and combine them to obtain:
>
> $$\mathbf{SF_u} = S_u / \sigma_{design} = AS_u / P_{design} = P_{fail} / P_{design} \qquad (a)$$
>
> Solving Eq. (a) for the design load $P_{design}$ gives:
>
> $$P_{design} = P_{fail} / \mathbf{SF_u} = 35,000/\,2.4 = 14,580\text{ lb}$$

## EXAMPLE 4.11

A single #20 gage steel wire with a solid cross section is 0.0348 in. in diameter. It is to carry a load of 100 N with a safety factor of 2.8. Specify the steel alloy from which the wire should be manufactured to avoid failure by rupture.

> **Solution:** Recall Eq. (4.11) and solve it to obtain an expression for $S_u$:
>
> $$\mathbf{SF_u} = AS_u / P_{design}$$
>
> $$S_u = (P_{design})\mathbf{SF_u}/A = 4(100\text{ N})(2.8)/[\pi\,(0.0348\text{ in.})^2\,(25.4\text{ mm/in.})^2] = 456.3\text{ MPa}$$
>
> It is necessary to refer to a materials handbook to select a suitable material for this application. Reference to Appendix B-2 indicates three steel alloys and four stainless steel alloys with ultimate tensile strengths exceeding the requirement for $S_u$ in this example.

## EXAMPLE 4.12

A cold drawn alloy steel wire, with a yield strength of 125,000 psi, is to carry a load of 600 lb while incorporating a safety factor of 2.25. You are to inform the purchasing representative regarding the size of the wire to be ordered.

**Solution:** To determine the size of the wire in this application, recall Eq. (4.12).

$$SF_y = S_y / \sigma_{design} = AS_y / P_{design}$$

Solve this relation for the area to obtain:

$$A = SF_y \, (P_{design})/S_y = \pi d^2/4$$

$$A = (2.25)(600)/125,000 = 0.0108 \text{ in}^2 = \pi d^2/4$$

$$d = 0.1173 \text{ in.}$$

Do you call the purchasing representative and order a cold drawn alloy steel wire with a diameter of 0.1173 in? **No!!!** If you make this mistake, you will soon learn that steel wire is available only with a standard diameter. Economic laws dictate that the most cost effective way to mass-produce materials is by limiting inventories to only those sizes and geometries that satisfy most customers. In Appendix A standard wire sizes and the corresponding gage numbers that are used to order them are listed. For this example, four entries from the listings in Appendix A for standard steel wire are presented below:

| Gage No.[5] | Diameter (in.) | Area (in²) |
|---|---|---|
| 10 | 0.1350 | 0.01431 |
| 11 | 0.1205 | 0.01140 |
| 12 | 0.1055 | 0.00874 |
| 13 | 0.0915 | 0.00658 |

The response to the purchasing representative is to procure Gage No. 11 wire. It is important for you to specify a standard size in ordering structural members. Standard sizes are available with minimum delay at the lowest cost. Analyses rarely give dimensions for structural members that correspond to the available standard sizes. The usual practice is to increase the size determined in the analysis to the next larger dimension available as a standard. This approach enhances safety while minimizing cost and delivery time.

## 4.3.2 Design Analysis of Rods and Bars

All of the equations [(4.1) to (4.12)], derived in Section 4.3.1 for cables, are also valid for rods and bars. If the external axial forces are tensile (tend to pull the bar apart), the internal forces and stresses are tensile and denoted with a positive sign. On the other hand, if the external axial forces are compressive (tend to squash the bar), the internal forces and stresses are compressive and denoted with a minus sign.

---

[5] There are several different standards that refer to gage numbers for wire and steel sheet. In this example we list the American Steel and Wire Co. standard that is commonly used for steel wire.

We must qualify the capability of a rod or bar to carry compressive loads. If the rod is very long and slender and the compressive force too high, the rod may buckle. **Buckling** is an unstable condition, and if the critical load is exceeded, the rod fails suddenly and catastrophically. In this chapter, we will assume that the rods or bars loaded in compression are sized to resist buckling; consequently, they fail due to excessive compressive stresses. However, the tendency for these structures to buckle cannot always be ignored. Theories describing the failure of **columns** due to buckling will be introduced in a later course titled Mechanics of Materials.

## *Centroids*

The relations derived for the stresses and deformations of cable assumed that the internal and external forces acted through the **centroid** of its cross sectional area. For cables or wire, with circular cross sections, the location of the centroid is obvious—it is at the center of the circle. However, for cross sections of more complex shapes, the location of the centroid is not obvious.

    A centroid is the point coinciding with the center of gravity of a two-dimensional area. For the circular cross section shown in Fig. 4.8, the center of the circle clearly locates the centroid. The center also serves as the origin for the centroidal axes $x_c$ and $y_c$. For cross sectional shapes such as ellipses, circles, squares and rectangles, the center may be located by inspection because these shapes are all symmetric about both horizontal and vertical axes. The centroid is located at the intersection of the two axes of symmetry. However, for non-symmetric figures, such as a triangle, a portion of a circular area, a parabolic area etc., locating the center of the gravity of the area by inspection is not possible.

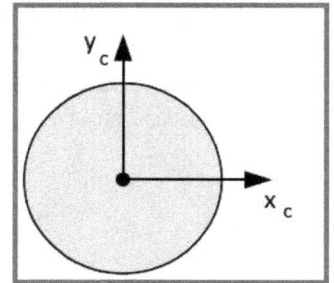

Fig. 4.8 A circular cross section with the center as the centroid and the centroidal axes $x_c$-$y_c$.

One approach for determining the location of a centroid is to consider the area as a flat plate positioned in the x-y plane with its outer normal vector in the z direction as illustrated in Fig. 4.9. Let's examine Fig. 4.9a and observe that the weight W of the plate acts downward through the **center of gravity** defined by the point $G_c$. The center of gravity and the centroid will be located at the same point for a plane body with uniform thickness and density. While the gravitational forces are distributed over the area of the plate, we represent the total weight of the plate as a concentrated force of magnitude W that acts downward though the center of gravity.

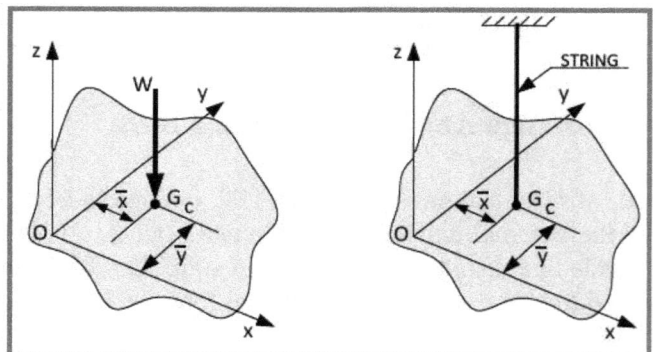

Fig. 4.9 A flat plate of arbitrary area with its weight acting though its center of gravity $G_c$.

(a)                    (b)

The center of gravity may be established experimentally or by analysis. The experimental approach is illustrated in Fig. 4.9b. Suppose we represent the plate with a sheet of cardboard cut to the shape of the arbitrary cross section. We scribe a Cartesian coordinate system on the cardboard to provide a reference for measuring the location of the center of gravity $G_c$. A string is fixed to the cardboard model at some point P near its center, and the model is suspended in space hanging from this string. In most cases, we cannot precisely estimate the location of the centroid of the model, and the model flops over onto its side. However, when the points P and $G_c$ coincide, the model will hang from the string with its plane horizontal and with its outer normal (the z axis) parallel to the string.

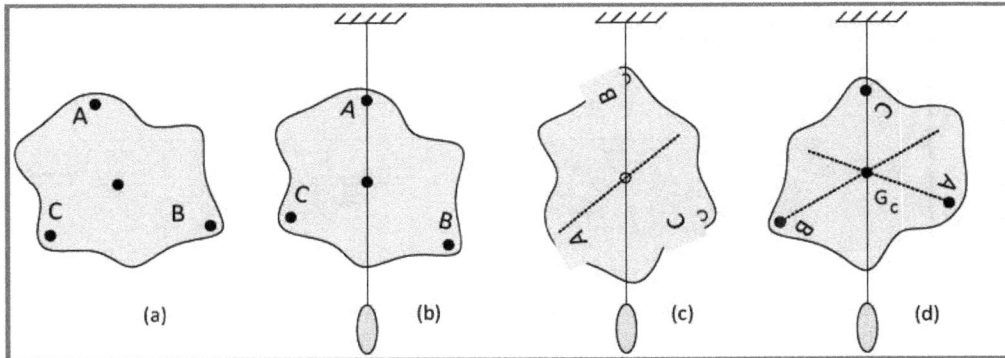

Fig. 4.10
Experimental
determination of
the centroid of an
arbitrary area.

Another experimental method for locating the center of gravity (also the centroid) is depicted in Fig. 4.10. Again we begin with a cardboard model of the area in question. With this approach we punch three small holes at points (A, B and C) located near the boundary of the model as illustrated in Fig. 4.10 a. We hang the model from a pin inserted through the hole at point A, and draw a straight line vertically downward from point A as shown in Fig. 4.10b. A plumb line or a level is used to aid in drawing a perfectly vertical line. We repeat this process with the model suspended from points B and then C as indicated in Figs. 4.10c and 4.10d. The three straight lines intersect; thus, locating the center of gravity or the centroid at a point $G_c$.

These experimental methods are effective in locating the position of the centroid and give an insight to the meaning of the centroid, center of gravity and the **first moment of an area**. There is one qualification that must be imposed on the fabrication of the model—the sheet of cardboard must be homogenous (of uniform thickness and density).

Analytical methods for locating the centroid of an arbitrary shaped area are covered in Appendix C.

### *Stresses in a Uniform Bar or Rod*

The procedure for determining the normal stresses $\sigma$ is the same as described previously for wires and cables except for the fact that the internal forces may be compressive or tensile. We begin by constructing a FBD to show the point of application and the directions of the internal and external forces. From the equilibrium relations, we determine the internal forces and their sign. Finally, Eq. (4.4) is employed to determine the normal stresses $\sigma$. We demonstrate this procedure with the example problems presented below.

## EXAMPLE 4.13

Determine the stress in the bar shown in Fig. E4.13 when subjected to a compressive axial force of F that acts through the centroid of the bar. The following numerical parameters define the bar and the applied load: F = 800 kN, L = 0.500 m, w = 100 mm and h = 50 mm.

Fig. E4.13

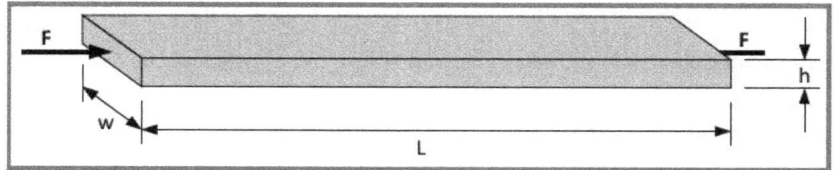

**Solution:** We make a section cut near the center of the bar, and construct the free body diagram shown below:

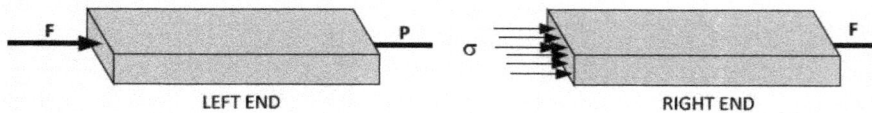

LEFT END          RIGHT END

The stresses $\sigma$ visible on the face exposed by the section cut are due to the internal force P. It is clear from $\Sigma F_x = 0$ that P = F; then from Eq. (4.4), we write:

$$\sigma = P/A = -800 \times 10^3 \text{ N}/(100)(50)\text{mm}^2$$

$$\sigma = -800/5 = -160 \text{ N/mm}^2 = -160 \text{ MPa}$$

The axial stress is a negative number (– 160 MPa) indicating that a compressive stress develops in the bar due to the compressive axial loading. Also observe that the value of the stress is independent of the material from which the bar is fabricated.

## EXAMPLE 4.14

Determine the stress in a two-foot long rod that has a diameter of 4 in. The bar is fabricated from mild steel with a yield strength $S_y = 30,000$ psi, and is subjected to an axial compressive load of 100,000 lb. Also, compute the safety factor of the bar against failure by yielding.

**Solution:** Again, we begin with a drawing of the rod and a free body diagram in Fig. E4.14 showing the forces F and P and the normal stresses $\sigma$.

Fig. E4.14

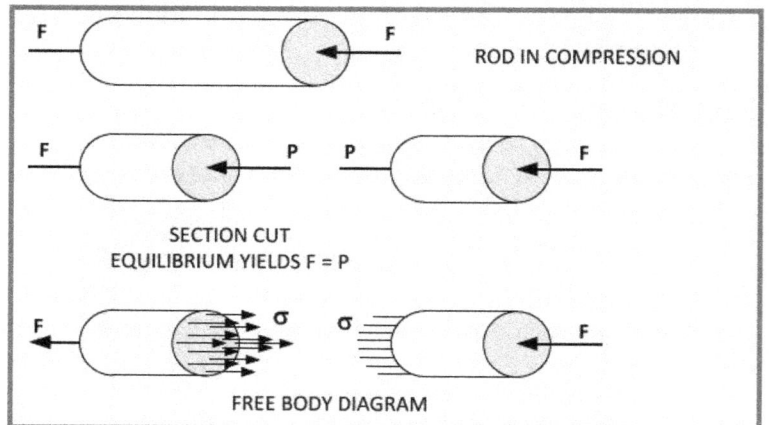

ROD IN COMPRESSION

SECTION CUT
EQUILIBRIUM YIELDS F = P

FREE BODY DIAGRAM

From Eq. (4.4), we write:

$$\sigma = P/A = -\,100{,}000 \text{ lb}/[\pi \times (2)^2] = -\,7{,}958 \text{ psi}$$

We determine the safety factor against yielding from Eq. (4.12).

$$\mathbf{SF_y} = S_y/S_{design} = -\,30{,}000/-\,7{,}958 = 3.77$$

Note, the safety factor is always a positive quantity. In this problem, we compared a compressive strength of $-\,30{,}000$ psi (taken as a negative quantity) with the design stress of $-7{,}958$ psi.

## *Deflection of Axially Loaded Bars and Rods*

The deformation of rods and bars may be described using Eqs. (4.1) and (4.2), as illustrated in the following examples.

## EXAMPLE 4.15

If the bar described in Example 4.13 is fabricated from steel, determine the length of the bar after the application of the compressive force.

**Solution:** An inspection of the free body diagram shown in Fig E4.13a indicates that the internal force P is constant from one end of the bar to the other and equal to the external force F. Using Eqs. (4.1) and (4.2), we may write:

$$\delta = L_f - L_0 = PL_0 /(AE) \qquad (a)$$

Solving Eq. (a) for $L_f$, gives:

$$L_f = L_0 \,[1 + P/(AE)] \qquad (b)$$

Substituting numerical values for the known parameters in Eq. (b) yields:

$$L_f = 0.5 \,\{1 - [(800 \times 10^3) /(0.05)(0.1)(207 \times 10^9)]\}$$

$$L_f = 0.5[1 - 0.7729 \times 10^{-3}] \text{ m} = (500 - 0.3864) \text{ mm} = 499.6 \text{ mm} \qquad (c)$$

Since the new length of the rod is 499.6 mm, it is apparent that it contracted by 0.4 mm under the action of the compressive load. The value of P in Eq. (4.2) is treated as a negative number when the axial load on the bar is compressive. Also, the value of the modulus of elasticity was taken as 207 GPa from Appendix B.

## EXAMPLE 4.16

Determine the axial deflection of the rod described in Example 4.14. Repeat the solution for the rod if it is fabricated from an aluminum alloy.

**Solution:** An inspection of the free body diagram in Fig. E4.14 indicates that the internal force P is constant over the entire length of the rod. From Eq. (4.2), we write:

$$\delta = PL/(AE) \tag{a}$$

Substituting the values for the known quantities for the steel rod in Eq. (a) gives:

$$\delta = - [(100,000)(2)(12)]/[(\pi)(2)^2 (30 \times 10^6)] = - 0.006366 \text{ in.}$$

Substituting the values for the known quantities for the aluminum rod in Eq. (a) gives:

$$\delta = - [(100,000)(2)(12)]/[(\pi)(2)^2 (10.4 \times 10^6)] = - 0.01836 \text{ in.}$$

Both solutions carry a negative sign indicating that the rod is compressed (shortened) by the action of the compressive force. It is interesting to observe that the rod fabricated from an aluminum alloy exhibited nearly three times the deflection of the steel rod. The reason for this difference is in the lower modulus of elasticity of aluminum relative to steel. The modulus of elasticity for aluminum and steel used in this calculation was $10.4 \times 10^6$ and $30 \times 10^6$ psi, respectively as cited in Appendix B.

## 4.4 SHEAR STRESSES

In the preceding discussions, the stresses created by internal forces were normal to the area exposed by the section cut. To emphasize this fact, we called these **normal stresses**. A second type of stress exists—a **shear stress**. As the name implies, the shear stress lies in the plane of the area exposed by the section cut. To show shear stresses in a more graphical manner, consider the stubby beam-like member loaded with the force F in Fig. 4.11.

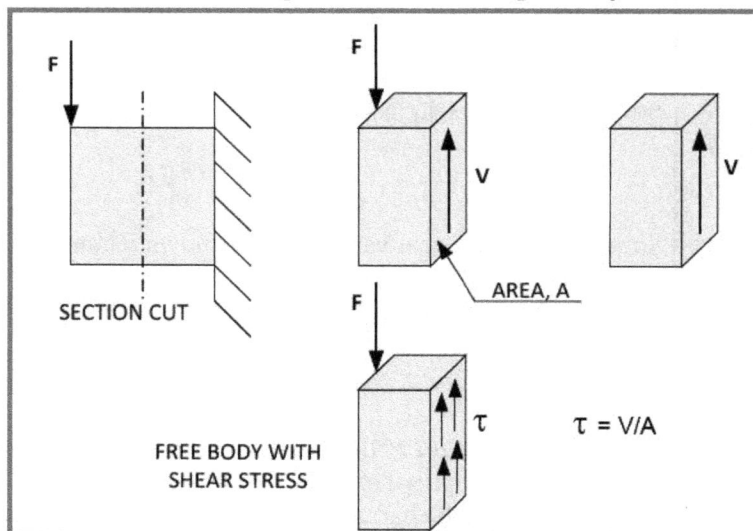

Fig. 4.11 A section cut made in a stubby beam-like member produces a free body.

First, cut the stubby member to create a free body of the left end, and then apply an internal shear force V in the plane of the area exposed by the section cut[6]. From $\sum F_y = 0$, we determine that F = V. The shear force V is produced by a shear stress $\tau$. The relation between shear stress $\tau$ and the shear force V is:

---

[6]To maintain the focus of the discussion on shear forces, we have not included the internal moment acting at the section cut on the FBD in Fig. 4.11.

$$\tau = V/A \tag{4.13}$$

We have assumed that the shear stress $\tau$ is uniformly distributed over the area of the stubby member. Later, in the discussion of beam theory found in a later course on Mechanics of Materials, we will find that the shear stress is not uniformly distributed over the cross sectional area of a beam. However, for many block-like members, the assumption of a uniform distribution of shear stresses is a reasonable approximation.

## EXAMPLE 4.17

A key is employed to keep a gear from slipping on a shaft when transmitting power. Forces F of 50 kN are created on the key at the locations shown in Fig. E4.17. Determine the shear stresses in the key if it is 8mm wide, 9 mm high and 50 mm long.

**Solution:** We construct a FBD of the key, presented in Fig. E4.17, showing the equal and opposite forces F and the shear plane between the gear and the shaft. We section the key along the shear plane, and draw another free body diagram of its lower portion. On this second free body diagram, we illustrate the shear stresses $\tau$ that occur on the shear plane.
The shear stresses on the key are given by Eq. (4.13) as:

$$\tau = V/A = (50 \times 10^3) /(0.008)(0.050) = 125 \text{ MPa}$$

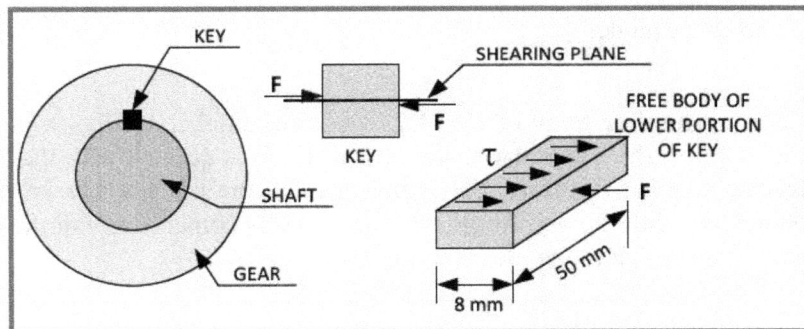

Fig. E4.17 A section cut showing shear forces acting on a key, which prevents the gear from slipping on the shaft

The average shear stress acting on the key is 125 MPa. To analyze the impact of this solution on the design of a structure, it is necessary to compare the imposed stress with the shear strength of the material from which the key is fabricated. The shear strength is usually lower than the yield or ultimate tensile strength of a material. A common practice is to consider either the yield or tensile strength of a material and multiply that value by 0.577 to estimate the shear strength. For example, if the tensile yield strength, $S_y$ of the steel used in manufacturing the key was 320 MPa, the yield strength in shear $S_{ys}$ is estimated as:

$$S_{ys} = 0.577 \, S_y = (0.557)(320) = 184.6 \text{ MPa}$$

The safety factor for the key is determined from:

$$SF = S_{sy} /\tau = 184.6/125 = 1.477$$

In interpreting the results for normal stresses, we recognize the difference between tensile (+) and compressive (−) values for the results. The reason for the careful distinction is the fact that the strength of material subjected to tensile or compressive stresses is often different. However, shear strength is not sensitive to the sign of the shear stress τ. The strength of a material to an imposed shear stress is not dependent on the direction of the imposed shear force. For this reason, we neglect the sign of the shear force in our analysis of stress on machine components.

## 4.5 STRESSES ON OBLIQUE PLANES

In making section cuts on bars and rods, we have limited our choices to sections perpendicular to the axis of the bar. This restriction was helpful because it simplified the state of stress that was examined. With these perpendicular section cuts, we considered only the normal stresses acting on the area exposed by the cut.

We showed that shear stresses exist in Section 4.4 and computed their magnitude in Example 4.17. In this case, we restricted the section cut to one parallel to the imposed system of equal and opposite forces. With this restriction on the section cut, only shear forces are developed. Clearly, by restricting the section cut we develop special cases where normal stresses exist in the absence of shear stresses and vice versa.

Let's treat the more general case with the bar cut at an arbitrary angle as shown in Fig 4.12.

Fig. 4.12 Bar with axial loading with a section cut at an arbitrary angle.

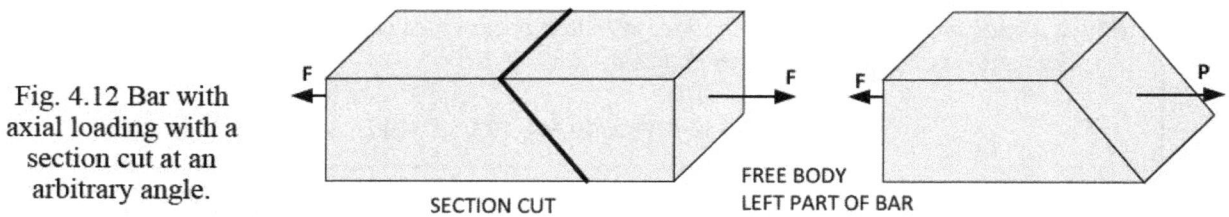

Let's cut the free body of Fig. 4.12 with still another section to produce a triangular shape as shown in Fig. 4.13a. The triangular shape in Fig. 4.13b is defined with the angle θ. The left side of this triangle has an area A; therefore, the hypotenuse of the triangle has an area = A/cos θ. Note, the x-axis is coincident with the axis of the bar; the n axis is coincident with the outer normal to the inclined cut; the t axis is tangential to the inclined cut.

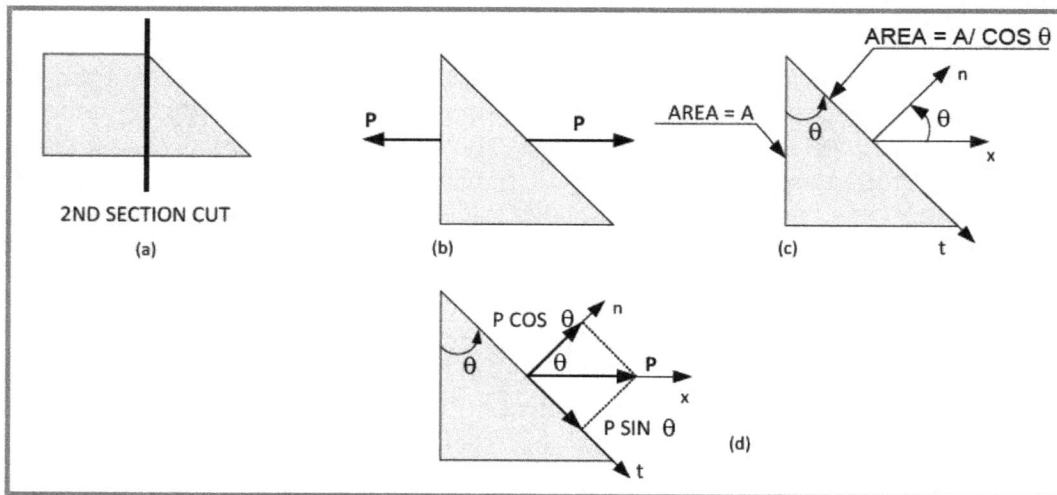

Fig. 4.13 Free body diagrams of the bar with an inclined cut: (a) The section cut is perpendicular to the axis of the bar to produce a triangle; (b) Internal forces acting on both faces of the element; (c) The area of the left side and the hypotenuse is illustrated; (d) The forces on the surface of the inclined cut are resolved into components.

As shown in Fig. 4.13c, the internal force P has been resolved into components along the n and t axes to yield:

$$P_n = P \cos \theta \qquad\qquad (a)$$

$$V_t = P \sin \theta \qquad\qquad (b)$$

To ascertain the stresses on the inclined surface, we simply divide either $P_n$ or $V_t$ by the area formed with the inclined section cut in accordance with Eq. (4.4) or Eq. (4.13). The stress $\sigma_\theta$ in the normal direction (n) is determined from Eq. (4.4) as:

$$\sigma_\theta = (P \cos \theta)/(A/\cos \theta) = (P/A) \cos^2 \theta \qquad\qquad (c)$$

If we define the axial stress $\sigma_x = P/A$, then we may write:

$$\sigma_\theta = (\sigma_x)\cos^2 \theta \qquad\qquad (4.14)$$

The shear stress $\tau_\theta$ that acts along the face of the inclined surface is given by Eq. (4.13) as:

$$\tau_\theta = P \sin \theta/(A/\cos \theta) = (P/A) \sin\theta\cos \theta \qquad\qquad (d)$$

This relation is rewritten as:

$$\tau_\theta = (\sigma_x/2) \sin2\theta \qquad\qquad (4.15)$$

Numerical results are presented in Fig. 4.14 for both $\sigma_\theta$ and $\tau_\theta$ as the angle of the inclined section varies from zero (a perpendicular cut) to 90° (a parallel cut). In this figure the axial stress $\sigma_x = 100$ units. From Fig. 4.14 it is evident that the normal stress $\sigma$ is a maximum when $\theta = 0°$ and the section cut is perpendicular to the axis of the bar. For this angle of the cut, $\sigma_\theta = \sigma_x = P/A$. Thus, when employing Eq. (4.4) to compute the normal stress, we determine the maximum possible value of $\sigma_\theta$. The shear stress is zero on the plane defined by $\theta = 0°$ where the normal stress is a maximum.

Both the normal and shear stress vanish when the section cut is made at 90°. The shear stress is a maximum when the section cut is defined by $\theta = 45°$. When $\theta = 45°$, the normal and shear stresses are equal to each other $\sigma_{45} = \tau_{45} = \sigma_x/2$.

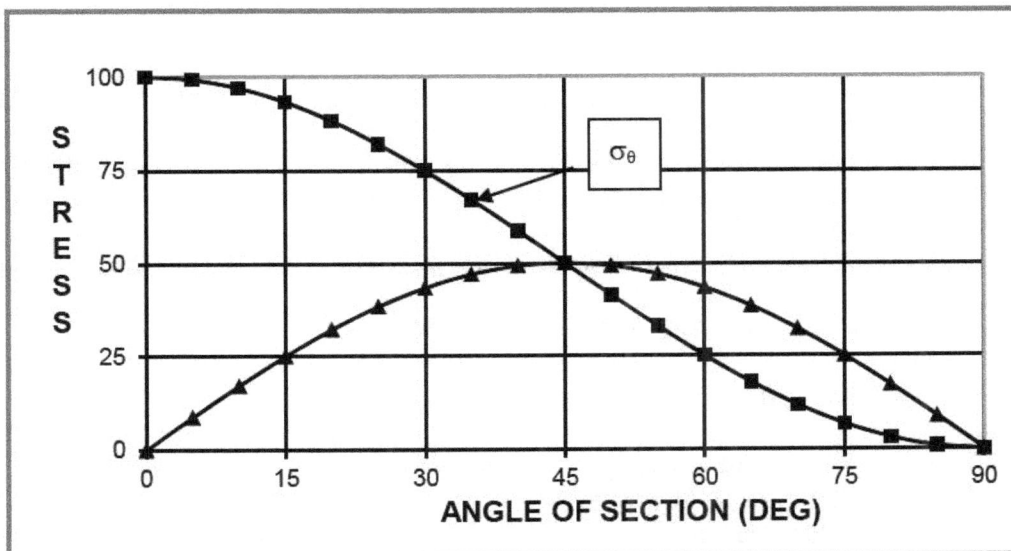

Fig. 4.14 Variation in $\sigma_\theta$ and $\tau_\theta$ with the angle of the inclined section.

The example of the axially loaded bar, with an inclined section cut, illustrates why stresses must be treated as tensor quantities. As we vary the angle of inclination of the section cut, two parameters are changing relative to either the normal or the shear stresses. First, the magnitude of the force components in the n and t direction changes with the angle $\theta$ because forces are vector quantities. Second, the area of the inclined surface exposed by the section cut increases with $\theta$. Both of these parameters affect the magnitude of the normal and shear stresses; therefore stresses must be treated as tensor quantities—not vectors.

## EXAMPLE 4.18

A circular rod 30 mm in diameter and 2 m long is subjected to an axial load of 120 kN. Determine:

- The maximum normal stress and the plane upon which it acts.
- The maximum shear stress and the plane upon which it acts.
- Draw a FBD of the section for the maximum normal stress.
- Draw a FBD of the section for the maximum shear stress.

---

**Solution:** The maximum normal stress occurs on a plane perpendicular to the axis of the rod where $\theta = 0°$. Equation (4.14) applies.

$$\sigma_{\theta=0} = \sigma_x \cos^2 \theta = (P/A) \cos^2 \theta = (120 \times 10^3)/[\pi (0.015)^2]\cos^2 0° = 169.8 \text{ MPa}$$

The maximum shear stress occurs when $\theta = 45°$ as indicated by the results depicted in Fig. E4.14. From Eq. (4.15), we write:

$$\tau_{\theta=45} = (\sigma_x/2) \sin 2\theta = (169.8/2) \sin 90° = 84.90 \text{ MPa}$$

The FBDs for the right portion of the bar showing the maximum normal and shear stresses are shown below:

Fig. E4.18 Section cuts for (a) maximum normal stress and (b) maximum shear stress.

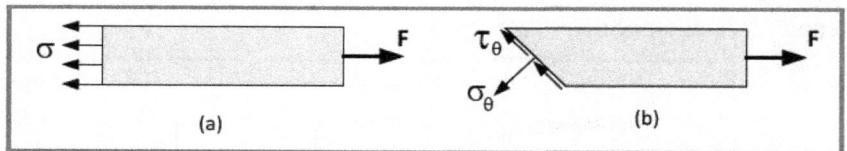

The FBD in Fig. E4.18a shows the normal stresses $\sigma$ acting on a plane area perpendicular to the axis of the rod. The shear stresses are shown in Fig. E4.18b where they act on a surface inclined at a 45° angle to the axis of the rod.

---

## 4.6 AXIAL LOADING OF A TAPERED BAR

The discussion of stress and deflection of a rod or bar has been limited to those members with a uniform cross sectional area. Of course, uniform bars and rods are most commonly employed in building structures because they are easy to design with and less expensive to manufacture; however, in some instances it may be more desirable to use members that are not uniform. In these cases, we must accommodate the effect of the changing cross sectional area over the length of the bar on both the stresses and displacement.

### 4.6.1 Normal Stresses in a Tapered Bar

The normal stresses that occur in a tapered bar or rod are determined using Eq. (4.4). The only consideration made to account for the taper in the bar is to adjust the area A to correspond with the position along the length of the bar. Let's consider the tapered bar presented in Fig. 4.15. The thickness of the bar, b, is constant along its entire length. The height, $h_x$, of the bar varies with position x according to the relation:

$$h_x = h_1 + (h_2 - h_1)(x/L)$$

The area, $A_x$, at any position x along the length of the bar is then:

$$A_x = bh_x = b[h_1 + (h_2 - h_1)(x/L)] \qquad (4.16)$$

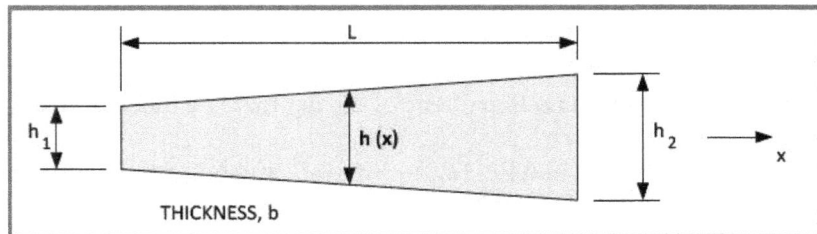

Fig. 4.15 Geometry of a tapered bar with a uniform thickness.

The normal stresses $\sigma_x$ due to an axial force are determined by substituting Eq. (4.16) into Eq. (4.4) to obtain:

$$\sigma_x = P/A_x = P/\{b[h_1 + (h_2 - h_1)(x/L)]\} \qquad (4.17)$$

where L is the length of the tapered bar.

## EXAMPLE 4.19

For a tapered bar similar to the one shown in Fig. 4.15, determine the normal stress as a function of position, x along the length of the bar. The geometry of the bar is given by $h_1 = 2$ in., $h_2 = 6$ in., $b = 3$ in. and $L = 36$ in. The axial load imposed on the tapered bar is 5,000 lb.

**Solution:** We employ Eq. (4.17) and write:

$$\sigma_x = P/\{b[h_1 + (h_2 - h_1)(x/L)]\} = 5,000/\{3[2 + (6 - 2)(x/36)]\} \qquad (a)$$

$$\sigma_x = 15,000/(18 + x)$$

The stress $\sigma_x = 833.3$ psi is a maximum at x = 0, and decreases to a minimum value of $\sigma_x = 277.8$ psi when x = 36 in.

### 4.6.2 Deflection of Tapered Bars

The deflection of a tapered rod may be determined from Eq. (4.2) although we must again modify the approach to accommodate for the changing cross sectional area over the length of the bar. To determine the axial deflection, this accommodation is more difficult since the total deformation of the bar is the sum of the incremental deflections at each position x along the entire length of the bar. We begin by considering an incremental length dx at some position x as shown in Fig. 4.16.

Fig. 4.16 An incremental length (slice) dx at position x along the length of the tapered bar.

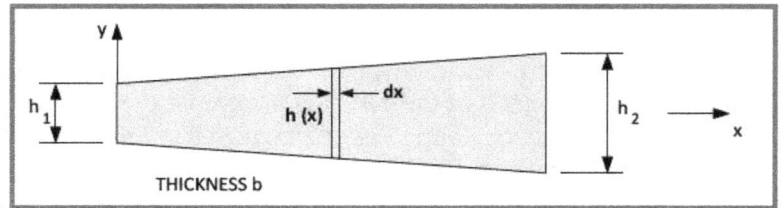

To accommodate for the taper in the bar, we determine the incremental deflection $d\delta$ of a bar of length dx. Since dx approaches zero, we treat the cross sectional area as a constant over the incremental length. Accordingly, we modify Eq. (4.2) to read:

$$d\delta = P\, dx/(A_x E) \tag{4.18}$$

where $A_x$ is the cross sectional area of the bar that is a function of x given in Eq. (4.16).

Substitute Eq. (4.16) into Eq. (4.18), simplify and integrate to obtain:

$$\delta = \frac{PL}{Eb} \int_0^L \frac{dx}{h_1 L + (h_2 - h_1)x} \tag{a}$$

$$\delta = \frac{PL}{Eb} \left( \frac{1}{h_2 - h_1} \right) \ln\left( \frac{h_2}{h_1} \right) \tag{4.19}$$

## EXAMPLE 4.20

Determine the deflection of the tapered bar described in Example 4.19 if the bar is fabricated from an aluminum alloy with a modulus of elasticity $E = 10.4 \times 10^6$ psi.

**Solution:** Let's solve this example problem by recalling Eq. (4.19).

$$\delta = \frac{PL}{Eb} \left( \frac{1}{h_2 - h_1} \right) \ln\left( \frac{h_2}{h_1} \right)$$

Substituting the parameters describing the geometry and the material constant for the bar into this relation yields:

$$\delta = [(5000)(36)/(10.4 \times 10^6)(3)][1/(6 - 2)]\ln(6/2) = 1.585 \times 10^{-3} \text{ in.}$$

Examine the magnitude of the deflection of the tapered bar. Is the deflection large or small? What reference do you use to judge? Clearly, the axial extension of the bar is small. For small quantities, we sometimes use the human hair as a reference. The diameter of a single strand of hair is slightly more than $2 \times 10^{-3}$ in.; hence, the extension is about 75% of a hair diameter.

## 4.7 AXIAL LOADING OF A STEPPED BAR

In some structures, bars of different cross sectional areas are employed where the area changes abruptly at some position along the length of the bar. An illustration of a stepped bar where the cross section undergoes an abrupt change is presented in Fig. 4.17.

The procedure for determining the normal stresses and deflection remains the same as for a bar with a uniform cross sectional area:

1. Draw a free body diagram of each section of the bar.
2. Use the equilibrium relations to establish the internal axial forces acting on each section.
3. Determine the stresses from Eq. (4.4) using the area of the section of interest.
4. Solve for the deflection of each segment of the bar from Eq. (4.2) and then add them to obtain the total deflection.

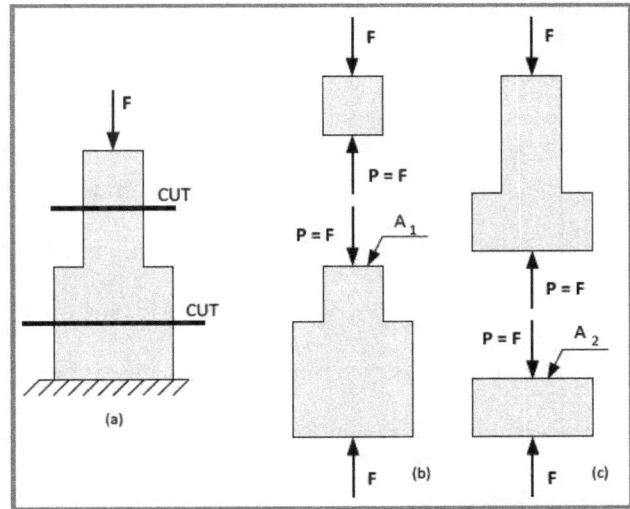

Fig. 4.17 A stepped bar with free body diagrams for each section of the bar.

### 4.7.1 Normal Stresses in Stepped Bars

The normal stress varies with the cross sectional area of the bar. For example, in Fig 4.17 the normal stress in the upper section of the bar is given by $\sigma = P/A_1$, and the stress in the lower section of the bar is $\sigma = P/A_2$.

## EXAMPLE 4.21

For the stepped bar, shown in Fig. E4.21, determine the normal stress in each of the two sections of the bar if $F_1 = 40$ kN; $F_2 = 80$ kN; $w_1 = 50$ mm; $b_1 = 60$ mm; $w_2 = 90$ mm; $b_2 = 60$ mm; $L_1 = 200$ mm and $L_2 = 300$ mm

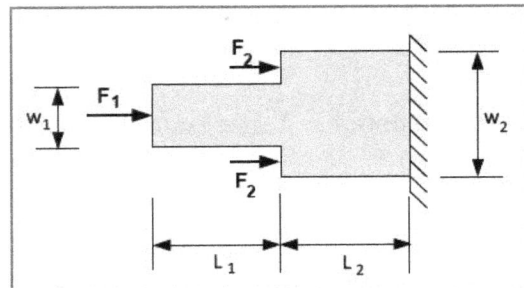

Fig. E4.21 The thickness of the bar is given by $b_1$ and $b_2$.

**Solution:**

Let's begin the solution by drawing free bodies of both sections of the bar. We make two section cuts A and B as illustrated in Fig. E4.21a. Then we draw free bodies associated with the portion of the bar to the left side of the section cut as shown below:

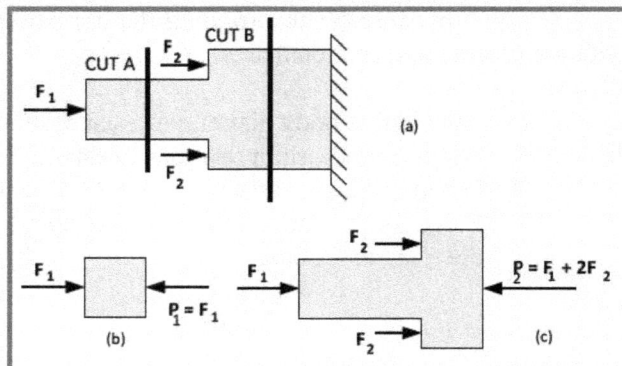

Fig. E4.21a FBDs of stepped sections.

From the equilibrium relations, it is clear that the internal force $P_1 = F_1$ in the smaller section of the stepped bar. Also, for the larger section of the bar, the internal force $P_2 = F_1 + 2F_2$.

Finally, from Eq. (4.4) for the smaller section of the bar, we write:

$$\sigma_1 = P_1/A_1 = -(40 \times 10^3)/[(50)(60)] = -13.33 \text{ N/mm}^2 = -13.33 \text{ MPa}$$

For the larger section of the bar, we write:

$$\sigma_2 = P_2/A_2 = -\{[40 + (2)(80)] \times 10^3\}/[(90)(60)] = -37.03 \text{ N/mm}^2 = -37.03 \text{ MPa}$$

Note, the minus sign indicates the stresses in the bar are compressive.

## 4.7.2 Deflection of Stepped Bars

To compute the axial extension or compression of the stepped bar, we consider each uniform section of the bar separately. Equation (4.2) is valid for each section because each has a uniform cross sectional area. For a stepped bar comprised of n uniform sections, we superimpose the individual deflections $\delta$ to obtain:

$$\delta_{total} = \delta_1 + \delta_2 + \delta_3 + \ldots + \delta_n \tag{4.20}$$

**EXAMPLE 4.22**

If the stepped bar defined in Example 4.21 is fabricated from a titanium alloy with a modulus of elasticity $E = 114$ GPa, determine the total deflection of the bar.

**Solution:** We recognize that the deflection of the stepped bar is determined from Eq. (4.20) and Eq. (4.2) as:

$$\delta_{total} = \delta_1 + \delta_2 = (P_1 L_1)/(A_1 E) + (P_2 L_2)/(A_2 E) \tag{a}$$

Since the modulus of elasticity is constant for both sections of the bar, Eq. (a) reduces to:

$$\delta_{total} = \delta_1 + \delta_2 = \frac{1}{E}\left[\frac{P_1 L_1}{A_1} + \frac{P_2 L_2}{A_2}\right] \qquad\qquad \text{(b)}$$

Substituting the numerical parameters for the unknown quantities in this relation yields:

$$\delta_{total} = \delta_1 + \delta_2 = \frac{1}{114 \times 10^9}\left[\frac{(-40 \times 10^3)(0.2)}{(0.05)(0.06)} + \frac{(-200 \times 10^3)(0.3)}{(0.09)(0.06)}\right] = -0.1209 \text{ mm}$$

Let's interpret this solution. The negative sign indicates the bar was compressed and the deformations reduced its length. The original length of the bar was $L_1 + L_2 = 500$ mm. When the total deformation of the bar is compared to this length, we find the deformation is very small— only 0.024%. This example again emphasizes that deformations of metallic members are usually very small. For this reason, we neglect these deformations and use the original lengths of structural members when substituting into the equilibrium equations.

## 4.8 STRESS CONCENTRATIONS

In our discussion of stresses in uniform, tapered or stepped bars, we have assumed that the stresses are uniformly distributed over the cross sectional area of the bar. For the uniform and the tapered bars, this is a valid assumption except near the ends of the bar where the external forces are applied. However, for the stepped bar, the stresses are not uniformly distributed across the section in the vicinity of the step. Indeed, whenever we encounter a discontinuity such as a step or a hole in a bar, the stresses tend to concentrate at that discontinuity. As a consequence, the use of Eq. (4.4) to compute the stresses seriously underestimates their actual value. We must account for the effect of the structural discontinuities by determining a suitable stress concentration factor. The topic of stress concentration factors will be introduced in the subsection below. However, since the topic is extremely important, the reader is referred to an excellent book by R. E. Peterson for more details [1].

### 4.8.1 Stress Concentration Factors Due to Circular Holes in Bars

Let's consider the bar with a centrally located circular hole subjected to an axial tension force as shown in Fig. 4.18. The stress distribution in a section removed three or more diameters from the hole is uniform with a magnitude given by Eq. (4.4) as $\sigma_o = P/(bw)$. However, on the section through the center of the hole the stress distribution shows significant variation. The stresses increase sharply adjacent to the discontinuity (the hole) and concentrate at this location. The maximum value of the normal stress occurs adjacent to the hole as indicated in Fig 4.19

We are interested in determining the maximum stress, $\sigma_{MAX}$ adjacent to the hole. It is convenient to express the maximum stresses in terms of a stress concentration factor by employing:

$$\sigma_{MAX} = K \sigma_{NOM} \qquad\qquad (4.21)$$

where K is the stress concentration factor and $\sigma_{NOM}$ is the nominal stress.

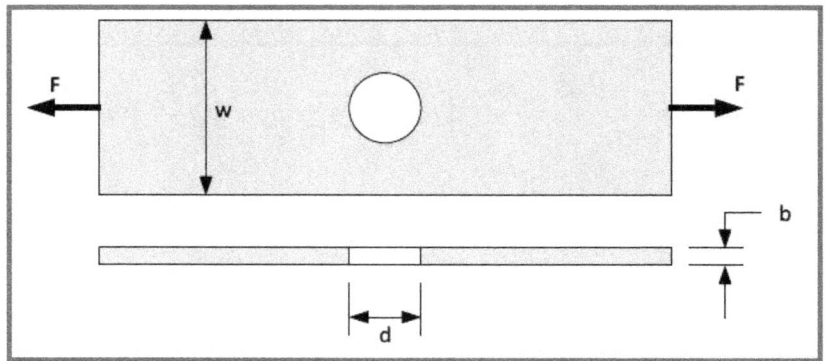

Fig. 4.18 A centrally located circular hole in an axially loaded bar.

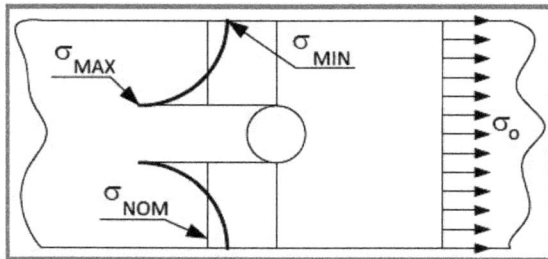

Fig. 4.19 Distribution of stress across a section through the center of the hole shows the concentration of stresses adjacent to the boundary of the hole.

The **stress concentration factor** K for a uniform thickness bar with a central circular hole subjected to axial loading is a function of the geometry depending on the ratio of d/w as shown in Fig. 4.20.

Fig. 4.20 Stress concentration factor for a central circular hole in an axially loaded bar.

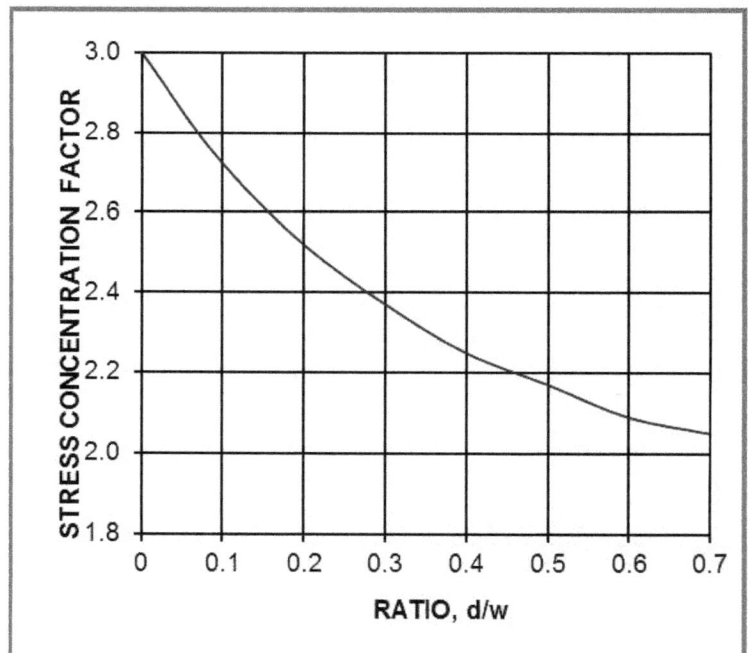

The **nominal stress** is the average stress across the net section containing the hole, and is given by:

$$\sigma_{NOM} = P/A_{NOM} = P/[(w - d)b] \qquad (4.22)$$

where b is the thickness of the bar, w is the bar width and d is the hole diameter.

The uniform stress $\sigma_o$ and the nominal stress $\sigma_{NOM}$ are related by:

$$P = \sigma_{NOM} A_{NOM} = \sigma_o A_{UNF} \qquad (a)$$

Substituting for the areas in Eq. (a) and simplifying yields:

$$\sigma_{NOM} = [w/(w - d)]\sigma_o \qquad (4.23)$$

The nominal stress is always greater than the uniform stress since the factor $w/(w - d)$ is always greater than one.

## EXAMPLE 4.23

A centrally located hole of diameter $d = 0.5$ in. is drilled in a long thin bar that is subjected to an axial load of 8,000lb. If the bar is defined by $w = 1.5$ in., and $b = 0.50$ in., determine the nominal stress, the stress concentration, and the maximum stress.

**Solution:** The nominal stress is given by Eq. (4.22) as:

$$\sigma_{NOM} = P/[b(w - d)] = 8,000/[(1.5 - 0.5)(0.5)] = 16,000 \text{ psi}$$

The stress concentration K is determined from Fig. 4.20, by locating the intercept of a vertical line originating at $d/w = 0.5/1.5 = 0.3333$ with the curve. This intercept gives $K = 2.33$.

Finally, the maximum stress is given by Eq. (4.21) as:

$$\sigma_{MAX} = K \sigma_{NOM} = (2.33)(16,000) = 37.28 \text{ ksi}$$

By drilling a hole in the bar, we increase the maximum stress significantly. The procedure using the stress concentration factor provides a simple yet effective approach for solving a very difficult stress analysis problem. Also, you should be aware that the uniform stress in the bar prior to drilling the hole was $\sigma_o = 10.67$ ksi; therefore, the presence of the hole increased the stresses by a factor of 3.50.

## 4.8.2 Stress Concentrations in Shouldered Bars

The stepped bar is another configuration with an abrupt change in the section of the bar. This geometric discontinuity produces a non-uniform distribution of stresses and a concentration of stress at the fillet used in transitioning from one section of the bar to the other. The geometric parameters involved in characterizing the stress concentration factor are $w_2/w_1$ and $r/w_1$ as defined in Fig. 4.21.

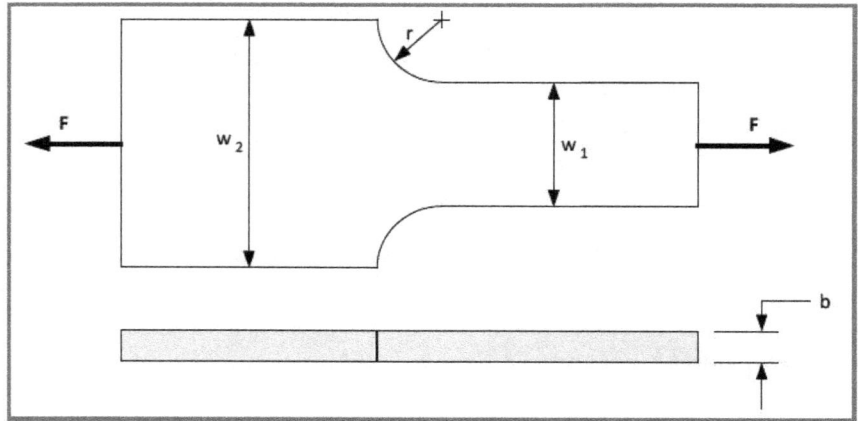

Fig. 4.21 Geometric parameters controlling the stress concentration in a stepped bar.

The maximum stress occurs at the fillet in the transition from the narrow section of the bar ($w_1$) to the wider section ($w_2$). The maximum stress is determined from Eq. (4.21). However, for the case of the stepped bar the nominal stress is defined as:

$$\sigma_{NOM} = P/(w_1 b) \tag{4.24}$$

The stress concentration factor K for the stepped bar depends on two different ratios $r/w_1$ and $w_2/w_1$. Curves showing the stress concentration factor K as a function of these ratios are presented in Fig. 4.22.

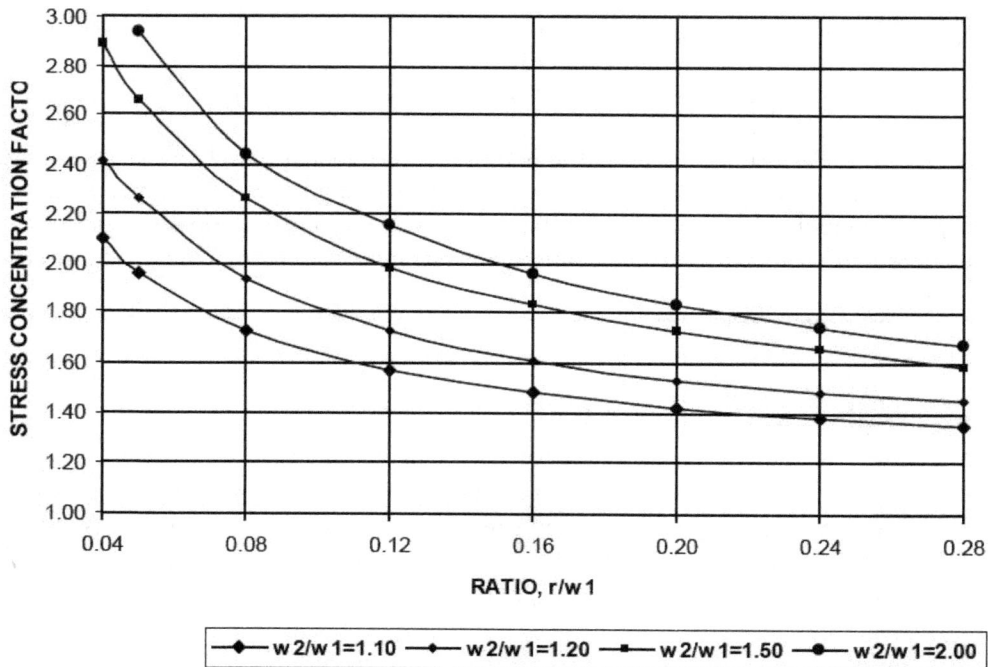

Fig. 4.22 Stress concentration factor for a stepped bar subjected to axial forces.

## EXAMPLE 4.24

Consider the axially loaded stepped bar shown in Fig. E4.24, and determine the maximum stress at the fillet in the transition region of the bar.

Fig. E4.24

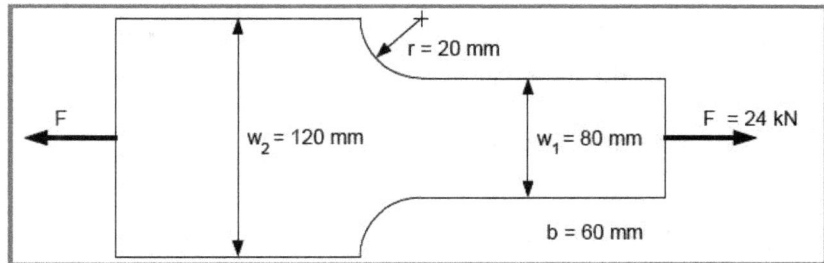

**Solution:** From Eq. (4.21), we write $\sigma_{MAX} = K\,\sigma_{NOM}$.

From Eq. (4.24), write:

$$\sigma_{NOM} = P/w_1 b = (24 \times 10^3)/[(80)(60)] = 5 \text{ MPa}$$

To determine the stress concentration factor K, we note that $r/w_1 = 20/80 = 0.25$, and $w_2/w_1 = 120/80 = 1.5$. Reading the chart for the stress concentration factor in Fig. 4.22, we find that K is:

$$K = 1.65 - (1/4)(1.65 - 1.58) = 1.63.$$

Finally, Eq. (4.21) yields:

$$\sigma_{MAX} = K\,\sigma_{NOM} = 1.63 \times 5 = 8.15 \text{ MPa}$$

## 4.9 SCALE MODELS

We have described methods for determining the stresses, strains and deflection of bars and rods subjected to axial tension or compression forces. The analysis required to determine these quantities is relatively simple when the bar can be isolated and the forces acting on it are known. However, if the bar is part of a complex structure, determining the forces acting on the bar may prove to be more difficult. In some cases, scale models of the structure under consideration are constructed and then subjected to various loads to verify structural integrity.

### Geometric Scale Factor

Scale models of structures are usually much smaller than the real thing. Let's call the real structure—a bridge, building or stadium—the prototype, and consider a scale model that is 100 times smaller than the prototype. In this case, the geometric scale factor $\mathbf{S} = 1/100$. If the same scale factor $\mathbf{S}$, is used for all three dimensions (i.e., length, width and thickness), we may write:

$$L_m = \mathbf{S}\,L_p$$

$$w_m = \mathbf{S}\,w_p \tag{4.25}$$

$$b_m = \textbf{S}\, b_p$$

where L, w and b represent length, width and thickness and subscripts m and p refer to the model and the prototype respectively.

We note from Eq. (4.25), that the scale model is exactly the same shape as the prototype except for its size. It is geometrically proportional.

### *Scaling Factor for Stresses*

Let's continue with the concept of scaling and consider the stresses occurring in both the model and the prototype. The stress $\sigma$ in a bar subjected to axial loading is given by Eq. (4.4) as:

$$\sigma = P/A = P/(w\,b) \tag{4.26}$$

where P is the axial load applied to the bar, and $A = w\,b$ is the cross sectional area of a bar of width w and depth b.

If the left side of Eq. (4.26) is divided by its right side, a unit dimensionless ratio involving the stress and its controlling parameters is formed:

$$\sigma(w\,b)/P = 1 \tag{4.27}$$

Let's designate a model and a prototype in Eq. (4.27) by writing:

$$\sigma_m\,(w_m\,b_m)/P_m = \sigma_p\,(w_p\,b_p\,)/P_p \tag{a}$$

Rearranging the terms in Eq. (a) gives:

$$\sigma_m = (P_m/\,P_p)(w_p\,/w_m\,)(b_p\,/\,b_m\,)\sigma_p \tag{b}$$

Substituting Eq. (4.25) into Eq. (b) yields:

$$\sigma_m = (P_m/\,P_p)(1/\textbf{S}^2)\sigma_p \tag{c}$$

Examining Eq. (c) shows that the stresses in the model and the prototype are related by the geometric scale factor $\textbf{S}$ and load scale factor $\textbf{L}$, which is defined by:

$$\textbf{L} = (P_m/\,P_p) \tag{4.28}$$

Finally, we substitute Eq. (4.28) into Eq. (c) and obtain:

$$\sigma_m = (\textbf{L}\,/\,\textbf{S}^2)\,\sigma_p \tag{4.29}$$

Clearly, the stresses produced in the model are related to the scale factors for the load and the geometry. It is also important to recognize that the geometric scale factor usually differs significantly from the load scale factor.

# EXAMPLE 4.25

Suppose we construct a scale model of the Golden Gate Bridge using a scale factor $\mathbf{S}$ of 1/500. Determine the size of the model if:

- Total bridge length$\Rightarrow$ 8981 feet
- Length of suspended structure$\Rightarrow$6450
- Length of main span$\Rightarrow$4200 feet
- Length of each side span$\Rightarrow$1125 feet
- Width of bridge$\Rightarrow$90 feet
- Diameter of main cable$\Rightarrow$36-3/8 inch
- Width of roadway between curbs$\Rightarrow$60 feet
- Lanes of vehicular traffic$\Rightarrow$6
- Weight of main span per lineal foot$\Rightarrow$21,300 lb
- Live load capacity per lineal foot$\Rightarrow$4,000 lb

**Solution:** The model bridge is constructed in three parts including the main span and the two side spans. The length of the main span of the model is given by Eq. (4.25) as:

$$L_m = \mathbf{S} \, L_p = (1/500)(4200) = 8.4 \text{ ft} = 100.8 \text{ in.}$$

The width of the model bridge is determined in the same manner as:

$$w_m = \mathbf{S} \, w_p = (1/500)( 90) = 0.18 \text{ ft} = 2.16 \text{ in.}$$

The diameter, D of the main cable on the model is:

$$D_m = \mathbf{S} \, D_p = (1/500)(36.375) = 0.073 \text{ in.}$$

All geometric features of the model of the bridge are determined in this manner.

# EXAMPLE 4.26

Let's suppose a very small strain gage is installed on a steel bar of a model used to simulate a critical structural member in a prototype. The gage provides a measurement of the strain equal to $4000 \times 10^{-6}$ when the model bridge was fully loaded. If the scaling factor for the load $\mathbf{L} = 1/100,000$, determine the stress in the main cable of the prototype.

**Solution:** The stress acting on the bar in the model is given by Hooke's law as:

$$\sigma = E\varepsilon \qquad\qquad (4.5 \text{ bis})$$

Substituting numerical values for the modulus of elasticity and the strain into Eq. (4.5) gives the stress in the main cable of the model as:

$$\sigma_m = E_m \, \varepsilon_m = (30 \times 10^6)(4000 \times 10^{-6}) = 120,000 \text{ psi}$$

where the modulus of elasticity $E = 30 \times 10^6$ psi for steel.

Solving Eq. (4.29) for the stress on the prototype gives:

$$\sigma_p = (S^2/ \ L) \ \sigma_m \tag{4.30}$$

Substituting the scale factors for **S** and **L** into Eq. (4.30) yields:

$$\sigma_p = (S^2/ \ L) \ \sigma_m = [(1/500)^2 \ /(1/100,000)][120,000] = (10/25)(120,000) = 48,000 \ psi$$

This example illustrates that the scale factors for the load and the geometry of the structure should be selected to limit the stresses induced in the model. In most instances, the scale factor for the load is much smaller than the scale factor for the geometry.

## *Scaling Factor for Displacements*

A model of the structure may also provide displacement measurements that may be used to predict displacements in the prototype. To develop the displacement relation between the model and the prototype, we again seek a unit dimensionless quantity that includes the variables controlling the displacement. Recalling Eq. (4.2) it may be shown that the displacement $\delta$ of a rod of length L subjected to an axial load P is given by:

$$\delta = PL/AE = PL/[(w \ b)E] \tag{4.31}$$

If the left side of Eq. (4.31) is divided by its right side, a unit dimensionless ratio is formed:

$$[\delta(w \ b)E]/(PL) = 1 \tag{4.32}$$

Let's identify the model and the prototype with appropriate subscripts in Eq. (4.32) and write:

$$\delta_m \ (w_m \ b_m) \ E_m \ /(P_m \ L_m \ ) = \delta_p \ (w_p \ b_p) \ E_p \ /(P_p \ L_p \ ) \tag{a}$$

Solving Eq. (a) for $\delta_p$ and substituting Eq. (4.25) and (4.28) into the result, yields:

$$\delta_p = [ \ (S \ E \ )/ \ L] \ \delta_m \tag{4.33}$$

where $E = E_m /E_p$ is the modulus scale factor between the model and the prototype.

If the same materials are employed in the manufacture of both the model and the prototype the modulus scale factor is one. However, selection of the model materials is not restricted. We may use a wide variety of materials to fabricate the model providing they respond in an elastic and linear manner.

## EXAMPLE 4.27

Suppose that a model of a truss type bridge structure is fabricated from members formed from sheet aluminum. The prototype structure is to be fabricated from steel with a span of 600 feet. The model is geometrically scaled so that its span is six feet. The capacity of the live load on the prototype is 15,000 lb/ft, and the model is loaded with 25 lb/ft. If the model deflects a distance of two inches under full load at the center of the span, determine the deflection of the prototype under the design load.

**Solution:** From Eq. (4.33), we write:

$$\delta_p = (\mathbf{S}\,\mathbf{E}\,/\mathbf{L}\,)\delta_m$$

Note the scale factors are given by:

$$\mathbf{S} = 6/600 = 1/100$$

$$\mathbf{L} = 25/15{,}000 = 1.667 \times 10^{-3}$$

$$\mathbf{E} = 10 \times 10^6 /30 \times 10^6 = 1/3$$

Substituting the scale factors into Eq. (4.33) yields:

$$\delta_p = (\mathbf{S}\,\mathbf{E}\,/\mathbf{L}\,)\delta_m = [(1/100)(1/3)/(1.667 \times 10^{-3})](2) = 4.0 \text{ in.}$$

In this instance, the deflection of the prototype is twice as large as the deflection of the model. The choice of materials for the model and the scaling factors for both the load and geometry influence the differences in the stress, and deflection between the model and the prototype.

## 4.10 STATICALLY INDETERMINATE AXIAL MEMBERS

Structures are considered to be statically determinate when the unknown forces, either internal forces or external reactions, can be determined using only the equilibrium relations. However, in some structures the number of unknown forces exceed the number of applicable equilibrium equations, and a solution is not possible without the introduction of additional equations. When the number of unknown forces exceeds the number of applicable equilibrium equations, the structure is **statically indeterminate.**

The approach for solving for the unknown forces in statically indeterminate structures is to use the equilibrium relations and then to introduce additional equations based on the deformation of the structure. The solution involves two separate but compatible parts:

1. Prepare the traditional FBD together with all of the applicable equations of equilibrium.
2. Prepare displacement diagrams together with deformation equations that are consistent with the forces shown on the FBD.

Let's consider an example to demonstrate this approach in solving for the unknown forces in a statically indeterminate structure.

### EXAMPLE 4.28

A 14 ft. long column supporting the corner of a very large building is fabricated from a steel pipe with an outside diameter of 8 in. and an inside diameter of 7 in. The pipe is filled with medium strength concrete to form a composite column that supports a 100-ton axial load as illustrated in Fig. E4.28. Determine the forces and stresses in the steel pipe and concrete plug.

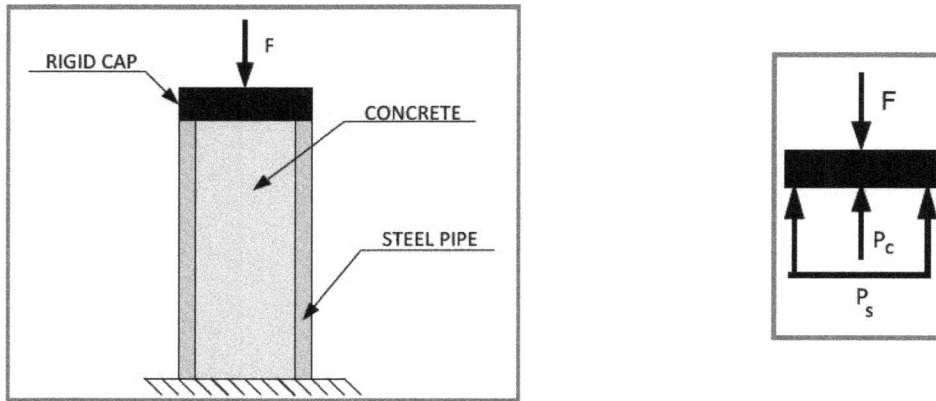

Fig. E4.28a

**Solution:**

**Step 1:** Prepare FBDs of the rigid cap, steel pipe and concrete plug as shown in Fig. E4.28a, and write the only applicable equilibrium equation, which is $\Sigma F_y = 0$.

$$\Sigma F_y = P_c + P_s - F = 0$$

$$P_c + P_s = 100 \text{ ton} \qquad\qquad (a)$$

where $P_c$ and $P_s$ are the internal forces in the concrete plug and steel pipe, respectively.

Because we have two unknown forces $P_c$ and $P_s$ and only a single applicable equation of equilibrium, the unknown forces are statically indeterminate.

**Step 2:** Prepare a drawing shown in Fig. E4.28b that depicts an exaggerated view of the deformations occurring in the concrete plug and the steel pipe when subjected to the internal forces $P_c$ and $P_s$. Because the two structural elements deform together under the action of the axial loads, we write the deformation equation.

$$\delta_c = \delta_s \qquad\qquad (b)$$

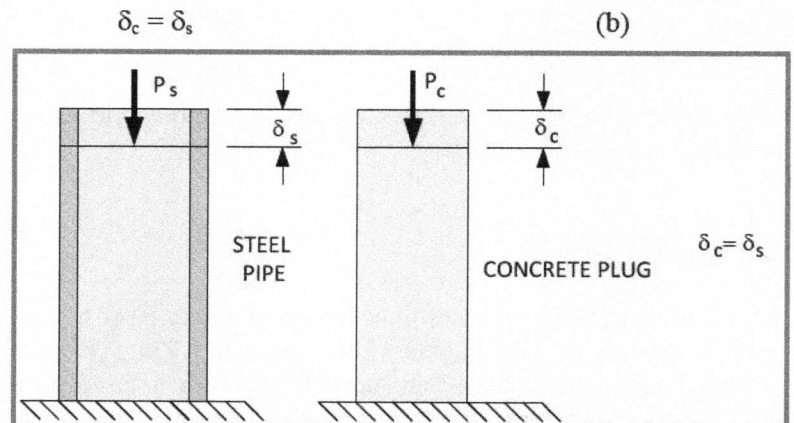

Fig. E4.28b

Recall Eq. (4.2) and write:

$$\delta_c = P_c L_c / A_c E_c = \delta_s = P_s L_s / A_s E_s \qquad\qquad (c)$$

Solve Eq. (c) for $P_c$ noting that $L_c = L_s$, to obtain:

$$P_c = (A_c/A_s)(E_c/E_s)P_s \qquad (d)$$

Equation (d) provides the second relation needed to solve for the unknown internal forces.

**Step 3:** Solve for the internal force $P_s$ by substituting Eq. (c) into Eq. (a) to obtain:

$$P_s = 100/[1 + (A_c/A_s)(E_c/E_s)] \qquad (e)$$

The area ratio is given by:

$$A_c/A_s = [d_i^2/(d_o^2 - d_i^2)] = (7.0)^2/[(8.0)^2 - (7.0)^2] = 3.267 \qquad (f)$$

Using values from Appendix B-1, the modulus ratio is given by:

$$E_c/E_s = 3.6/30 = 0.12 \qquad (g)$$

Substituting the results from Eqs. (f) and (g) into Eq. (e) gives:

$$P_s = 71.84 \text{ ton} \qquad (h)$$

Then from Eqs. (a) and Eq. (h), it is clear that:

$$P_c = 28.16 \text{ ton} \qquad (i)$$

**Step 4:** Solve for the stresses in the concrete plug and steel pipe as:

$$\sigma_c = P_c/A_c = [4(28.16)(2000)]/[\pi(7.0)^2] = 1,463 \text{ psi}$$

$$\sigma_s = P_s/A_s = [4(71.84)(2000)]/\{\pi[(8.0)^2 - (7.0)^2]\} = 12,200 \text{ psi}$$

Note that both of these stresses are compressive.

## EXAMPLE 4.29

A grade 8 steel bolt ¾ in. in diameter (10 threads/in.) is employed to clamp an aluminum bushing between two rigid platens as illustrated in Fig. E4.29. The aluminum bushing has a 2.0 in. outside diameter and 1.0 in. inside diameter and is 10.00 in. long. After the unit is assembled with a snug fit, the nut is tightened by ½ of a turn. Determine the axial stresses in the bolt and the bushing. Also determine the deflection of the aluminum bushing.

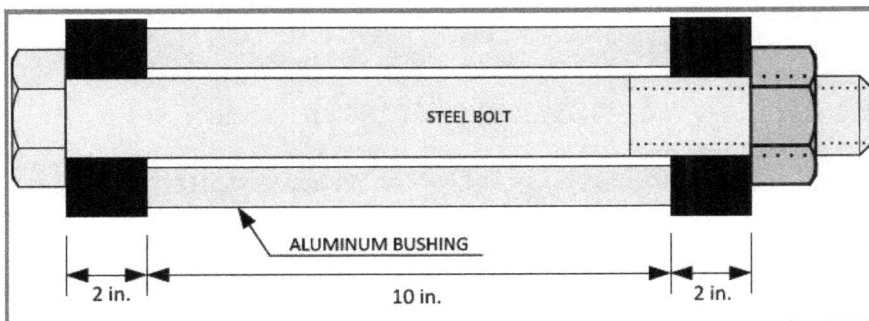

STEEL BOLT

ALUMINUM BUSHING

2 in.    10 in.    2 in.

Fig. E4.29

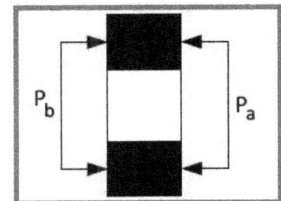

$P_b$    $P_a$

Fig. E4.29a

**Solution:**

**Step 1:** Prepare FBDs of one of the rigid platens as shown in Fig. E4.29a, and write the only applicable equilibrium equation, which is $\Sigma F_x = 0$.

$$\Sigma F_x = -P_a + P_b = 0$$

$$P_a = P_b \qquad\qquad \text{(a)}$$

where $P_a$ and $P_b$ are the internal forces in the aluminum bushing and the steel bolt, respectively. They are equal in magnitude and opposite in sign. Because we have two unknown forces $P_a$ and $P_b$ and only a single applicable equation of equilibrium, the unknown forces are statically indeterminate.

**Step 2:** Prepare a drawing shown in Fig. E4.29b that depicts an exaggerated view of the deformations occurring in the bolt and the aluminum bushing when subjected to the internal forces $P_a$ and $P_b$. When the nut is tightened the steel bolt tends to stretch and the aluminum bushing is compressed (shortened). The total of the two deformations must equal the amount of displacement induced by rotating the nut by ½ of a turn, which is $(1/2)(1/10) = 0.050$ in.

Fig. E4.29b

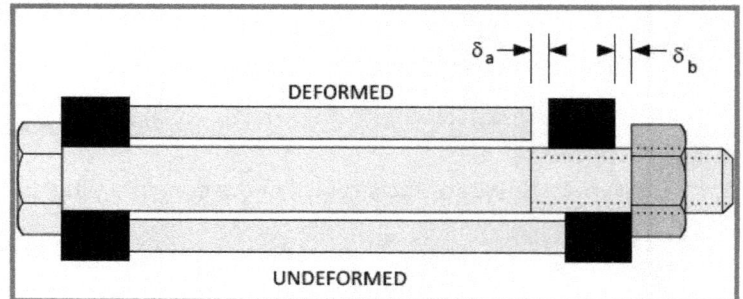

$$\delta_a + \delta_b = .050 \text{ in.} \qquad\qquad \text{(b)}$$

**Step 3:** Recall Eq. (4.2) and Eq. (a) and write:

$$\delta_a = P_a L_a / A_a E_a = P_b L_a / A_a E_a = [(4)(10)/\{\pi[(2)^2 - (1)^2](10.4 \times 10^6)\}] P_b = 0.4081 \times 10^{-6} P_b$$

$$\delta_b = P_b L_b / A_b E_b = \{(4)(14)/[\pi(0.75)^2 (30 \times 10^6)]\} P_b = 1.0563 \times 10^{-6} P_b \qquad \text{(c)}$$

Substitute Eq. (c) into Eq. (b) and solve for $P_b$.

$$P_b [0.4081 + 1.0563] = 0.050 \times 10^6$$

$$P_a = P_b = 34,140 \text{ lb} \qquad\qquad \text{(d)}$$

**Step 4:** The stresses are given by Eq. (4.4) as:

$$\sigma_b = P_b/A_b = (4)(34,140)/[\pi(0.75)^2] = 77.290 \text{ ksi (tension)}$$

$$\sigma_a = P_a/A_a = (4)(34,140)/[3\pi] = 14.490 \text{ ksi (compression)}$$

**Step 5:** The amount of the compression of the aluminum bushing is given by Eq. (4.2) as:

$$\delta_a = P_a L_a / A_a E_a = [(34,140)(10)(4)]/[(3\pi)(10.4 \times 10^6)] = 0.01393 \text{ in. (shorter)}$$

These two examples demonstrate the approach employed to solve statically indeterminate problems. This approach involves two important steps—writing the applicable equations of equilibrium based on an accurate FBD and writing a deformation equation based on a drawing showing the deformation of the structural elements involved.

## 4.10.1 Thermal Stresses

Thermal stresses are produced in structural elements by the constraint of the free expansion of a material subjected to a temperature change. To illustrate this fact, consider a long thin rod of length $L_o$ at a temperature $T_o$. When the temperature of the rod increases to say $T_1$, the rod undergoes a free expansion with an attendant increase in length $\delta_T$ given by:

$$\delta_T = L_o\,\alpha(T_1 - T_o) = L_o\,\alpha\Delta T \qquad (4.34)$$

where $\alpha$ is the thermal coefficient of expansion.

It is clear from the definition of strain and Eq. (4.3) that a thermal strain $\varepsilon_T$ accompanies a free expansion.

$$\varepsilon_T = \delta_T\,/L_o = \alpha\Delta T \qquad (4.35)$$

Thermal stresses in a free expansion process are zero because the temperature change does not generate the internal forces P required for thermal stresses. Thermal strains induce thermal stresses if, and only if, the structural element is constrained in some manner. When the structural element is totally constrained and free expansion is not permitted, the thermal stresses in the uniaxial rod are given by:

$$\sigma_T = E\alpha\Delta T \qquad (4.36)$$

To derive Eq. (4.36), consider the free expansion of a long thin rod due to a temperature change $\Delta T$ as shown in Fig. 4.23a.

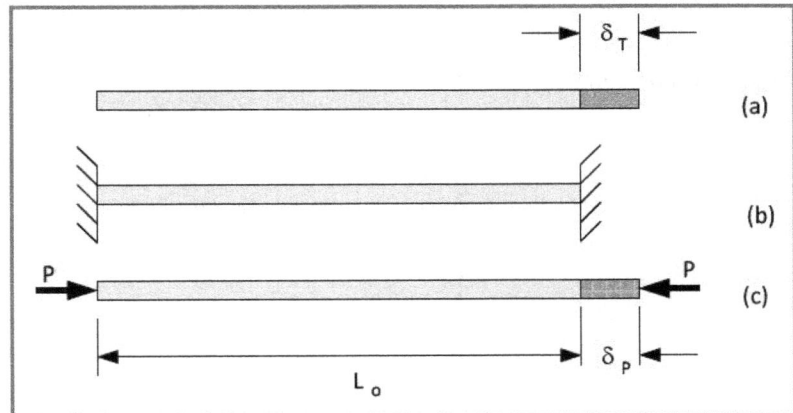

Fig. 4.23 Rod with free expansion, constraint and internal forces P.

If the rod is constrained within rigid walls as illustrated in Fig. 4.23b, the length remains at $L_o$ after the temperature change $\Delta T$. As the rod attempts to expand against the constraint offered by the rigid walls, an internal force P is produced as shown in Fig. 4.23c. The internal force P is sufficiently large to produce an axial displacement (contraction) equal to the thermal expansion. Hence, we can write:

$$\delta_T = \delta_P \qquad (4.37)$$

Substituting Eqs. (4.2) and (4.34) into Eq. (4.37) and simplifying yields:

$$\sigma_T = P/A = E\alpha\Delta T \qquad (4.36bis)$$

The results obtained by applying Eq. (4.36) should be considered as an upper bound on the thermal stresses generated by a temperature change $\Delta T$. The derivation of this relation assumed total constraint of the free expansion of the bar. In practice, total constraint is difficult to achieve and because the displacements are small, even slight relief in the constraint markedly reduces the magnitude of the thermal stresses.

We have included a table showing the coefficient of thermal expansion, modulus of elasticity and thermal stresses for a $\Delta T = 100°$ F or $100°$ C in Table 4.1.

**Table 4.1**
**Coefficient of thermal expansion, modulus of elasticity and thermal stress induced by $\Delta T = 100°$ F or $100°$ C in various engineering materials.**

| Material | $E \times 10^6$ psi | $\alpha \times 10^{-6}/°F$ | $\sigma_T$ (ksi) | $E \times 10^9$ (Pa) | $\alpha \times 10^{-6}/°C$ | $\sigma_T$ (MPa) |
|---|---|---|---|---|---|---|
| Steel | 30 | 6.3 | 18.90 | 207 | 11.3 | 233.9 |
| Aluminum | 10.4 | 12.9 | 13.42 | 72 | 23.2 | 167.0 |
| Brass | 16 | 11.1 | 17.76 | 110 | 20.0 | 220.0 |
| Stainless Steel | 27.5 | 9.6 | 26.40 | 190 | 17.3 | 328.7 |
| Titanium | 16.5 | 4.9 | 8.09 | 114 | 8.8 | 100.3 |

## EXAMPLE 4.30

Two heavy blocks of steel are permanently fastened together by a steel shrink link that is fitted into pockets machined into the two blocks as illustrated in Fig. E4.30. If a shrink link has a body length of 100.0 mm at a temperature of 275° C, determine the clamping force provided by the shrink link at an ambient temperature of 20° C. Also determine the stress in the link after it has cooled to ambient temperature. The body of the link has a square cross section with a side dimension of 50 mm.

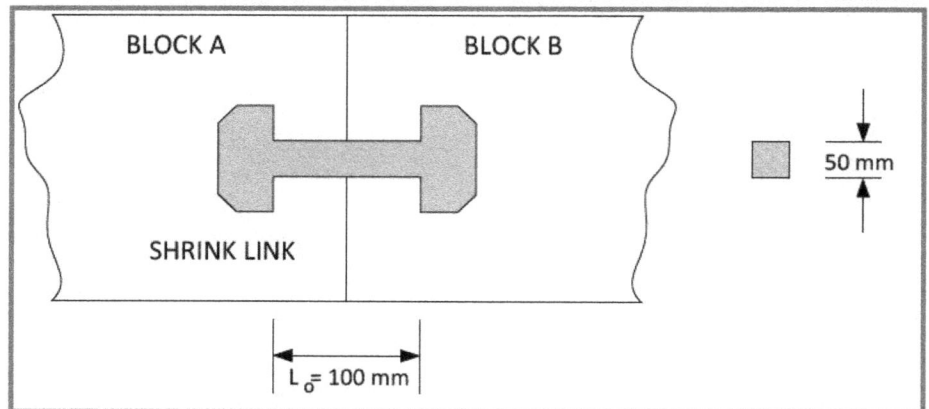

Fig. E4.30

---

**Solution:**

**Step 1:** Upon cooling to ambient temperature the link attempts to shrink an amount $\delta_T$ given by Eq. (4.34) as:

$$\delta_T = L_o\, \alpha\Delta T = 100(11.3 \times 10^{-6})(275 - 20) = 0.2882 \text{ mm} \qquad (a)$$

**Step 2:** We assume that the free contraction of the link is totally constrained by the two steel blocks. This constraint condition yields:

$$\delta_T = \delta_P \qquad\qquad (b)$$

Substituting Eq. (4.2) into Eq. (b) gives the internal force P developed in the shrink link as:

$$P = (\delta_P AE)/L = [(0.2882)(2500)(207 \times 10^3)]/100 = 1491 \text{ kN} \qquad (c)$$

**Step 3:** The thermal stress induced in the link by the $\Delta T = 255°$ C is given by Eq. (4.36) as:

$$\sigma_T = P/A = E\alpha\Delta T = (1491 \times 10^3)/(0.050)^2 = 596.4 \text{ MPa} \qquad d)$$

## EXAMPLE 4.31

A thin retaining ring, fabricated from stainless steel is fitted over the polished steel journal of a solid shaft 7.000 inches in diameter. The outside diameter of the retaining ring is 7.375 in. and its length is 0.50 in. To fit the ring over the journal it is heated to 350 °F and slid on the shaft to its designated position. The ring becomes snug at a temperature of 295 °F. Determine the stress in the retaining ring and the clamping pressure between the ring and the journal.

**Solution:**

**Step 1:** Upon cooling from the snug temperature of 295° F to the ambient temperature of 75° F, the ring attempts to shrink diametrically by an amount $\delta_T$ given by Eq. (4.34) as:

$$\delta_T = d_o\, \alpha\Delta T = (7.000)(9.6 \times 10^{-6})(295 - 75) = 14.78 \times 10^{-3} \text{ in.} \qquad (a)$$

**Step 2:** We assume that the free contraction of the retaining ring is totally constrained by the polished journal of the solid steel shaft. This constraint condition yields:

$$\delta_T = \delta_P \qquad\qquad (b)$$

Consider the circumference $C = \pi d_o$ of the retaining ring. Then the free contraction $\Delta C$ of the circumference of the ring is given by:

$$\Delta C = \pi\Delta d = \pi\delta_T = \pi\delta_P \qquad (c)$$

The increase in the circumference of the ring due to the generation of an internal force P is obtained by substituting into Eq. (4.2) as:

$$\Delta C = (P\pi d)/(AE) = \pi\delta_T \qquad (d)$$

Solving Eq. (d) for the internal force P in the ring gives:

$$P = (\delta_T AE)/d_o = (14.78 \times 10^{-3})(0.375/2)(0.5)(27.5 \times 10^6)/7.000 = 5.444 \text{ kip} \qquad (e)$$

**Step 3:** The thermal stress induced in the retaining ring due to $\Delta T = 220°$ F is given by Eq. (4.36) as:

$$\sigma_T = P/A = E\alpha\Delta T = (27.5 \times 10^6)(9.6 \times 10^{-6})(220) = 58.08 \text{ ksi} \qquad (f)$$

**Step 4:** To determine the interfacial pressure between the ring and the journal, consider the FBD of the ring shown in Fig. E4.31.

Fig. E4.31

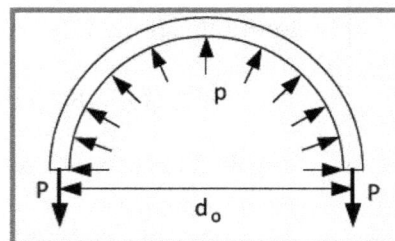

Writing $\Sigma F_y = 0$ gives an expression containing the interfacial pressure p as:

$$\Sigma F = p(db) - 2P = 0 \qquad\qquad p = 2P/(db) \qquad\qquad (g)$$

where $b = 0.5$ in. is the width of the ring

The interfacial pressure is then given by:

$$p = 2(5.444 \times 10^3)/(7.0)(0.5) = 3,111 \text{ psi} \qquad\qquad (h)$$

## 4.11 SUMMARY

This chapter treats structural members that are long and thin. Flexible members, such as wire and cable, only support axial tensile loads. They cannot support compressive forces because they buckle under very low loads. In addition, they cannot support transverse loads because their transverse stiffness is also negligible. Under increasing tension forces applied along the axis of these long, thin, and flexible members, they stretch, yield and finally rupture.

Rods and bars are sufficiently stiff to carry compressive forces. In this chapter, we consider only axial loading of the bars and rods. The case of transverse loading is deferred until Chapter 10. The equations derived for stresses and deflections in wire and cable are also applicable to bars and rods. When dealing with compressive forces and stresses, a minus sign is used to indicate the direction of the loading. We also assume compressively loaded bars and rods are sized so they will not fail by buckling. Examples are presented for determining stresses and deflections of uniform axial bars.

Relations have been derived to determine the stretch (axial deformation). Stress and strain have been defined and the stress-strain relation (Hooke's law) has been discussed. Several examples have been provided to guide you in determining, stress, strain, and axial extension.

We have shown that the internal force P is equal to the external force F for these uniaxial members. The internal force is produced by normal stresses that are uniformly distributed over the cross section of the uniaxial member.

We describe failure by yielding and rupture and introduce the stress-strain curve for different types of materials in the process. Ultimate tensile strength and yield strength are defined. Safe design philosophy is discussed together with the use of factors of safety in design. Examples are provided to demonstrate problem-solving techniques dealing with strength and safety factors. The importance of specifying standard sizes of structural members in design is illustrated in one of the examples.

The concept of models that are geometrically similar to actual structures is introduced. The stresses and displacements produced by loading or deforming a model are related to those developed in the prototype (structure). Relations involving scaling factors are employed to determine the stresses and displacements in the structure based on measurements made on a geometrically scaled model.

The equations developed for uniform members subjected to axial forces are summarized below:

$$\delta = L_f - L_o \qquad (4.1)$$

$$\delta = (PL)/(AE) \qquad (4.2)$$

$$\varepsilon = \delta/L \qquad (4.3)$$

$$\sigma = P/A \qquad (4.4)$$

$$\sigma = E\varepsilon \qquad (4.5)$$

$$\varepsilon = P/(AE) \qquad (4.6)$$

$$P_{INT} = \int \sigma \, dA \qquad (4.7)$$

$$P_{INT} = \sigma \int dA = \sigma A \qquad (4.8)$$

$$S_u = \sigma_{fail} = P_{fail}/A \qquad (4.9)$$

$$S_y = \sigma_{yield} = P_{yield}/A \qquad (4.10)$$

$$SF_u = S_u/\sigma_{design} \qquad (4.11)$$

$$SF_y = S_y/\sigma_{design} \qquad (4.12)$$

$$\tau = V/A \qquad (4.13)$$

Stresses on a section with an inclined cut at an angle $\theta$ were determined using the equilibrium relations and the simple definition for normal and shear stresses as:

$$\sigma_\theta = (\sigma_x)\cos^2 \theta \qquad (4.14)$$

$$\tau_\theta = (\sigma_x/2) \sin 2\theta \qquad (4.15)$$

In Fig. 4.14, we note that both the normal stress $\sigma_\theta$ and the shear stress $\tau_\theta$ vary with the angle $\theta$. The normal stress is a maximum when $\theta = 0°$, and the section cut is perpendicular to the axis of the bar. The shear stress vanishes on this section. The shear stress is a maximum when $\theta = 45°$. On this plane, $\sigma_{45} = \tau_{45} = \sigma_x/2$.

Tapered bars subjected to axial loading were described. For the stresses, Eq. (4.4) was modified to account for the variable cross sectional area as a function of position x along the length of the tapered bar.

$$\sigma_x = P/A_x \qquad (4.17)$$

For deflection of the tapered bar, we considered an incremental length dx along the length of the bar and modified Eq. (4.2) to read as:

$$d\delta = P \, dx/(A_x \, E) \qquad (4.18)$$

To determine the deflection $\delta$ of the tapered bar, we write an expression for $A_x$ and integrate Eq. (4.18) from zero to L. Examples are provided to demonstrate the computation technique for tapered bars.

We have also described techniques for determining the stresses in stepped bars. The stresses in individual sections of the bar are computed from Eq. (4.4) using the appropriate cross sectional area for the section under consideration. Displacements of each section of the bar are determined from Eq. (4.2) and then superimposed to give the total deflection as shown in Eq. (4.20).

$$\delta_{total} = \delta_1 + \delta_2 + \delta_3 + \ldots + \delta_n \qquad (4.20)$$

Stress concentrations that develop at discontinuities in bars and rods have been discussed. These stress concentrations occur for all types of loading: axial, transverse and torsion. The stress concentration increases the maximum stresses by a significant amount. We determine the maximum stresses from:

$$\sigma_{MAX} = K \, \sigma_{NOM} \qquad (4.21)$$

The stress concentration factor K is defined for a bar with a central circular hole in Fig. 4.20 and for a stepped bar in Fig. 4.22. The nominal stresses for these two discontinuous bars are given by:

$$\sigma_{NOM} = P/(w - d)b \qquad (4.22)$$

$$\sigma_{NOM} = P/w_1 b \qquad (4.24)$$

Examples demonstrating the method for determining the stress concentration factor, and the nominal and maximum stresses are provided.

Modeling, an approach to experimentally determine stresses and displacements in large complex structures, was introduced. The relations for the stresses and displacement between the model and the structure (prototype) are given by:

$$\sigma_m = (L \, / \, S^2) \, \sigma_p \qquad (4.29)$$

$$\delta_p = [ \, (S \, E )/ \, L] \, \delta_m \qquad (4.33)$$

Structures are considered to be statically determinate when the unknown forces, either internal forces or external reactions, can be determined using only the equilibrium relations. However, in some structures the number of unknown forces exceed the number of applicable equilibrium equations, and a solution is not possible without the introduction of additional equations. When the number of unknown forces exceeds the number of applicable equilibrium equations, the structure is **statically indeterminate.** The approach for solving for the unknown forces in statically indeterminate structures is to use the equilibrium relations and then to introduce additional equations based on the deformation of the structure. The solution involves two separate but compatible parts:

1. Prepare the traditional FBD together with all of the applicable equations of equilibrium.
2. Prepare displacement diagrams together with deformation equations that are consistent with the forces shown on the FBD.

Thermal stresses are produced in structural elements by the constraint of the free expansion of a material subjected to a temperature change. When the free expansion or contraction is totally constrained the thermal stresses are given by:

$$\sigma_T = E\alpha\Delta T \qquad\qquad (4.36)$$

# REFERENCES

1. Peterson, R. E., <u>Stress Concentration Design Factors</u>, 2nd Ed., Wiley & Sons, New York, 1990.

# PROBLEMS

4.1   A steel music wire (Gage No. 10 with a diameter of 0.024 in.) is employed to support an axial load P of 8.64 lb. If the wire was initially 22 ft in length, determine its final length and the stress and strain in the wire.

4.2   A steel music wire is tightened on a guitar by rotating the wire supporting post through an angle of 11°. If the post has a diameter $d = 0.25$ in. and the length of the wire is 32 in., determine the strain induced in the guitar string.

4.3   Referring to Problem 4.2, determine the additional stress induced in the guitar string as it is tightened.

4.4   A hemp rope 15 m long and 45 mm in diameter is supported from one of the roof beams in a gymnasium. Three gymnasts climb this rope together. The lead gymnast has a mass of 55 kg, the next 72-kg and the trailing gymnast 52 kg.

   (a) Determine the maximum stress in this rope.
   (b) Determine the stress in the rope at a location between the lead and intermediate gymnast.
   (c) Determine the stress in the rope at a location between the intermediate and trailing gymnast.

4.5   A suspension bridge is to carry a roadway and traffic that may weigh up to 5,000 ton. If the safe load that can be imposed on the wire rope to be used in the construction is 175 ton, specify the number of cables to be employed. Justify your answer.

4.6   Describe the constraints on the relation $\sigma = E\varepsilon$.

4.7   Determine the design load that can be specified for Gage No. 00 steel wire with a strength of 530 MPa, if the design specification calls for the safety factor of 2.5.

4.8   Discuss the factors to consider in establishing the safety factor in a design specification.

4.9   For the cable-pulley-mass arrangement shown in the figure to the right, determine the stresses in the cable. The effective cable diameter d is 0.500 in.

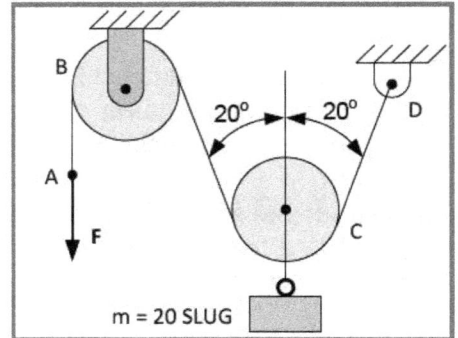

B

20°   20°   D

A

F

C

m = 20 SLUG

4.10  Cold drawn steel alloy wire exhibits an ultimate tensile strength of 600 MPa and a yield strength of 500 MPa. If this alloy is employed, specify the gage of the wire if it is to support a design load of 7.5 kN with a safety factor of 3.2. Specify the gage number required based on both yielding and rupturing as modes of failure.

4.11  Determine the safety factor for the wire CD in the cable-weight arrangement shown in the figure to the right. The steel cables are each 10 m long, exhibit a yield strength of 500 MPa and have an effective cross sectional area of 8 mm$^2$. Assume that the forces in the horizontal wires are equal in magnitude and opposite in direction ($F_{CB} = - F_{CA}$). Also assume $F_{CB} = (0.18)F_{CD}$.

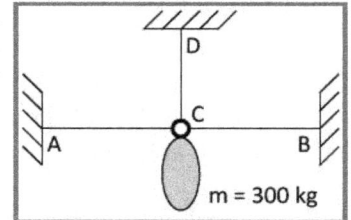

D

C

A      B

m = 300 kg

4.12  Determine the mass m, shown in the diagram to the right, that is required to yield the cable CD if the cables are each 22 ft long, exhibit a yield strength of 80ksi and have an effective cross sectional area of 0.015 in.$^2$ Assume that the forces in the horizontal wires are equal in magnitude and opposite in direction ($F_{CB} = - F_{CA}$). Also assume $F_{CB} = (0.25)F_{CD}$. Is the cable-weight arrangement stable after yield? Why?

4.13   The wire-mass system, shown in the figure to the left, is constructed using steel wire. Specify a suitable alloy and the gage number for the wires if the criterion for failure is based on yielding. Use the same diameter for both wires. The safety factor imposed on the design is 2.9.

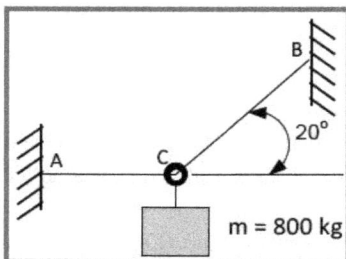

B

20°

A   C

m = 800 kg

4.14  The wire-mass system, shown in the figure to the right, is constructed using aluminum wire. Specify a suitable alloy and the gage number for the wires if the criterion for failure is based on ultimate tensile strength. Use the same diameter for both wires. The safety factor imposed on the design is 1.8.

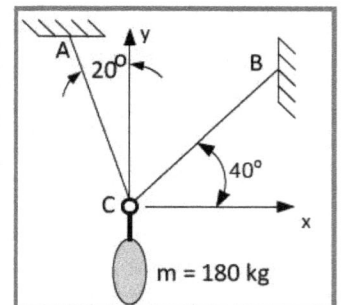

y

A      B

20°

40°

C      x

m = 180 kg

4.15  The wire-mass system ,shown in the figure to the left, is constructed using steel wire fabricated from 1010 A.  A safety factor of 2.9, based on the ultimate tensile strength, is to be employed for both wires.  Find the maximum weight that can be lifted (in pounds) and specify the required gage numbers for each wire.

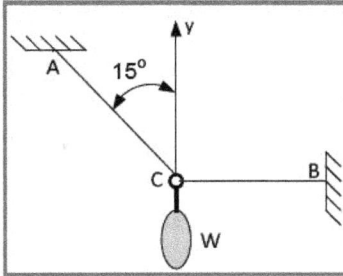

4.16  Describe in an engineering brief the differences between a wire rope and a rod.

4.17  Describe in an engineering brief the similarities between a wire rope and a rod.

4.18  A long thin steel bar with a length of 1.8 m, a width of 50 mm, and a thickness of 20 mm is subjected to an axial force of 22 kN.  Determine the tensile stress and the axial deformation of the bar.

4.19  A bar fabricated from steel with a tensile strength of 54.0 ksi is subjected to an axial tensile force of 20 kip.  The bar is designed with a safety factor of 3.2.  Determine the design stress for the bar and the required cross sectional area.

4.20  A bar fabricated from an aluminum alloy with a tensile strength of 400 MPa is subjected to an axial tensile force of 75 kN.  The bar is designed with a safety factor of 2.5.  Determine the design stress for the bar and the required cross sectional area.

4.21  A bar fabricated from brass with a tensile strength of 247 MPa is subjected to an axial compressive force of 76 kN.  The bar is designed with a safety factor of 2.8.  Determine the design stress for the bar and the required cross sectional area.  Assume the tensile and compressive stresses are equal.

4.22  A key used to lock a gear onto a shaft is subjected to a shear force of 8.0 kip.  If the key is 0.25 in. wide by 0.375 in. high and 1.5 in. long, determine the shear stress acting on the key.  Draw a free body diagram showing this shear stress.

4.23  If the key in Problem 4.22 is machined from a steel alloy with a tensile yield strength of 48,000 psi, determine the safety factor for the key.

4.24  A rectangular bar, shown in the figure to the right, is subjected to an axial tensile force F = 30 kN.  Determine the normal stress and the shear stress on an inclined plane if the angle of the section cut is $\theta = 30°$.

4.25  A rectangular bar, similar to the one shown in the figure to the right, is adhesively bonded along a 20° section cut.  The normal stress in the bond line is limited to 2.5 ksi and the shear stress to 1.5 ksi.  If the bar has a cross sectional area A = 1.75 in.$^2$ and a safety factor of 3.0 is specified, determine the largest axial force that can be applied to the bar.

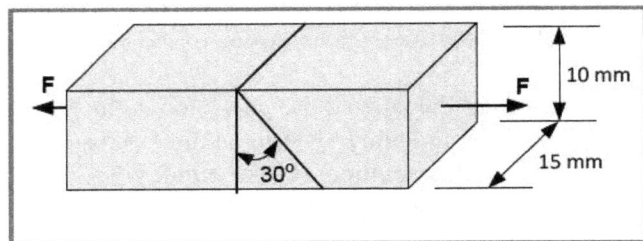

4.26 A rod with a circular cross section with a diameter d is fabricated from two pieces as illustrated in the figure to the right. If a force F is applied to the bar, derive the expression for the normal and shear stress acting in the plane of the adhesive joint as a function of angle $\phi$.

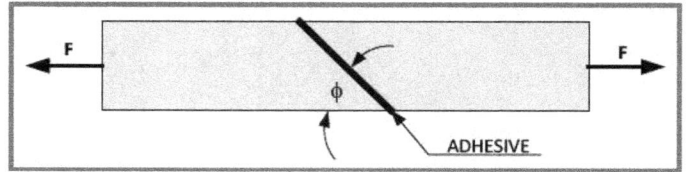

4.27 For the bar illustrated in the figure to the right, determine the shear stresses in the adhesive joint if the angle $\phi$ is varied from 0° to 90°. The axial force applied to the bar is 2.0 kN and its diameter is 20 mm. Also compute the normal stresses acting on the adhesive joint. Hint: Use a spreadsheet to determine the shear and tensile stresses in the adhesive joint for the range of $\phi$ specified.

4.28 A tensile bar, defined in the figure to the right, has a cross sectional area of 175 mm$^2$ and is subjected to a tensile force F. The stresses on the inclined plane A—B are $\sigma_\theta = 81$ MPa and $\tau_\theta = -27$ MPa. Determine the stress $\sigma_x$, the angle $\theta$ and the force F.

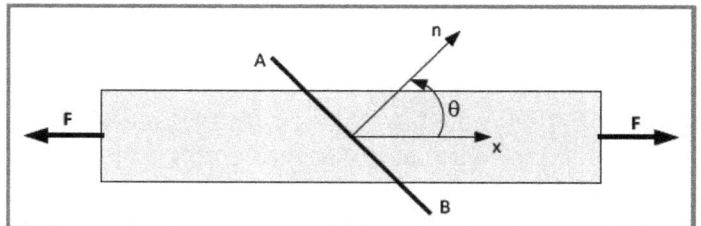

4.29 The tensile bar, defined above and in the figure to the right, has a cross sectional area of 175 mm$^2$ and is subjected to a tensile force F. The stresses on the inclined plane A—B are $\sigma_\theta = 81$ MPa and $\tau_\theta = -27$ MPa. Determine the shear and normal stresses acting on an inclined plane with $\theta = 40°$.

4.30 For the tensile bar in Problem 4.29, prepare a graph that shows the stresses $\sigma_\theta$ and $\tau_\theta$ as a function of the angle of the inclined cut as $\theta$ varies from 0 to 90°.

4.31 Consider the tapered bar presented in the figure below and to the right . The tapered bar is subjected to an axial force of 500 kN. Prepare a graph of the axial stress $\sigma_x$ as a function of x as it varies from zero to 4 m.

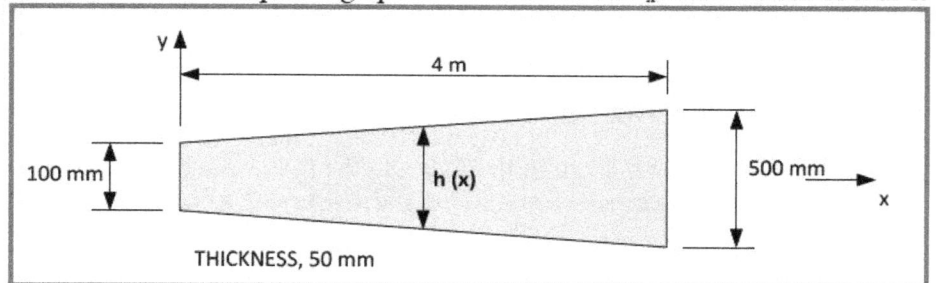

4.32 Consider the tapered bar presented in Problem 4.31. If the bar is fabricated from an aluminum alloy, determine the extension of the bar when it is subjected to an axial tensile force of 500 kN.

4.33 Consider the tapered bar presented in the figure below and to the right. If the bar is fabricated from an aluminum alloy, determine the extension of the bar when it is subjected to a tensile force of 120 kN. In this solution, do not employ Eq. (4.19). Instead, use Eq. (4.18) and perform a numerical integration on a spreadsheet.

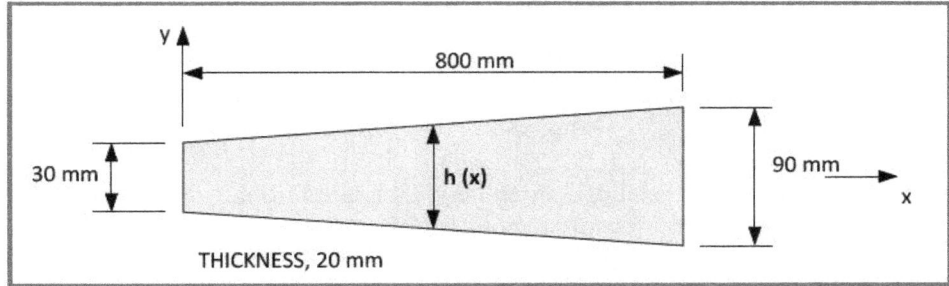

**4.34** Describe the procedure employed to solve for the stresses and deflection in a stepped bar subjected to axial loading.

**4.35** For the stepped bar, illustrated in the figure to the right, determine the stresses in both portions of the bar and its total deflection. The bar is fabricated from steel.

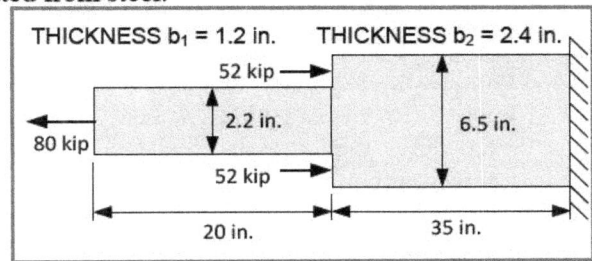

**4.36** A tensile bar 40 in. long, shown in the figure below, is subjected to an axial force of 10 kip. Determine: (a) the nominal stress, (b) the stress concentration factor K, and (c) the maximum stress.

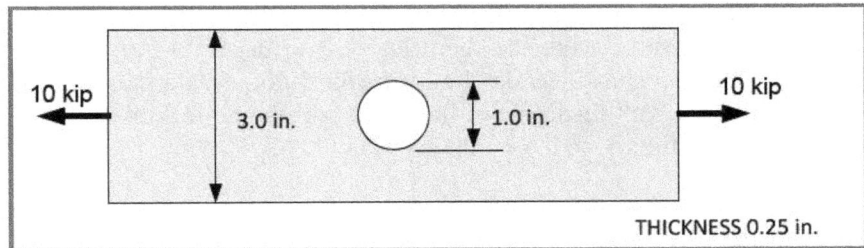

**4.37** A stepped tensile bar, illustrated in the figure below, with fillets at the transition between the small and the large section is subjected to an axial tensile force of 75 kN. Determine the maximum stress at the fillets.

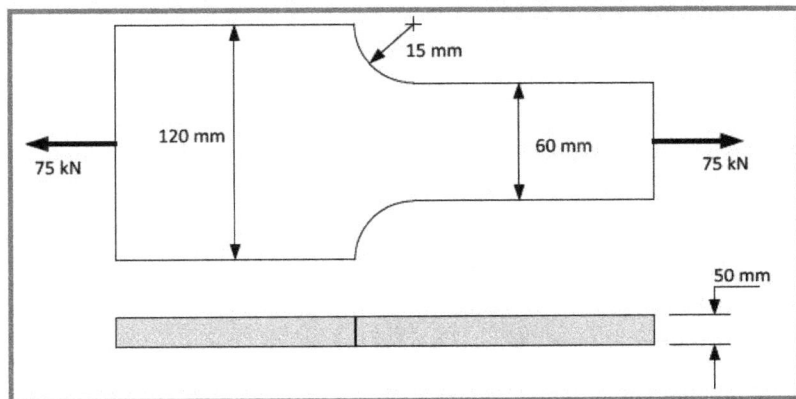

4.38  A scale model of a large structure has been fabricated from steel and tested. A strain gage on one member of the structure indicated an axial strain of $\varepsilon = 1{,}450$ $\mu$m/m. Determine the stress in the corresponding member of the prototype. The numerical parameters defining the scaling factors for the loads and the size of the structural member are $w_m = 1.6$ mm, $w_p = 80$ mm, $b_m = 2.0$ mm, $b_p = 100$ mm, $L_m = 14$ mm and $L_p = 7.0$ m. The scaling factor $L = 1/10{,}000$.

4.39  Suppose that a model of a structure is fabricated from members formed from sheet aluminum. The prototype structure is to be fabricated from steel with an open span of 200 feet. The model is geometrically scaled so that its span is four feet. The capacity of the live load on the prototype is 150 lb/ft, and the model is loaded with 2.5 lb/ft. If the model deflects a distance of 0.220 in. under full load at the center of the span, determine the deflection of the prototype under the design load.

4.40  The tensile bar, presented in the figure to the right, is fabricated with an adhesive joint inclined at an angle $\phi$ to the axis of the bar. Determine the optimum angle $\phi$ for the inclined plane if the stresses in the adhesive are not to exceed either 3.5 ksi in tension or 2.8 ksi in shear. Also determine the maximum force that can be applied to the member without exceeding the stresses in the adhesive joint.

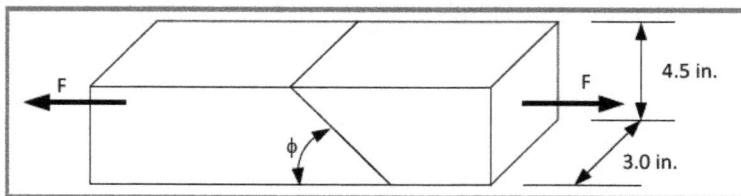

4.41  A structural member is fabricated from a solid round bar of steel with a diameter $d = 50$ mm. If the member is 6.0 m long, determine the maximum axial force that can be applied if the axial stress is not to exceed 175 MPa and the total elongation is not to exceed 0.14% of its length.

4.42  A steel tie rod with a diameter of 1.0 in. and length of 34.0 in. is employed to compress a brass bushing with an outside diameter of 3.0 in. and a length of 24.0 in. as shown in the figure to the right. Determine the minimum wall thickness of the bushing if the deflection of the tie rod is limited to $\delta = 0.020$ in. Note that $F = 10$ kip.

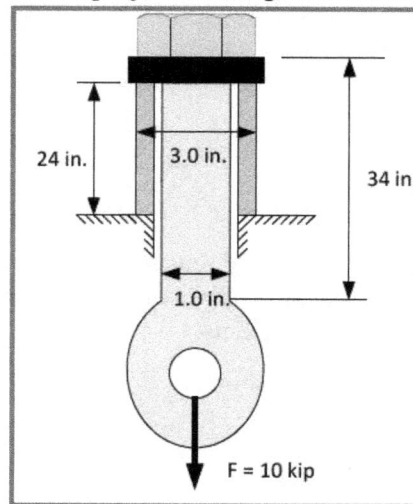

4.43  A short column with a height of 2.0 m is fabricated by adhesively bonding aluminum faceplates to a core of plastic foam as shown in the figure below and to the left. The foam plastic core has a square cross section with an elastic modulus of 1200 psi. Determine the stresses in the aluminum plates and the plastic foam. Also, determine the displacement of the column under the action of the applied force $F = 35$ kN.

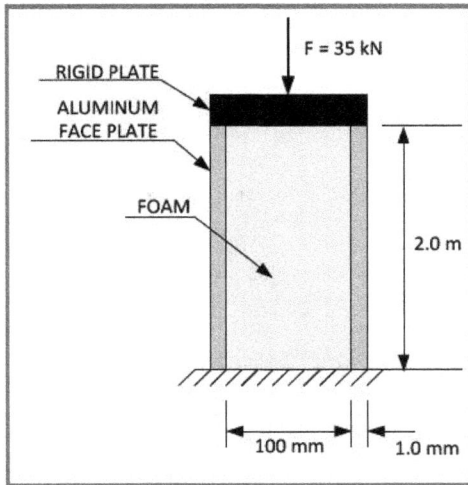

4.44 A grade 8 steel bolt, 1.0 in. in diameter (14 threads/in.) is employed to clamp a brass bushing between two rigid platens as illustrated in the figure below. The brass bushing has a 2.5 in. outside diameter and 1.5 in. inside diameter and is 12.00 in. long. After the unit is assembled with a snug fit, the nut is tightened by 1/3 of a turn. Determine the axial stresses in the bolt and the bushing. Also determine the deflection of the brass bushing.

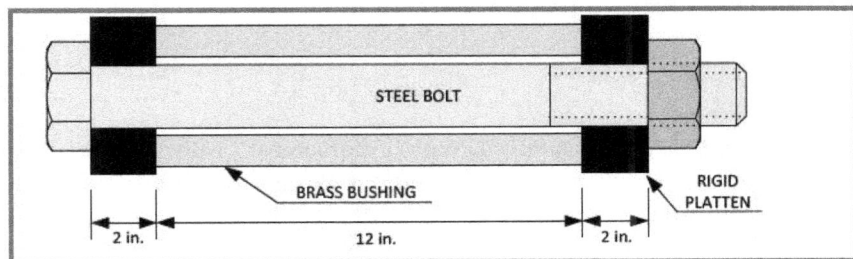

4.45 Twenty-five steel reinforcing rods with a ¼ in. diameter are placed within a high strength concrete column with a square cross section that supports an applied force of 50 kip. Determine the stresses in the steel reinforcing bars and the concrete. Also determine the amount of deflection of the column. Reference the figure to the right.

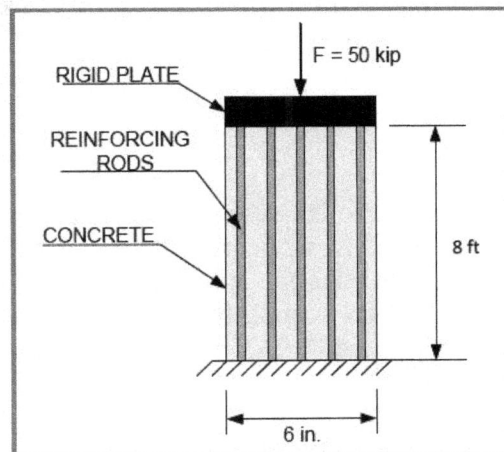

4.46 A long aluminum rod is connected to a shorter brass cylinder with a bolted flange as shown in the figure to the left. Prior to assembly a gap of δ = 0.20 mm occurred between the flange plates. Bolts were inserted and tightened bringing the flange plates together. Determine the stresses induced in both the rod and the cylinder by the assembly operation. Also determine the displacement of the face of each flange.

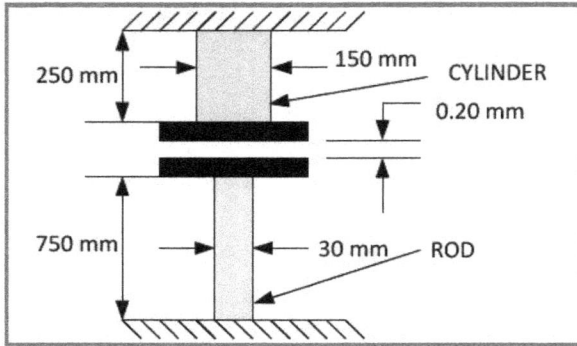

**4.47** A copper and stainless steel rod are assembled between two rigid walls at an ambient temperature of 20° C as shown in the figure to the right. If the temperature is increased by an amount ΔT = 120 °C, determine the thermal stresses induced in each rod. Also determine the change in length of each member.

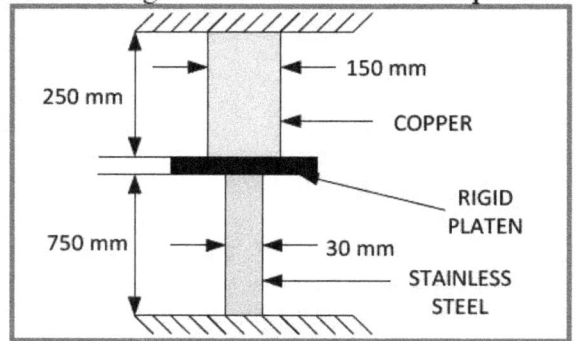

**4.48** A high strength steel bolt with a diameter of 1.0 in. passes through an aluminum bushing with a cross sectional area of 3.0 in.$^2$ as shown in the figure below. The unit is assembled with a snug fit at an ambient temperature 70 °F. If the temperature of the assembly increases by ΔT = 60 °F, determine the thermal stresses induced in the bolt and the bushing.

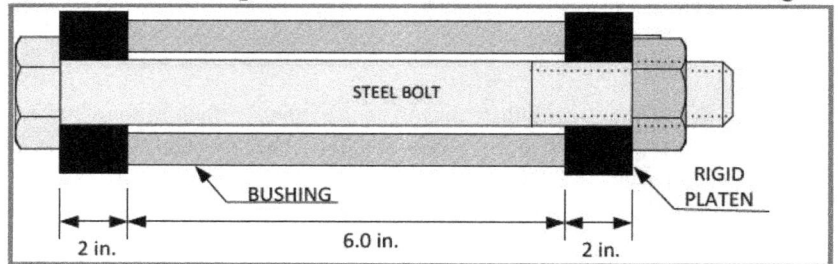

**4.49** A three bar suspension system, shown in the figure to the right, is assembled at an ambient temperature of 20° C. Determine the axial stresses in each bar after a force F of 100 kN is applied to the rigid platen and the temperature is increased by ΔT = 80 °C. The bars labeled A are fabricated from aluminum with a cross sectional area of 750 mm$^2$ and the bar labeled B is fabricated from stainless steel with a cross sectional area of 1,250 mm$^2$.

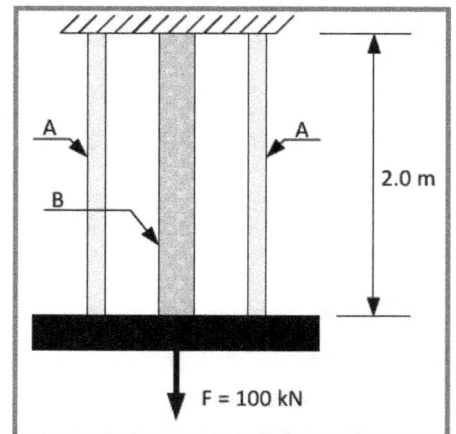

# CHAPTER 5

# MATERIAL PROPERTIES

## 5.1 INTRODUCTION

Today structures are usually fabricated with steel or steel reinforced concrete. These materials are subjected to stresses due to gravitational loads and due to live loads. The gravitational loads (weight of the structure itself) are constant with respect to time. The live loads are due to vehicular traffic for bridges, placement of furniture or equipment, or storage of inventory, which varies with time. To ensure the safety of the structure, the strength of the materials (steel, wood, concrete, etc.) must be adequate to support the applied stresses with a comfortable, but not excessive margin of safety. This premise seems simple, but on careful study, we find several strengths are needed to characterize the behavior of structural materials—yield, tensile or ultimate, and fatigue. To perform design analysis, it is also necessary to describe elastic constants that define the stiffness of these materials. To judge the adequacy of these materials for large structures, we examine their ductility (i.e. elongation and percent reduction in area). In this chapter, we define these different strengths, and demonstrate methods for accounting for the strength of construction materials in the design of structural components.

## 5.2 TYPES OF STRUCTURAL FAILURES

Structures fail while in service for three different reasons. First, by fracturing where the structure breaks into two or more pieces. This is the most catastrophic mode of failure since the structure collapses and usually falls. Often life and property is endangered or lost, and careers are ruined.

The second type of failure is due to yielding. When a structural member is stressed beyond its capabilities, the material, from which the member is fabricated, yields and redistributes the excessive stresses. In yielding, the structure undergoes excessive deformation, but usually it does not break or become unstable. The structure fails because of excessive plastic deformations that alter the geometry of one or more of the structural members. No one is satisfied with a beam that is permanently bowed or bent. Usually these deformed structures have to be straightened or replaced in the repair of the structure.

The third mode of failure is due to excessive elastic deformation. In this case, the structure remains intact, and after the load is removed, it returns to its original shape. The structure fails because small elastic deformations change its shape and alter its performance. For example, suppose the structure is a tool holder for a lathe that is used in a cutting operation removing material from a shaft. In order to maintain the tolerances on the diameter of the shaft, it is necessary to limit the deformations of the tool holder to the specified tolerances. If the tool holder deflects by a significant amount, it fails because it is too flexible and cannot maintain the tool in a relatively fixed location.

In analyzing a structure to ensure that it does not fracture or yield, we compare the stress developed in the structure with the strength of the material used in its fabrication. In making the comparison, we employ one or more strength criteria, which characterize the capability of the material. These include:

- Yield strength
- Ultimate tensile strength
- Fatigue strength
- Compressive strength

A table giving some of the properties characterizing commonly employed engineering materials is shown in Appendix B.

## 5.3 THE TENSILE TEST

We determine several material properties, as shown in Appendix B, by conducting standardized tensile tests[1]. For example, the yield and ultimate tensile strength are determined in a tensile test. To begin, we prepare **tensile specimens** of the materials under investigation. The standard size of the specimen is shown in Fig. 5.1.

Fig. 5.1 Tensile test specimen (dimensions in inches).

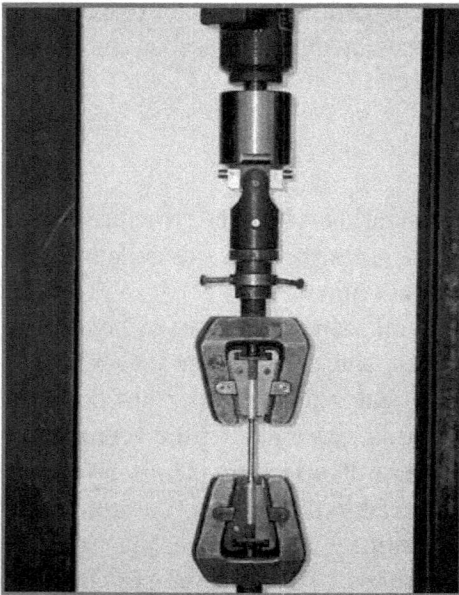

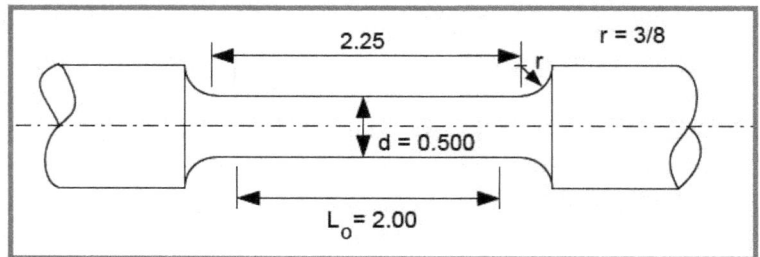

Fig. 5.2 Tensile specimen, wedge-grips, flex-joints and load cell in a testing machine.

The tensile specimen is mounted in a **universal testing machine** similar to the one illustrated in Fig. 5.2. The machine may be a mechanical type with one head fixed and the other driven by screws, or it may be a hydraulic machine where the movable head is driven by a hydraulic cylinder. In either instance, the testing machine applies a load F along the axis of the tensile specimen. The load is applied slowly until the specimen yields and/or fails by fracturing.

---

[1] The American Society for Testing Materials (ASTM) publishes standards that define the test specimen and procedures for measuring the material properties described in this chapter. The standard for tension testing of metallic materials is E8 – 99, (see the *Annual Book of ASTM Standards* volume 03.01).

During the tension test, we measure the applied load F and the stretch δ over the gage length $L_o$ of the specimen. A **load cell** on the universal testing machine indicates the applied load F and an **extensometer** mounted on the specimen, shown in Fig. 5.3, measures the stretch δ. The electrical signals from the load cell and the extensometer are recorded together on a x-y chart to provide a load-deformation curve that is proportional to the stress-strain curve. The load-deflection (stretch) curves recorded during the tension tests are converted into stress-strain curves that characterize the tensile behavior of the metallic material by utilizing:

$$\sigma = F/A_o \qquad (4.4)$$

$$\varepsilon = \delta/L_o \qquad (4.3)$$

where $L_o$ and $A_o$ are the initial gage length and cross sectional area of the tensile specimen.

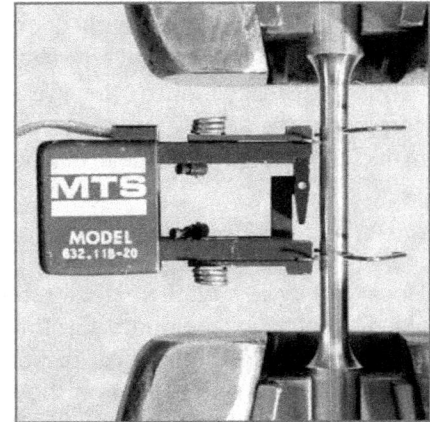

Fig. 5.3 A tension specimen with an extensometer for measuring the stretch δ.

This procedure gives the **engineering stress** and the **engineering strain,** which differ from the **true stress** and the **true strain**. The distinction between the "engineering" and "true" quantities is described in Section 5.6.

For brittle materials, which do not exhibit significant plastic deformation before fracture, the stress-strain curve is nearly linear until failure, as indicated in Fig. 5.4. The stress $\sigma_f$ producing the failure of the brittle specimen is recorded during the tensile test. It is this particular value of the stress that defines the material property known as the ultimate tensile strength $S_u$ for a brittle material.

$$S_u = \sigma_f \qquad (5.1)$$

A photograph of a tension specimen that exhibited brittle failure is depicted in Fig. 5.5. Note the absence of any necking or extensive plastic deformation in the uniform section of the specimen. Brittle failure is dangerous because it is sudden and catastrophic. The fracture initiates without warning and the cracks propagate across the specimen (structure) in microseconds. Of course, in the selection of materials for our designs, we avoid the use of brittle materials in structures to preclude the possibility of a catastrophic failure. Brittle materials such as gray cast iron are sometimes used because of its excellent casting properties and for its ability to damp vibrations. In the usage of brittle materials, we are always careful to maintain a state of compressive stress in the structure.

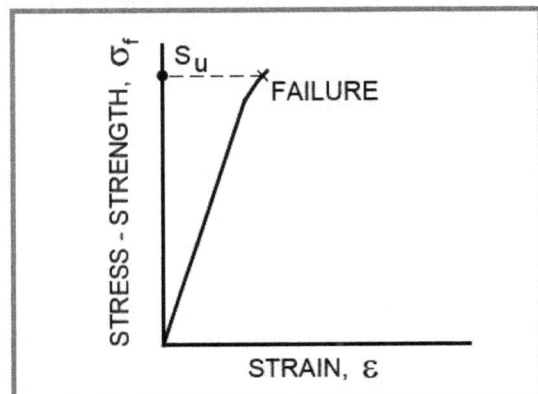

Fig. 5.4 Stress-strain curve for a brittle metallic material.

Fig. 5.5 Failure of a brittle material occurs suddenly with little plastic deformation.

Structures are designed with ductile materials, which yield and undergo extensive plastic deformation prior to rupture. A typical stress-strain curve for a ductile material (mild steel) is presented in Fig. 5.6.

An inspection of the stress-strain diagram shows that there are four different regions with each corresponding to a different behavior of the ductile steel. The first is called the **elastic region** where the material responds in a linear manner. In this region, Hooke's law applies and we write:

$$\sigma = E\,\varepsilon \qquad\qquad (4.5)$$

Hooke's law is a mathematical model of material behavior, but it is valid only in the elastic region. The elastic region extends until the low carbon steel (or some other ductile material) begins to yield. When the material yields, the linear response of the material ceases and Eq. (4.5) is no longer valid.

The second region on the stress-strain diagram depicts the initiation of yielding where the stress in the specimen exceeds the **elastic limit**. Slip planes have developed in the specimen, and deformation by slip along these planes is occurring with small or negligible increases in the stress. The stress-strain relationship is non-linear in this region. In some ductile materials, a small decrease in the applied stress is observed as the tensile specimen yields and continues to elongate by slip.

The third region describes a material behavior known as **strain hardening**. The easy slip that occurred during the initial yielding phase becomes more difficult to induce. As a consequence, higher stresses are required to continue deforming the tensile specimen. The stresses increase with increases in the strain; however, the relationship is not linear and Eq. (4.5) is not valid in this region.

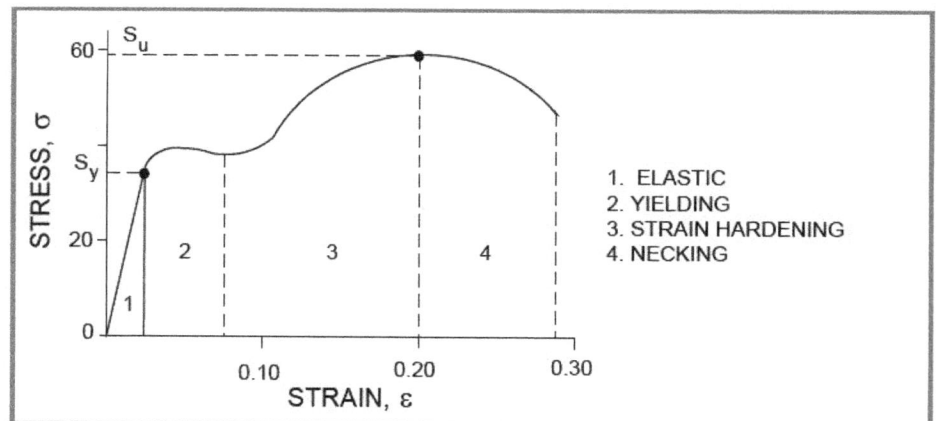

Fig. 5.6 Stress-strain diagram for low carbon steel.

1. ELASTIC
2. YIELDING
3. STRAIN HARDENING
4. NECKING

In the final region, the tensile specimen undergoes a dramatic change in appearance: it begins to **neck**. The deformation becomes localized to a small area along the length of the bar. During this phase of the deformation, the specimen resembles an hourglass. The neck decreases in diameter with increasing deformation until the specimen fails by rupturing. The axial deformation, which occurs as the tensile specimen necks, requires no increase in the applied loads. Indeed, the load may actually decrease significantly during the necking phase of the deformation processes. The appearance of a ruptured tension specimen fabricated from low carbon steel is illustrated in Fig. 5.7.

Regions 2, 3 and 4 are often combined and called the **plastic regime** for a ductile material.

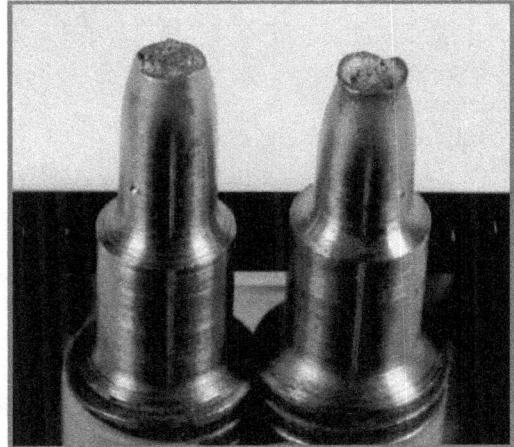

Fig. 5.7 Ductile failure of a tensile specimen.

## 5.4 MATERIAL PROPERTIES

The tensile test provides several material properties that are important in the analysis of engineering components and structures. These properties include two measures of strength, two measures of ductility, and two elastic constants. Let's first discuss strength.

### 5.4.1 Measures of Strength

The two measures of strength determined in a tensile test of a ductile material are the yield strength and the ultimate tensile strength. The yield strength, as the name implies, is the stress required to induce yielding:

$$S_y = \sigma_y \qquad (5.2)$$

To establish the yield stress $\sigma_y$, we examine the stress-strain diagram and attempt to identify the stress when yielding initiated. For some ductile materials with stress-strain diagrams similar to that shown in Fig. 5.6, the precise identification of $\sigma_y$ is clear. However, the yield behavior of other materials is much less well defined. For example, suppose a material exhibits the stress-strain behavior as indicated in Fig. 5.8 a. Where is the yield point?

It is evident in Fig. 5.8a that the stress-strain curve is non-linear, but we might differ in defining the point where slip and yield initiated. To eliminate the ambiguity in the definition of the yield point from a stress-strain diagram, the offset method is employed. We construct a line parallel to the linear portion of the $\sigma$-$\varepsilon$ curve in the elastic region. This line is offset along the strain axis by 0.2% or $\varepsilon = 0.002$. The intersection of the offset line with the $\sigma$-$\varepsilon$ curve defines the yield point as illustrated in Fig. 5.8 b. With the **0.2% yield stress** $\sigma_y$ defined, we use Eq. (5.2) to establish the yield strength $S_y$ for the material.

The ultimate tensile strength is established from the maximum stress on the specimen that occurs at the onset of necking. We have defined this point on Fig. 5.8 b, and have established the ultimate stress $\sigma_u$. The ultimate tensile strength $S_u$ for a ductile material is given by:

$$S_u = \sigma_u \qquad (5.3)$$

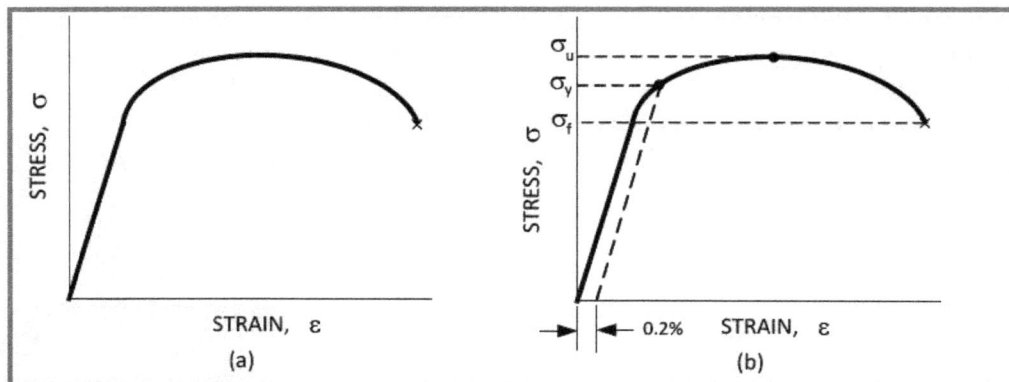

Fig. 5.8 The offset method for determining the stress $\sigma_y$ for yielding.

The stress at failure, $\sigma_f$ for a ductile material, shown in Fig. 5.8 b, is only of academic interest. If the structure or machine component is properly designed, the deformation state will probably be limited to the elastic region. In some instances, plastic deformations are tolerated in design, but these exceptions are limited to yielding in small local regions under conditions of constant loading. Global yielding is not tolerated, and yielding under the action of cyclic stresses is a recipe for disaster. For this reason, the stresses imposed on a structure are usually less than $\sigma_y$ and certainly less than $\sigma_u$.

### 5.4.2 Measures of Ductility

There are two common measures of ductility for metallic materials. The first is the **percent elongation** given by:

$$\% \, e \; = \; \frac{L_f - L_0}{L_0} \times 100 \qquad\qquad (5.4)$$

where $L_0$ is the gage length of the tensile specimen — usually 2.0 in. or 50 mm and $L_f$ is the final deformed length of the specimen between the marks for the gage length.

For low carbon steels, the percent elongation is in the range from 20 to 35%; however, the percent elongation decreases with increasing carbon content in the steel and increasing strength.

The second measure of ductility is the **percent reduction in area** that is given by:

$$\% \, A = \frac{A_0 - A_f}{A_0} \times 100 = \left[\left(\frac{d_0^2 - d_f^2}{d_0^2}\right)\right] \times 100 \qquad\qquad (5.5)$$

where $A_0$ is the initial cross sectional area of the tensile specimen given by $\pi d_0^2/4$ and $A_f = \pi d_f^2/4$ is the final cross sectional area of one of the ruptured ends of the specimen.

The percent reduction in area for low carbon steel is typically in the range of 60 to 70%. As was the case with percent elongation, the percent reduction in area decreases as the carbon content of steel is increased to enhance its strength.

There is a trade-off between ductility and strength for metallic alloys. To illustrate the loss in ductility with increasing strength, let's examine typical stress-strain curves for three different types of steels that are shown in Fig. 5.9. As the strength increases the strain to failure decreases with an accompanying decrease in the ductility. For low carbon steels, the strain to failure usually exceeds 60 to

70%. This value decreases to 30 to 40% for higher carbon steels. The very high strength steel alloys fail at strains ranging from 10 to 20%.

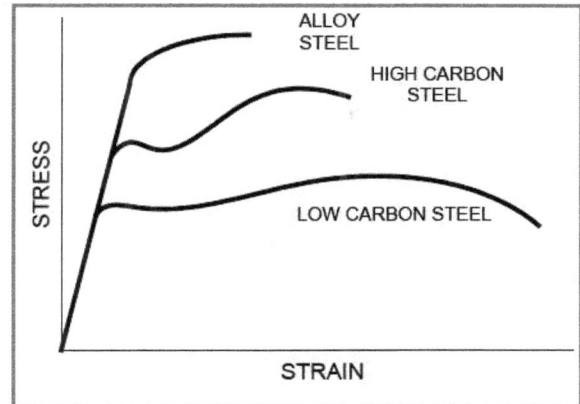

Fig. 5.9 Stress-strain curves for three different types of steels.

## EXAMPLE 5.1

A steel supplier provides you with data from a recent series of tensile tests of two different steels that they sell to your corporation. Your manager questions the ductility of both materials, and asks you to determine it. An examination of the supplier's data indicates:

Lower cost steel at $9.00/100lb
$L_f = 66$ mm and $d_f = 9.1$ mm

Higher cost steel at $9.74/100lb
$L_f = 61$ mm and $d_f = 10.8$ mm

The gage length $L_o$ and diameter $d_o$ for specimens from both types of steel was 50 mm and 12.5 mm respectively.

**Solution:** For the measure of ductility known as the percent elongation, Eq. (5.4) gives:

$$\%e = 100(66 - 50)/50 = 32\% \text{ for the lower cost steel.}$$

$$\%e = 100(61 - 50)/50 = 22\% \text{ for the higher cost steel.}$$

For the measure of ductility known as the percent reduction in area, Eq. (5.5) gives:

$$\%A = 100(d_o^2 - d_f^2)/d_o^2 = 100[(12.5)^2 - (9.1)^2]/(12.5)^2 = 47.0\% \Rightarrow \text{lower cost steel.}$$

$$\%A = 100(d_o^2 - d_f^2)/d_o^2 = 100[(12.5)^2 - (10.8)^2]/(12.5)^2 = 25.3\% \Rightarrow \text{higher cost steel.}$$

When your manager asks which steel has the higher ductility, what is your response?

### 5.4.3 Elastic Constants

*Modulus of Elasticity*

In this discussion of the elastic constants, we limit the application of these constants to characterize deformations only in the linear elastic region of the elastic-plastic regime. In this region, the material is elastic and recovers completely when the load or stress is removed from the specimen. The slope of the stress-strain curve is defined as the **modulus of elasticity** or Young's modulus. The slope illustrated in Fig. 5.10 is given by:

$$E = \sigma/\varepsilon \tag{4.5}$$

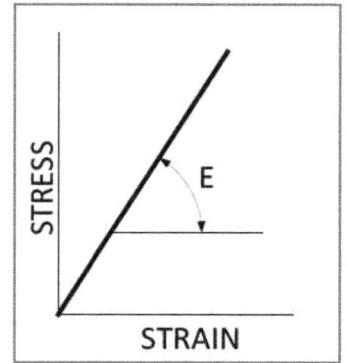

Fig. 5.10 The slope of the stress-strain curve determines the modulus of elasticity.

## EXAMPLE 5.2

After conducting a tensile test with a mild steel specimen, you measure the slope $\Delta P/\Delta \delta$ of the trace on the load-deflection curve (P, $\delta$) and find:

$$\Delta P/\Delta \delta = 7850 \text{ lb} / 2.68 \times 10^{-3} \text{ in.}$$

If the specimen diameter is 0.504 in. and the gage length of the extensometer is 2.00 in., determine the modulus of elasticity E.

> **Solution:** From Eq. 5.5, we write:
>
> $$E = \sigma/\varepsilon = (P/A) / (\delta/L) = PL / \delta A = \Delta PL/\Delta \delta A$$
>
> $$E = [(7850\text{lb})(2 \text{ in.})]/[(2.68 \times 10^{-3} \text{ in.})(\pi (0.252)^2 \text{ in.}^2)] = 29.36 \times 10^6 \text{ psi}$$

### *Poisson's Ratio*

Another elastic constant, **Poisson's ratio**, may be measured in the tensile test; however, strain gages must be attached to the specimen to measure the strain in both the axial and the transverse directions. Poisson's ratio is defined as:

$$\nu = - \varepsilon_t/\varepsilon_a \tag{5.6}$$

where $\varepsilon_t$ is the strain in the transverse direction, and $\varepsilon_a$ is the strain in the axial direction.

When a specimen is pulled in tension, its length extends and its diameter contracts. This contraction is usually 1/10 to 1/2 of the amount of the specimen's axial stretch. Extension of the specimen in the elastic region produces a contraction that is usually too small to be observed during the tensile test.

Poisson's ratio is a measure of the relative contraction occurring in the diameter of the tensile specimen as it is loaded. Although too small to be observed, the Poisson effect is a very important phenomenon because it significantly affects the analysis of structures subjected to multiaxial states of stress, where stresses are imposed in more than one direction.

To describe the Poisson contraction in more detail, consider the drawing of a rectangular bar of an elastic material with dimensions L, W, and D as shown in Fig. 5.11a. Next, apply an axial strain $\varepsilon_a$ and write:

$$\varepsilon_a = \Delta L/L \tag{a}$$

In the deformed state, the dimensions of the bar change to L + $\Delta$L, W + $\Delta$W, and D + $\Delta$D as shown in Fig. 5.11b. The strains $\varepsilon_t$ in the x and y directions (both x and y are transverse directions) are both due to the Poisson effect and are equal. They are given by:

$$\varepsilon_{tx} = \Delta W/W = \quad \varepsilon_{ty} = \Delta D/D \qquad\qquad\qquad (b)$$

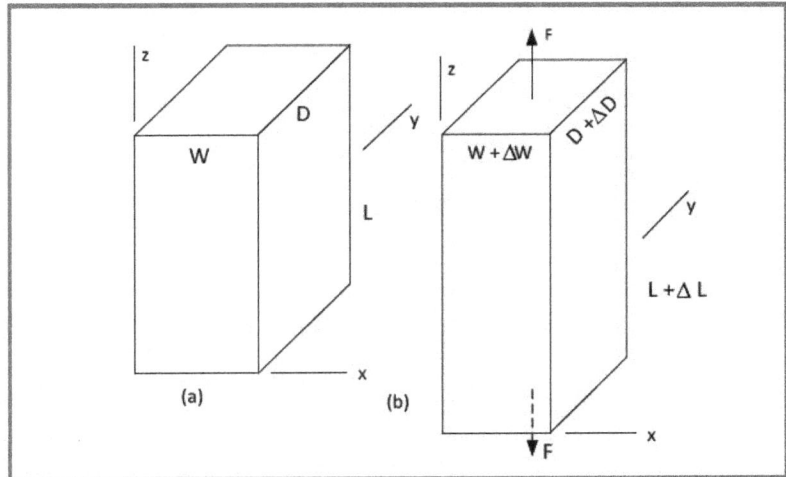

Fig. 5.11 A rectangular bar before and after imposing an axial deformation.

If we substitute Eq. (5.6) into Eq. (b), we obtain:

$$\varepsilon_t = -\,v\varepsilon_a = -\,v\Delta L/L = \Delta W/W = \Delta D/D \qquad\qquad (c)$$

From Eq. (c) it is clear that the changes in the transverse dimensions of the bar due to the Poisson effect are given by:

$$\Delta W = -\,v\Delta L(W/L); \qquad\qquad \Delta D = -\,v\Delta L(D/L) \qquad\qquad (d)$$

From this elementary analysis of the deformed geometry of a rectangular bar, we note that the material property **Poisson's ratio, v** gives the transverse deformation in terms of the axial deformation when a structural member is subjected to a uniaxial loading. The quantities $\Delta W$ and $\Delta D$ are negative when $\Delta L$ is positive because the Poisson effect produces a contraction in the transverse (x and y) directions when the body is extended in the axial (z) direction.


## EXAMPLE 5.3

You cut a circular hole with a 100-mm diameter in a large sheet of dental dam (very thin rubber sheet). If the sheet, originally 1,000 mm long by 500 mm wide, is stretched until it is 1,300 mm long, determine the new width of the sheet and the dimensions of the deformed hole. Assume that rubber is perfectly elastic with a Poisson's ratio, $v = \frac{1}{2}$.

**Solution:** Recall Eq. (5.6) and write:

$$\varepsilon_t = -\,v\varepsilon_a = -\,v\,\Delta L/L_o = -\,(1/2)\,(1300 - 1000)/1000 = -\,0.150$$

$$\varepsilon_a = \Delta L/L_o = (1300 - 1000)/1000 = 0.300$$

Since these strains are imposed over the entire sheet, determine the deformed width $W_{NEW}$ from:

$$W_{NEW} = W_o + \Delta W = W_o + \varepsilon_t W_o = (1 + \varepsilon_t)W_o = (1 - 0.150)500 = 425 \text{ mm}$$

The dimensions of the deformed hole are given by:

$$D_a = D_o + \Delta D_a = D_o + \varepsilon_a D_o = (1 + \varepsilon_a)D_o = (1 + 0.300)100 = 130 \text{ mm}$$

$$D_t = D_o + \Delta D_t = D_o + \varepsilon_t D_o = (1 + \varepsilon_t)D_o = (1 - 0.150)100 = 85 \text{ mm}$$

where $D_a$ and $D_t$ are the axial and transverse diameters of the hole after deformation.

What is the geometric form of the deformed hole in the rubber sheet?

## EXAMPLE 5.4

Determine the change in volume of a rectangular bar subjected to an axial strain of $\varepsilon_a = 2 \times 10^{-3}$. The dimensions of the bar before deformation were $L = 4W = 3D = 15$ in. The material from which the bar is fabricated has a Poisson's ratio of $v = \frac{1}{4}$.

**Solution:** The original volume, **V** of the bar is given by:

$$\mathbf{V} = L \times W \times D = L^3/12 = (15)^3/12 = 281.25 \text{ in}^3 \tag{a}$$

From Eq. (5.6), we determine the new dimensions of the bar as:

$$W_{NEW} = (1 - \varepsilon_a/4)W; \quad D_{NEW} = (1 - \varepsilon_a/4)D,; \quad L_{NEW} = (1 + \varepsilon_a)L \tag{b}$$

We rewrite Eq. (b) as:

$$W_{NEW} = (1 - \varepsilon_a/4)L/4; \quad D_{NEW} = (1 - \varepsilon_a/4)L/3; \quad L_{NEW} = (1 + \varepsilon_a)L \tag{c}$$

The new volume is given by:

$$V_{NEW} = L_{NEW} \times W_{NEW} \times D_{NEW} = (1 + \varepsilon_a)(1 - \varepsilon_a/4)(1 - \varepsilon_a/4)(L^3/12) \tag{d}$$

From Eq. (a) and (d), it is evident that:

$$\Delta V = [(1 + \varepsilon_a)(1 - \varepsilon_a/4)(1 - \varepsilon_a/4) - 1](L^3/12) \tag{e}$$

Substituting $\varepsilon = 2 \times 10^{-3}$ into Eq. (e) yields:

$$\Delta V = [(1 + 2 \times 10^{-3})(1 - 0.5 \times 10^{-3})(1 - 0.5 \times 10^{-3}) - 1] \text{ V} \tag{f}$$

Performing the calculation gives:

$$\Delta V = 0.00099825 \text{ V} = (0.00099825)(281.25) = 0.2808 \text{ in}^3$$

The percentage change in the volume is given by:

$$\Delta V/V = [(1 + \varepsilon_a)(1 - v\varepsilon_a)(1 - v\varepsilon_a) - 1] = 0.000998 = 0.0998\%$$

From these results it is evident that the change in the volume is extremely small for strains in the elastic region.

### *Shear Modulus*

Another elastic constant that we often use in the analysis of structures subjected to shear stress is the **shear modulus**, G. The shear modulus relates the shear stress to the shear strain by:

$$\tau = G\,\gamma \qquad\qquad (5.7)$$

where $\tau$ is the shear stress and $\gamma$ is the shear strain.

The shear strain $\gamma$ is defined as the change in angle of two perpendicular lines when a body is deformed. For instance, suppose we have a Cartesian coordinate system scribed on a body with the x and y axes serving as the two perpendicular lines. When the body deforms under the action of a shear stress $\tau$, these two lines rotate, and the angle at their intersection changes from 90° to some other angle 90° ± $\gamma$ as illustrated in Fig. 5.12.

Fig. 5.12 Shear strain is the change in a right angle when a body deforms.

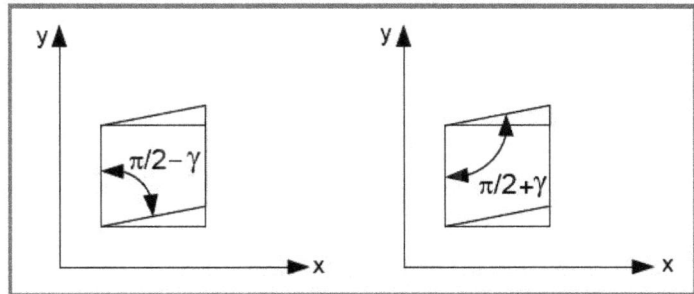

The shear modulus G is related to the modulus of elasticity and Poisson's ratio by:

$$G = \frac{E}{2(1 + \nu)} \qquad\qquad (5.8)$$

## EXAMPLE 5.5

Determine the shear modulus G for:

(a) Steel with the modulus of elasticity $E = 30 \times 10^6$ psi and Poisson's ratio $\nu = 0.30$.
(b) An aluminum alloy with $E = 73$ GPa and Poisson's ratio $\nu = 0.33$.

**Solution:** From Eq. (5.8) we write for steel the following relation:

$G = (30 \times 10^6) / [2(1 + 0.30)] = 11.54 \times 10^6$ psi for steel.

For the aluminum alloy, the shear modulus is given by:

$G = (73 \times 10^9) / [2(1 + 0.33)] = 27.44$ GPa for an aluminum alloy.

## 5.5 HOOKE'S LAW

In the elastic region, there is a linear relationship between stress and strain. In the simple case of uniaxial loading, which produces a uniaxial state of stress in long thin structural members such as bars and beams, Hooke's law gives the relationship:

$$\sigma = E\,\varepsilon \qquad\qquad (4.5)$$

However, some structural members are not long and thin. For example, tubes or cylindrical pressure vessels are subjected to biaxial stresses. Stresses, due to an applied pressure p, develop in both the axial (x) and the circumferential (y) directions as shown in Fig. 5.13.

Fig. 5.13 Biaxial stresses induced by pressure loading a cylindrical vessel.

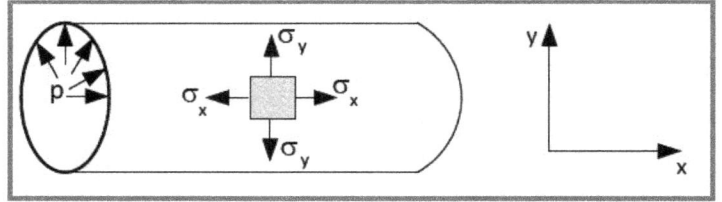

When a biaxial state of stress occurs in a structural member, the simple form of Hooke's law given in Eq. (4.5) is not valid. The Poisson effect produces an interaction between the stresses and strains in the two orthogonal directions. To accommodate the Poisson effect, we express the strains in terms of the stresses as:

$$\varepsilon_x = \frac{1}{E}\left(\sigma_x - v\sigma_y\right)$$

$$\varepsilon_y = \frac{1}{E}\left(\sigma_y - v\sigma_x\right)$$

(5.9)

The stresses in terms of the strains are:

$$\sigma_x = \frac{E}{\left(1-v^2\right)}\left(\varepsilon_x + v\varepsilon_y\right)$$

$$\sigma_y = \frac{E}{\left(1-v^2\right)}\left(\varepsilon_y + v\varepsilon_x\right)$$

(5.10)

We employ Eqs. (5.9) and (5.10) with two-dimensional structural members such as tubes, plates and shells.

## EXAMPLE 5.6

The stresses in the wall of a cylindrical pressure vessel are:

$$\sigma_x = 7,500 \text{ psi} \qquad \text{axial direction}$$
$$\sigma_y = 15,000 \text{ psi} \qquad \text{hoop direction}$$

If the pressure vessel is fabricated from steel plates with E = 30 × 10$^6$ psi and v = 0.30, determine the strain in the hoop and axial directions.

**Solution:** Recall Eq. (5.9) and write:

$$\varepsilon_x = \frac{1}{E}\left(\sigma_x - v\sigma_y\right) = \frac{1}{30}[7,500 - (0.30)(15,000)] \times 10^{-6} = 100 \times 10^{-6}$$

$$\varepsilon_y = \frac{1}{E}\left(\sigma_y - v\sigma_x\right) = \frac{1}{30}[15,000 - (0.30)(7,500)] \times 10^{-6} = 425 \times 10^{-6}$$

It is interesting to observe that the pressure vessel develops biaxial tensile stresses with a ratio of 2 hoop to 1 axial; however, the ratio of strain is 4.25 hoop to 1 axial. The change in the ratio of strains is due to the Poisson effect.

## EXAMPLE 5.7

Electrical resistance strain gages are employed to measure the strain in the hoop and the axial directions on a high-pressure pipeline that runs close to a residential neighborhood. The strains are:

$$\varepsilon_a = \varepsilon_x = 600 \times 10^{-6} \qquad \text{and} \qquad \varepsilon_h = \varepsilon_y = 2650 \times 10^{-6}$$

If the pipe in the line is fabricated from a high strength alloy steel with E = 200 GPa and Poisson's ratio $v$ = 0.29, determine the axial and hoop stresses and comment on the safety of the pipeline.

**Solution:** Recall Eqs. (5.10) and write:

$$\sigma_a = \sigma_x = \frac{E}{(1-v^2)}\left(\varepsilon_x + v\varepsilon_y\right) = \frac{200}{1-(0.29)^2}\left[600 + (0.29)(2650)\right] \times 10^{9-6} = 298.7 \text{ MPa}$$

$$\sigma_h = \sigma_y = \frac{E}{(1-v^2)}\left(\varepsilon_y + v\varepsilon_x\right) = \frac{200}{1-(0.29)^2}\left[2650 + (0.29)(600)\right] \times 10^{9-6} = 616.6 \text{ MPa}$$

**Comment on the safety of the pipeline:**

Although the specific type of steel has not been defined, an examination of the strength of the steels listed in Appendix B indicates that the hoop stress is very high. Material samples should be obtained and tensile tests conducted to determine the yield and tensile strength of the actual steel used in constructing the pipeline. Preliminary analysis indicates that the safety of the pipeline is suspect.

## 5.6 TRUE STRESS AND TRUE STRAIN

In the standard tensile test, we define the (engineering) stress as:

$$\sigma = P/A_o \qquad\qquad (4.4)$$

where $A_o$ is the cross sectional area of the tensile specimen.

As we load the tensile specimen, the diameter of the bar decreases due to Poisson contraction. Clearly, the cross sectional area of the specimen is changing during the standard tensile test. How do we handle this situation?

There are two approaches. First, with the engineering approach, we ignore the changes in the cross sectional area and compute the stress (engineering) using Eq. (4.4). We also determine the strain (engineering) by:

$$\varepsilon = \delta/L_o \tag{4.3}$$

where $L_o$ is the initial gage length.

Obviously the gage length of the tensile specimen is increasing as we load the tensile bar. However, if we treat the strain as an "engineering" quantity, it is computed using Eq. (4.3). The errors resulting by neglecting the changes in the area A and the gage length L are small when the analysis is limited to stresses and strains in the elastic region

The second approach is to accommodate both the changing cross sectional area A and the gage length L in the analysis of the results from a tensile test. With this approach the "true" stress $\sigma_T$ is given by:

$$\sigma_T = F/A_F \tag{5.11}[2]$$

where $A_F$ is the cross sectional area of the tensile specimen at any applied force F.

The true strain[3] is also defined by considering an incremental change in length dL over a gage length L, which is increasing with the applied load. With this definition the true strain is given by:

$$d\varepsilon_T = dL/L \tag{5.12}$$

We integrate Eq. (5.12) as the gage length increases from $L_o$ to $L_F$ and obtain:

$$\varepsilon_T = \int_{L_0}^{L_F} \frac{dL}{L} = \ln\frac{L_F}{L_0} \tag{5.13}$$

The engineering strain $\varepsilon$ is related to the true strain by:

$$\varepsilon_T = \ln (L_F/L_o) = \ln [(L_o + \Delta L)/L_o] = \ln (1 + \varepsilon) \tag{5.14}$$

If the results of a tensile test are interpreted using the definitions of true stress and true strain, the stress-strain curve in the plastic regime is changed markedly. A typical true stress-true strain diagram is illustrated in Fig. 5.14. Inspection of this figure shows that the true stress-strain curve differs dramatically from the engineering stress-strain curve. For the elastic region where the strains are very small, the two curves are nearly identical. However, for larger strains significant differences are evident. The true stress is larger than the engineering stress for a given load and the true strain is smaller than the engineering strain. The true stress does not decrease with the onset of necking because the area at the neck is decreasing more rapidly than the load. The engineering stress, which is based on the initial area $A_0$ of the specimen, decreases because the load decreases as the neck develops.

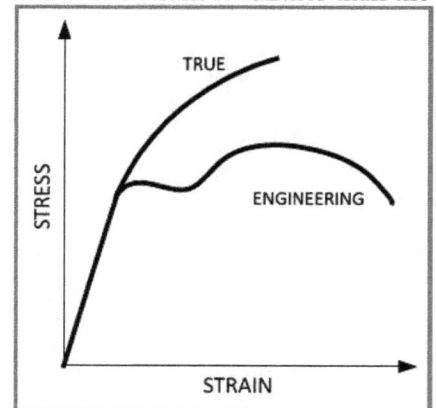

Fig. 5.14 Illustration of a true-stress—true-strain diagram.

---

[2] We interchange the external force F and the internal force P in the equation for stress in the specimen because clearly P = F in a tensile specimen.

[3] True strain is sometimes referred to as "natural" or "logarithmic" strain.

# EXAMPLE 5.8

The rupture load on a standard tensile specimen with an initial diameter $d_o = 0.500$ in. and initial gage length $L_o = 2.00$ in. was 12,000 lb. If the diameter $d_F$ at the neck of the ruptured specimen was 0.415 in. and the gage length after failure was $L_F = 2.650$ in., determine:

    a. The true stress at failure.
    b. The true strain at failure.
    c. The engineering stress at failure.
    d. The engineering strain at failure.

---

**Solution:** The true stress at failure is given by Eq. (5.11) as:

$$\sigma_T = F/A = 4F/\pi d_F^2 = [(4)(12000)]/[\pi(0.415)^2] = 88{,}710 \text{ psi}$$

The true strain at failure is given by Eq. (5.14) as:

$$\varepsilon_T = \ln(L_F/L_o) = \ln(2.650/2.000) = 0.281 = 28.1\%$$

The engineering stress at failure is given by Eq. (4.4) as:

$$\sigma = F/A_o = 4F/\pi d_o^2 = [(4)(12000)]/[\pi(0.500)^2] = 61{,}120 \text{ psi}$$

The engineering strain at failure is given by Eq. (4.3) as:

$$\varepsilon = (L_F - L_o)/L_o = (2.650 - 2.000)/2.000 = 0.325 = 32.5\%$$

---

## 5.7 FATIGUE STRENGTH

Many structures are loaded only once. For example, the structural members in the Hancock tower in Chicago were loaded by the dead weight of all of the structural components as the tower was constructed. The people, furnishings and equipment going into the tower added some weight, but this additional load was small when compared to the weight of the basic structure. For structures and machine components loaded only once, either the yield strength or the ultimate tensile strength is adequate to predict the safety of the structure.

However, many structures and machine components are subjected to repeated loading. Very heavy tractor-trailer trucks, for example, traverse a bridge. With the passage of each truck every bridge member is subjected to another cycle of loading. Consider the engine of your automobile. If you are driving along a highway with the tachometer registering 3600 RPM, you are subjecting the crankshaft in your engine to 3600 cycles of load for each minute of operation.

There are two detrimental effects due to the cyclic loading of structures and machine components. First, the design strength of the material from which the structure is fabricated is lowered. We design to a fatigue strength $S_f$ that is a function of the cyclic stresses and the number of cycles of load imposed onto the structure.

Second, fatigue failures in **ductile** materials are of a **brittle** nature (i.e., structural failure and collapse occur catastrophically). The failure mechanism in fatigue is markedly different from that observed in yielding or rupture. In fatigue, microscopic cracks are initiated due to accumulated irreversible slip in a very thin layer of material adjacent to the surface. These cracks grow larger and extend into the material until reaching critical size. At this point, the cracks become unstable and extend at very high speed across the structural member producing sudden and catastrophic collapse. Examination of a surface of a fatigue failure shows the brittle nature of the phenomena. Although the material may be classified as ductile based on the results of a tensile test, there is no visible sign of coarse slip or necking associated with failure in fatigue.

## 5.7.1 The $S_f$ - N Diagram

To accommodate for the degrading effects of cyclic loading, we compare the applied stresses with the **fatigue strength, $S_f$** of the material from which the structure is fabricated. A typical example of the fatigue strength for low carbon steel is presented in the $S_f$-N diagram of Fig. 5.15.

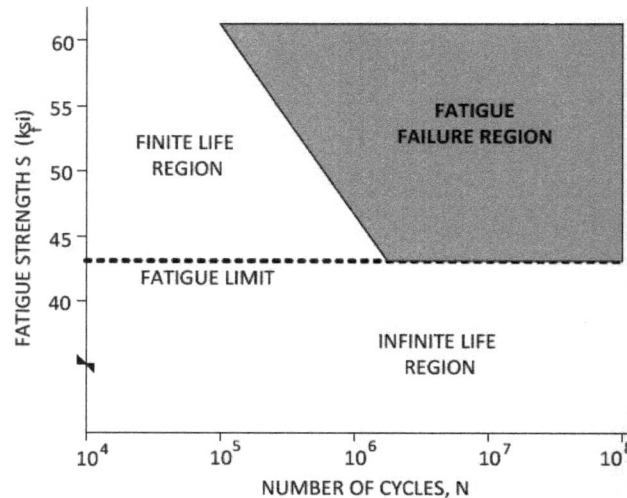

Fig. 5.15 $S_f$ – N curve for high strength low alloy steel.

The $S_f$ - N diagram is a graphical representation of both safe and critical states of cyclic stresses. The $S_f$ - N diagram is divided into three different regions:

1. A failure region (shaded) where many cycles of high stresses produce a fatigue crack leading to fracture.
2. Safe regions where component stresses lower than the fatigue (endurance) limit ensure infinite cyclic life.
3. A finite-life region where a specified number of cycles can be endured at a specified stress level that is greater than the fatigue limit.

For infinite life, the fatigue strength is often called the endurance limit $S_e$ where:

$$S_e = S_f \qquad \text{for N} > 10^6 \text{ cycles} \qquad (5.15)$$

For finite life, with the number of cycles less than $10^6$, the fatigue strength $S_f$ is larger than $S_e$. The value of $S_f$ used in a design analysis is determined from the $S_f$-N diagram for the material used to fabricate the structural elements. Example 5.10 demonstrates the techniques used to determine the strength $S_f$ associated with finite life.

In comparing the cyclic stresses with the strength as defined in Fig. 5.15, we use the alternating portion of the applied stresses. The alternating stress, $\sigma_a$, and the mean stress, $\sigma_m$, for different types of cyclic loading are defined in the stress-time diagrams presented in Fig. 5.16. We compute the mean stress for cyclic loading by identifying the maximum and minimum stress in a given cycle of applied loading. The mean stress is the average of the maximum and minimum stress. The relation used to determine the mean cyclic stress is given by:

$$\sigma_m = (\sigma_{max} + \sigma_{min})/2 \qquad\qquad (5.16)$$

The alternating stress is determined from the difference in the maximum and minimum stress during a typical load cycle. The relation used to determine the alternating cyclic stress is given by:

$$\sigma_a = (\sigma_{max} - \sigma_{min})/2 \qquad\qquad (5.17)$$

The method used to determine the safety of a structural member when subjected to cyclic loading is described in Example 5.9.

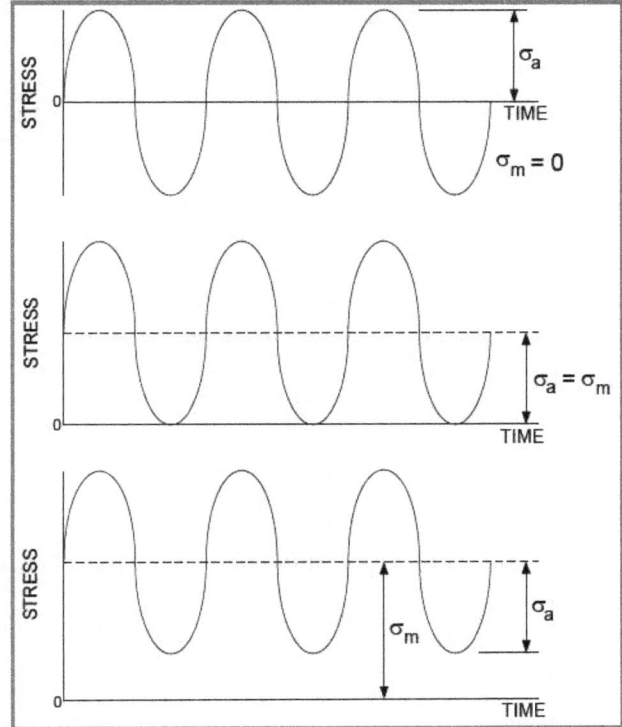

Fig. 5.16 Different types of cyclic stresses imposed on a structure.

## EXAMPLE 5.9

Consider a cross beam used in constructing a bridge. The dead weight of the bridge structure produces a stress of 22,000 psi in this beam. However, when a fully loaded tractor-trailer crosses over the beam the stress increases to 65,000 psi. The beam is fabricated from a material with the fatigue properties described in Fig. 5.15. The strengths, yield and ultimate tensile, are 54 ksi and 88 ksi respectively. Will the beam fail in fatigue? If failure occurs, predict the cyclic life of the structure. Neglect the effect of the mean stress on fatigue strength in this example.

**Solution:** Determine the alternating stresses imposed on the bridge beam from Eq. (5.17).

$$\sigma_a = (\sigma_{max} - \sigma_{min})/2 = (65,000 - 22,000)/2 = 21.5 \text{ ksi}$$

Next, compare this value to the fatigue limit of 43 ksi that is obtained by reading the $S_f$ - N curve in Fig. 5.15. Clearly $S_f = 43$ ksi $> \sigma_a = 21.5$ ksi; the alternating stresses are less than the endurance limit and the bridge beam is safe. Indeed, the safety factor is determined from a modified form of Eq. (4.11) as:

$$\mathbf{SF_f} = S_f/\sigma_a \qquad\qquad (5.18)$$

$$SF_f = 43/21.5 = 2.00$$

The mean stress imposed on the bridge beam during the cyclic loading is determined from Eq. (5.16) as:

$$\sigma_m = (\sigma_{max} + \sigma_{min})/2 = (65,000 + 22,000)/2 = 43.5 \text{ ksi}$$

This is a relatively large mean stress to impose on a beam. As a consequence, the fatigue strength will be lower than that obtained from the $S_f$–N curve to account for the effects of the large tensile mean stress. This solution is modified in Example 5.11 to account for the effects of mean stresses in reducing the allowable alternating stresses.

## EXAMPLE 5.10

Consider the crankshaft of a high performance racecar which is subjected to maximum and minimum stresses of + 345 MPa and – 345 MPa. If the crankshaft is fabricated from the high strength low alloy steel described in Fig. 5.15, determine its fatigue life.

**Solution:** Let's first determine the alternating stress $\sigma_a$ for the crankshaft from Eq. (5.17).

$$\sigma_a = (\sigma_{max} - \sigma_{min})/2 = [345 - (-345)]/2 = 345 \text{ MPa}$$

Next convert this result from MPa units to ksi units so that we may employ a graphical approach using the fatigue properties of the high strength low alloy steel shown in Fig. 5.15.

$$\sigma_a = 345 \text{ MPa} \times 1 \text{ ksi} /6.895 \text{ MPa} = 50.0 \text{ ksi}$$

Plot $\sigma_a = 50.0$ ksi at point A on the ordinate of the $S_f$-N diagram shown in Fig. E5.10. Extend a line parallel to the abscissa until it intersects the shaded (fatigue failure) region and plot point B. Drop a vertical line from point B until it intersects the axis defining the cyclic life to establish point C. Point C gives the anticipated cyclic life, which in this example is less than $10^6$ cycles. For a crankshaft operating at 6000 RPM in a high-performance racing car, the anticipated life is less than three hours. Clearly, this life is too short by several orders of magnitude and the design of the crankshaft is inadequate. Either the crankshaft size must be increased to lower the stress level or a different material with a higher endurance limit $S_e$ must be selected.

Fig. E5.10

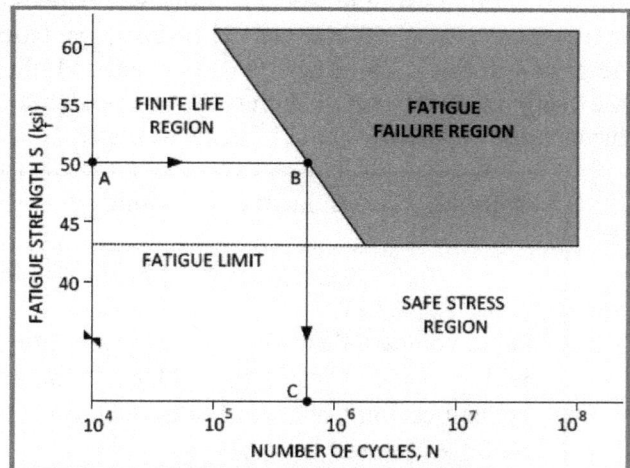

## 5.7.2 The Goodman Diagram

In some instances, the cyclic loading on a structural member produces a combination of alternating and mean stresses. When mean stresses are superimposed on the alternating stresses, the effect is to decrease the fatigue strength of the material. Goodman and Gerber have developed empirical methods to determine a modified fatigue strength that accounts for the detrimental effects of combined alternating and mean stresses. The modified fatigue strength $S_a$ is presented as a function of the cyclic mean stress in Fig. 5.17. Examination of Fig. 5.17 reveals that the modified fatigue strength $S_a$ is equal to $S_e$ when $\sigma_m = 0$ for all the empirical relations. The decrease in the modified fatigue strength is linear with the increase in $\sigma_m$, with the Goodman method. The fatigue strength $S_a$ becomes zero when $\sigma_m = S_u$, for the Goodman method and Gerber methods.

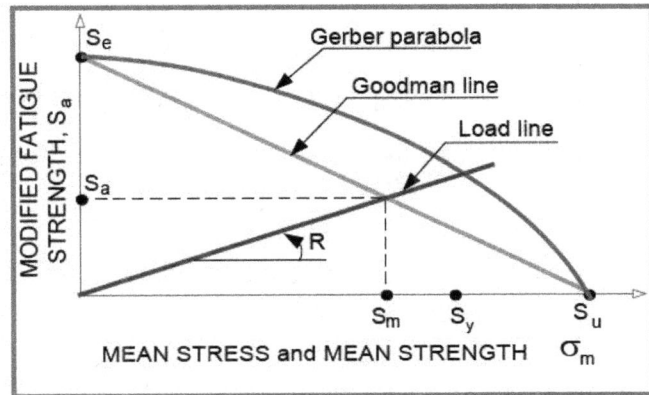

Fig. 5.17 Modified fatigue strength $S_a$ decreases as the cyclic mean stresses increase.

Let's consider the Goodman method, because it is the most commonly used technique and it is conservative compared to the Gerber method. We show a load line with a slope R in Fig. 5.17. The slope R is given by the ratio:

$$R = \sigma_a/\sigma_m = S_a/S_m \qquad (5.19)$$

The load line extends from the origin and intersects the Goodman line. At this point, vertical and horizontal lines are drawn to intersect the axes, defining the allowable alternating strength $S_a$ and the allowable alternating mean strength $S_m$. We may also characterize the modified fatigue strength $S_a$ in equation format for the Goodman and Gerber methods as:

$$S_a = S_e\,[1 - (S_m/S_u)] \qquad \text{(Goodman)} \qquad (5.20a)$$

$$S_a = S_e\,[1 - (S_m/S_u)^2] \qquad \text{(Gerber)} \qquad (5.20b)$$

where $S_e$ is the endurance limit, $S_u$ is the ultimate tensile strength, and $S_m$ is the mean strength.

A graph of $S_a$ as a function of $S_m$ for Eqs. (5.20) is shown in Fig. 5.17. The result shown as the blue curve is a parabola and the red curve is linear. Fatigue test data for both steels and aluminum alloys are usually located in the region between the red and the blue curves. The red line due to Goodman's method is more conservative than the Gerber method.

# EXAMPLE 5.11

Let's reconsider the bridge described in Example 5.9. In the previous solution, we neglected the influence of the mean stress on the fatigue strength. In this solution, let's account for the degrading effect of the cyclic mean stress on the fatigue strength by using the Goodman method for adjusting the fatigue strength. Determine the safety factor using the values determined for the allowable alternating and mean fatigue strengths. Recall from Example 5.9 that the fatigue strength was 43 ksi.

**Solution:**

Recall the results for the alternating and mean stresses from Example 5.9.

$$\sigma_a = (\sigma_{Max} - \sigma_{Min})/(2) = (65,000 - 22,000)/(2) = 21.5 \text{ ksi} \tag{a}$$

$$\sigma_m = (\sigma_{Max} + \sigma_{Min})/(2) = (65,000 + 22,000)/(2) = 43.5 \text{ ksi} \tag{b}$$

Determine the slope R of the load line from Eq. (5.19) as:

$$R = \sigma_a/\sigma_m = 21.5/43.5 = 0.4943 \tag{c}$$

Next, let's determine the allowable alternating and mean strengths. Substituting Eq. (5.10) into Eq. (5.20a) yields:

$$S_a = R \, S_e \, S_u/(R \, S_u + S_e) \tag{5.21a}$$

and

$$S_m = S_a /R \tag{5.21b}$$

Solving for $S_a$ and $S_m$ gives:

$$S_a = \frac{(0.4943)(43)(88)}{(0.4943)(88) + 43} = 21.62 \text{ ksi}$$

and $\hspace{12cm}$ (d)

$$S_m = S_a/R = 21.62/0.4943 = 43.74 \text{ ksi}$$

The safety factor is given by setting $S_a = S_f$ in Eq. (5.18) as:

$$SF_f = S_a/\sigma_a = 21.62/21.5 = 1.006 \tag{e}$$

When accounting for the mean stresses on fatigue strength, the safety factor was reduced slightly more than 1.0. In the previous solution for Example 5.9, the safety factor $SF_f = 2.0$; whereas, in this analysis the safety factor is reduced to 1.006. Clearly, the effect of the mean stress is to reduce the allowable fatigue strengths $S_a$ and $S_m$. With a safety factor of only 1.006, the beam size must be increased or a substitute material employed with higher fatigue strength.

## 5.8 SUMMARY

Structures and machine components fail because of fracturing, yielding or excessive elastic deformation. In predicting either fracture or yielding, we compare the stress imposed on the structure with the strength of the material from which it is fabricated. In some analyses, we employ a handbook to provide the strength of the materials, and in more critical designs we conduct standardized tensile tests to establish the strength of the specific materials to be employed in construction.

The standard ASTM tensile test is introduced in Section 5.3. Using a standard tensile specimen and standard test procedures a force-deflection curve is generated. We convert this force-deflection data to a stress-strain curve, and then interpret the curve to give the strength, ductility and the elastic constants. Care is exercised in the interpretation to distinguish between "brittle" and "ductile" materials because the yield strength cannot be measured for brittle materials. For ductile materials, procedures are described for measuring both the yield strength and the ultimate tensile strength.

The ductility of a material is indicated by two measurements — the percent elongation and the percent reduction in area.

$$\% \, e \, = \, \frac{L_f - L_0}{L_0} \times 100 \qquad (5.4)$$

$$\% \, A = \frac{A_0 - A_f}{A_0} \times 100 \qquad (5.5)$$

Methods for determining both of these quantities have been given. We noted that for most types of steel the ductility decreased with increasing carbon content used for strength enhancement.

Two elastic constants, the modulus of elasticity E and Poisson's ratio $\nu$, are measured in a tensile test. The modulus of elasticity is determined from the slope of the stress-strain curve in the elastic region. Poisson's ratio is determined from the ratio of strain in both the axial and transverse directions measured during a uniaxial tension test.

$$\nu = - \, \varepsilon_t / \varepsilon_a \qquad (5.6)$$

The shear modulus, another elastic constant often used in analysis of shear stresses and torsion loading of circular shafts, is introduced.

$$\tau = G \, \gamma \qquad (5.7)$$

$$G \, = \, \frac{E}{2(1 \, + \, \nu)} \qquad (5.8)$$

Clearly, the shear modulus G is not independent as it is a function of E and $\nu$.

Two versions of Hooke's law were discussed. The historical version, due to Robert Hooke, where $\sigma = E\varepsilon$, is valid only when the state of stress is uniaxial. However, when the structure is subjected to biaxial stresses, the uniaxial form of Hooke's law is not valid. Instead we must represent the relations between stress and strain with the biaxial equations given by:

$$\varepsilon_x = \frac{1}{E}\left(\sigma_x - v\sigma_y\right); \quad \varepsilon_y = \frac{1}{E}\left(\sigma_y - v\sigma_x\right) \qquad (5.9)$$

$$\sigma_x = \frac{E}{\left(1-v^2\right)}\left(\varepsilon_x + v\varepsilon_y\right); \quad \sigma_y = \frac{E}{\left(1-v^2\right)}\left(\varepsilon_y + v\varepsilon_x\right) \qquad (5.10)$$

The difference between engineering and true stress and strain is described. Definitions of true stress and true strain are given, and a relation between true strain and engineering strain is derived.

$$\varepsilon_T = \int_{L_0}^{L_F}\frac{dL}{L} = \ln\frac{L_F}{L_0} \qquad (5.13)$$

$$\varepsilon_T = \ln(L_F/L_0) = \ln\left[(L_0 + \Delta L)/L_0\right] = \ln(1 + \varepsilon) \qquad (5.14)$$

The difference between engineering and true strains is not significant if the strains are small, but the difference is significant for large values. For this reason, we usually employ engineering stress and strain if we are performing analyses of structures subjected to stresses and strains in the elastic region. Structures subjected to stresses beyond the yield conditions deform significantly, and we usually perform plastic analysis using the definitions of true stress and true strain.

The concept of failure by cyclic loading is introduced. The repeated loading induces fatigue damage in structural members, which produces a brittle failure at stresses that are often lower than the measured yield stress. The $S_f - N$ curve that describes the relation between the number of cycles and the fatigue strength of a material is presented. Examples are provided that illustrate an approach for determining the safety of a structure or a machine component subjected to fatigue loading.

## REFERENCES

1. Hertzberg, R. W., Deformation and Fracture Mechanics of Engineering Materials, 4th Edition, John Wiley, New York, NY 1996.
2. Madayag, A. F. Metal Fatigue, John Wiley, New York, NY, 1969.
3. Forrest, P. G., Fatigue of Metals, Pergamon Press, Oxford, UK, 1962.

## PROBLEMS

5.1 Sketch a stress-strain diagram for a relatively brittle material and identify the ultimate tensile strength on the diagram.

5.2 Sketch a stress-strain diagram for a relatively ductile material and identify the yield strength and ultimate tensile strength on the diagram.

5.3 Sketch a stress-strain diagram for a ductile material that does not exhibit a well-defined yield point. Identify on the diagram the yield strength and the ultimate tensile strength.

5.4 Write a test plan for conducting a standard tensile test.

5.5 Determine the yield strength and the ultimate tensile strength for the material represented by the stress-strain diagram shown in the figure to the right.

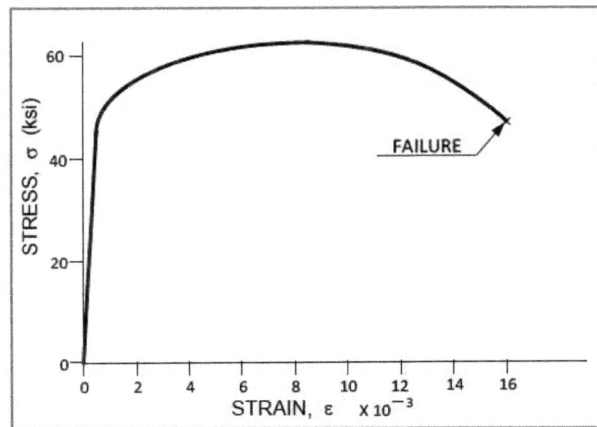

5.6 Determine the yield strength and the ultimate tensile strength for the material represented by the stress-strain diagram shown in the figure to the left.

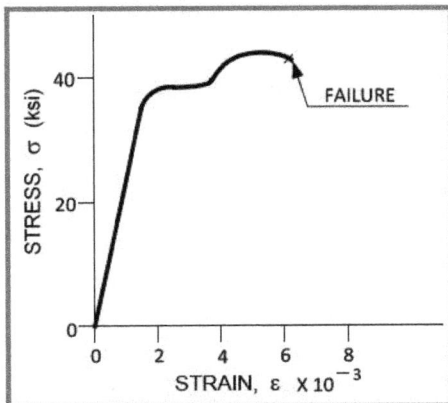

5.7 Determine the percent elongation and the percent reduction in area if measurements of $L_f = 66.4$ mm and $d_f = 10.52$ mm were made during a standard tensile test. The initial length and diameter of the standard tensile specimen is $L_o = 50$ mm and $d_o = 12.7$ mm.

5.8 In conducting a tensile test of an aluminum alloy, you adjust the signals from the load cell and the extensometer to zero and then load the specimen. You dwell at several loads in the elastic region of the stress-strain response, and record readings of the load F and the stretch $\delta$. The values of F and $\delta$ are plotted on an F-$\delta$ diagram to obtain the slope $\Delta F/\Delta\delta = (16,000 \text{ N})/(100 \times 10^{-3} \text{ mm})$. Determine the elastic modulus of the aluminum alloy if the initial length and diameter of the tensile specimen is $L_o = 50$ mm and $d_o = 12.7$ mm.

5.9 If a steel alloy exhibits a Poisson's ratio of $v = 0.30$, determine the diameter of a tensile specimen when it is subjected to an elastic stress of $\sigma = 52.6$ ksi. The dimensions $L_o$ and $d_o$ for the specimen are 2.00 in. and 0.500 in., respectively.

5.10 A long (3 m) sheet of an aluminum alloy is stretched in a manufacturing process until it is 3.8 m in length. If the sheet was initially 0.6 m wide and 2 mm thick, determine its new width and thickness. In the plastic regime, Poisson's ratio is 0.5 for all metallic materials because volume is conserved in the plastic deformation process.

5.11 A long (3.8 m) sheet of an aluminum alloy is stretched in a manufacturing process until it is 4.6 m in length. If a central hole initially 75 mm in diameter was drilled in the sheet prior to stretching, determine its new dimensions. In the plastic regime, Poisson's ratio is 0.5 for all metallic materials because volume is conserved in the plastic deformation process.

5.12 Prove that volume of an object is conserved in a uniaxial stress state if $v = \frac{1}{2}$.

5.13 If a steel alloy is linearly elastic until the applied axial stress equals the yield strength, determine the strain at yield for a structural steel with a yield strength of 38 ksi and a modulus of elasticity of $30 \times 10^6$ psi.

5.14 A long, thin, aluminum-alloy bar is subjected to an axial stress of 227 MPa. Determine the strain in the axial and transverse directions of the bar.

5.15 A spherical pressure vessel, fabricated from a steel alloy, is subjected to a biaxial stress field with $\sigma_x = \sigma_y = 20.6$ ksi. Determine the strains $\varepsilon_x$ and $\varepsilon_y$.

5.16 An electrical strain gage is placed with an arbitrary orientation on a spherical pressure vessel fabricated from a titanium alloy. The gage provides a strain measurement $\varepsilon = 1,140 \times 10^{-6}$. Determine the stresses $\sigma_x$ and $\sigma_y$ in terms of MPa.

5.17 Prove that the ratio of hoop strain to axial strain for a cylindrical pressure vessel is given by:

$$\varepsilon_h/\varepsilon_a = (2 - v)/(1 - 2v).$$

5.18 An orthogonal pair of electrical strain gages is placed on a cylindrical pressure vessel fabricated from an aluminum alloy. The strain gages provide measurements of the axial strain $\varepsilon_a = 264 \times 10^{-6}$ and hoop strain $\varepsilon_h = 1,232 \times 10^{-6}$. Determine the axial stress $\sigma_a$ and the hoop stress $\sigma_h$ in terms of MPa.

5.19 A standard tensile specimen ruptured with an applied force $F = 11.3$ kip. Your measurement of the diameter at the neck region after failure gives $d_f = 0.366$ in. Determine the engineering stress and the true stress at failure.

5.20 Determine the engineering strain and the true strain for a tensile specimen with $L_o = 2.00$ in. if the final length, $L_f$ is given in the table shown below:

| Final Length $L_f$ (in.) | Engineering Strain | True Strain |
|---|---|---|
| 2.1 | | |
| 2.2 | | |
| 2.3 | | |
| 2.4 | | |
| 2.5 | | |
| 2.6 | | |
| 2.7 | | |
| 2.8 | | |
| 2.9 | | |

5.21 Determine the engineering strain and the true strain for a tensile specimen with $L_o = 50$ mm if the final length $L_f$ is given in the table below:

| Final Length $L_f$ (mm) | Engineering Strain | True Strain |
|---|---|---|
| 53 | | |
| 56 | | |
| 59 | | |
| 62 | | |
| 65 | | |
| 68 | | |
| 71 | | |
| 74 | | |
| 77 | | |

5.22 Complete the conversion table that relates engineering strain and true strain.

| Engineering Strain | True Strain |
|---|---|
| 0.001 | |
| 0.002 | |
| 0.005 | |
| 0.010 | |
| 0.020 | |
| 0.050 | |
| 0.100 | |
| 0.200 | |
| 0.500 | |
| 1.000 | |

5.23 The maximum stress imposed at the root of a tooth on a spur gear is $\sigma_{max} = 78.9$ ksi and the minimum stress $\sigma_{min} = 0$. Neglecting the influence of the cyclic mean stress on the fatigue life, determine the anticipated life of the gear, if the $S_f - N$ curve presented to the right characterizes its fatigue properties.

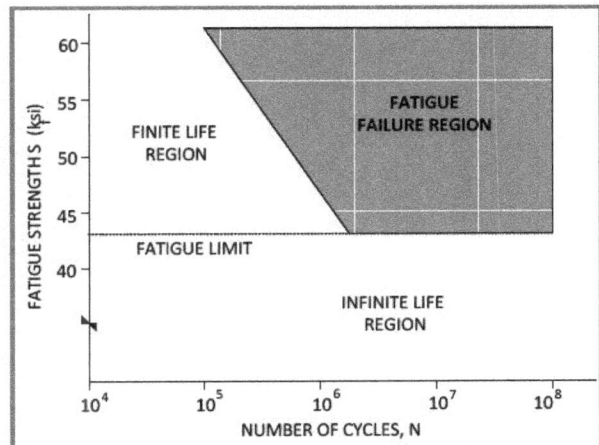

5.24 A rule for approximating the fatigue strength is $S_f = S_u/2$. Using this rule, determine the maximum stress that can be imposed on a structural member with a dynamic load that is twice the static load as illustrated in the figure below. The structural member is fabricated from steel with $S_u = 340$ MPa. Neglect the influence of the cyclic mean stresses on the fatigue behavior.

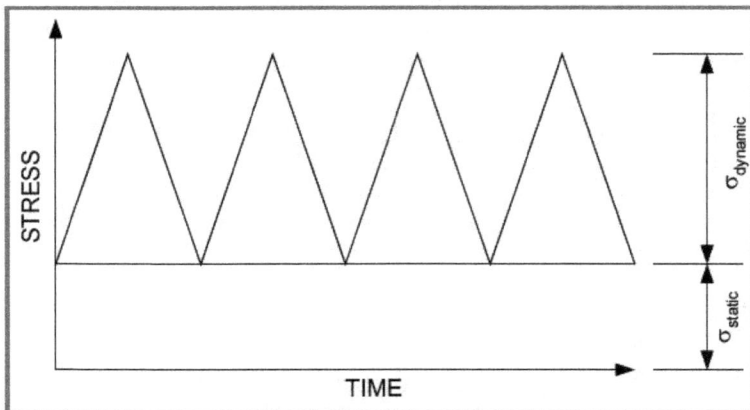

5.25 Reconsider Problem 5.23 taking into account the effect of the cyclic mean stress.

5.26 Reconsider Problem 5.24 taking into account the effect of the cyclic mean stress.

5.27 Repeat Problem 5.24 with the dynamic load equal to the static load.

5.28 Repeat Problem 5.24 with the dynamic load equal to one half of the static load.

5.29 Beginning with Eq. (5.9) derive Eq. (5.10).

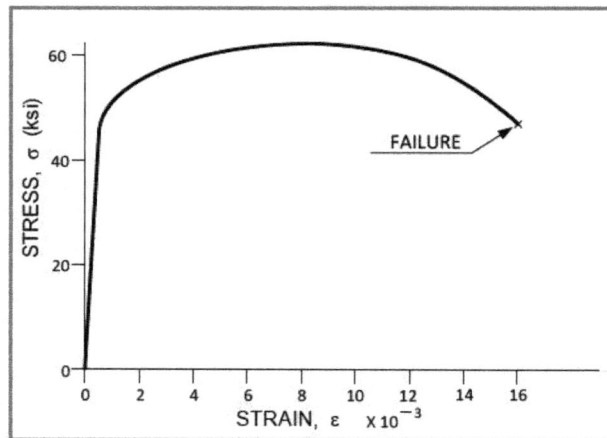

5.30 Convert the stress-strain curve shown in the figure to the right to a true stress-true strain curve.

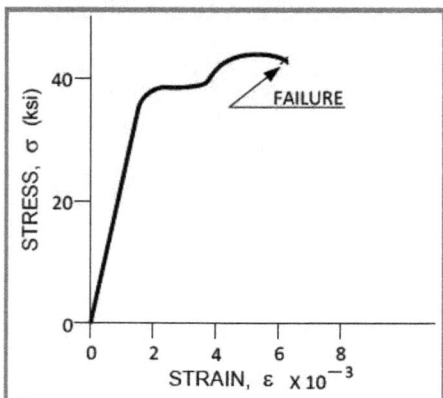

5.31 Convert the stress-strain curve shown in the figure to the left to a true stress-true strain curve.

5.32 At the proportional limit in a tensile test, a 2 in. gage length had elongated by 0.003 in. and the diameter of the standard specimen was smaller by 0.00026 in. The load measured by a load cell fitted to the tensile machine at the proportional limit was 5,000 lb. Determine the modulus of elasticity, Poisson's ratio and the proportional limit for this material.

5.33 A rod with a diameter of 1.00 in. and length of 8.0 ft undergoes an extension of 0.220 in. when subjected to an axial force of 54.0 kip. The diameter of the rod decreases by 0.0007 in. at this load. Determine the modulus of elasticity, Poisson's ratio and the shear modulus for the rod's material.

5.34 A standard tensile specimen was fitted with an extensometer with a 2.00-inch gage length and tested until failure. The force and extension measured during the test is presented in the table below. The diameter of the specimen at the fracture neck was 0.413 in. Analyze the data using a spreadsheet and determine the following quantities.

a. Modulus of elasticity
b. Yield strength (0.2% offset)
c. Ultimate strength
d. Fracture stress
e. Strain at fracture
f. Percent elongation
g. Percent reduction in area.

| Load (kip) | Extension (in.) | Load (kip) | Extension (in.) |
|---|---|---|---|
| 0 | 0 | 10.8 | 0.251 |
| 2.0 | 0.0021 | 11.2 | 0.282 |
| 4.0 | 0.0039 | 11.6 | 0.303 |
| 6.0 | 0.0058 | 11.6 | 0.318 |
| 7.0 | 0.0072 | 11.5 | 0.327 |
| 7.8 | 0.0084 | 11.2 | 0.351 |
| 8.3 | 0.0280 | 10.8 | 0.367 |
| 8.7 | 0.0530 | 10.1 | 0.394 |
| 9.2 | 0.0860 | 9.7 | 0.402 |
| 9.8 | 0.1450 | 9.6 | 0.420 |
| 10.4 | 0.2220 | | |

5.35 A standard tensile specimen 12.7 mm in diameter was fitted with an extensometer with a 50 mm gage length and tested until failure. The force and extension measured during the test is presented in the table below. The diameter of the specimen at the fracture neck was 9.0 mm. Analyze the data using a spreadsheet and determine the following quantities.

a. Modulus of elasticity
b. Yield strength (0.2% offset)
c. Ultimate strength
d. Fracture stress
e. Strain at fracture
f. Percent elongation
g. Percent reduction in area.

| Load (kN) | Extension (mm.) | Load (kN) | Extension (mm) |
|---|---|---|---|
| 0 | 0 | 35.0 | 3.54 |
| 5.0 | 0.010 | 37.0 | 4.52 |
| 10.0 | 0.019 | 38.0 | 5.53 |
| 15.0 | 0.029 | 38.0 | 6.18 |
| 20.0 | 0.042 | 36.0 | 6.66 |
| 25.0 | 0.048 | 33.0 | 7.35 |
| 26.0 | 0.50 | 30.0 | 7.86 |
| 27.0 | 0.95 | 27.0 | 8.42 |
| 29.0 | 1.50 | 23.0 | 8.78 |
| 31.5 | 2.21 | 19.0 | 9.32 |

# CHAPTER 6

# TRUSSES

## 6.1 INTRODUCTION

Trusses are structures commonly used to efficiently span relatively long distances. As such, they are employed in the design of bridges or the support for roofs of stadiums and both large and small buildings. A **truss** is defined as a large structure made of many smaller uniaxial members (bars, rods or cables) that are connected together to form a strong and stiff arrangement. The term truss is derived from the Middle English word *trusse*, which means, "bundle". Several of the geometries used in truss design are presented in Fig. 6.1.

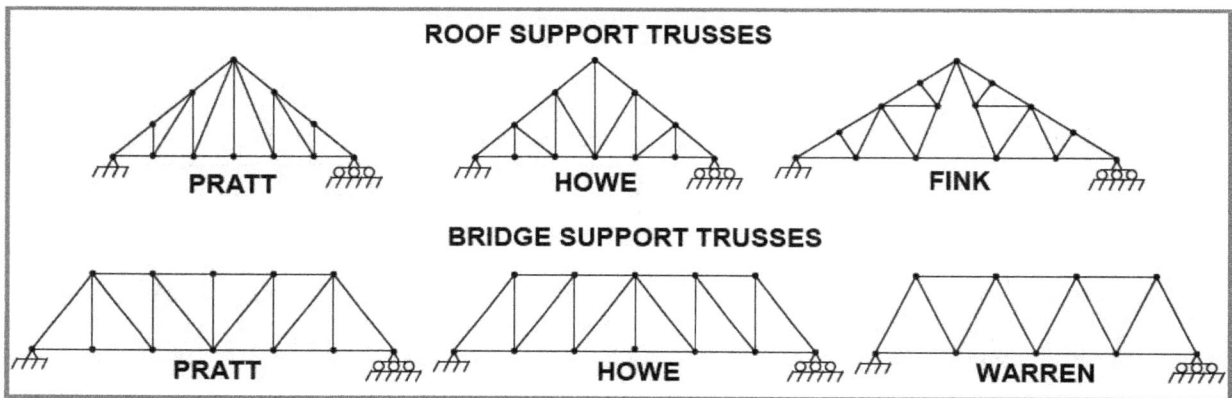

**ROOF SUPPORT TRUSSES**

PRATT     HOWE     FINK

**BRIDGE SUPPORT TRUSSES**

PRATT     HOWE     WARREN

Fig. 6.1 Different types of trusses used in the design of bridges and roof supports.

### 6.1.1 Stability

If we examine the geometry of the trusses presented in Fig. 6.1, it is clear that they are all made from straight members arranged to form an array of triangles. The triangular form is very important because it insures stability of the truss regardless of the direction of the applied forces. Let's examine the stability of a four-member frame with a rectangular arrangement of the members as defined in Fig. 6.2. The rectangular frame in Fig. 6.2a is fabricated with joints at the four corners A, B, C and D. If a horizontal force F is applied to this frame at point C, the joints tend to act as hinges[1], and the frame rotates as shown in Fig. 6.2b. If the force F is maintained as the frame rotates, the structure collapses.

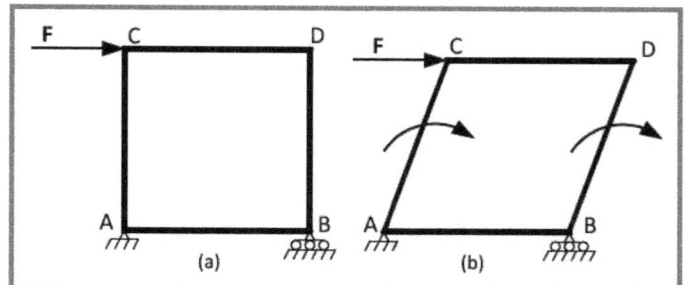

Fig. 6.2 A four-bar frame with a rectangular arrangement is unstable.

---

[1]Joints can be designed to resist rotation; however, it is usually more efficient and less costly to insure stability by adding an additional member to the structure that prevents rotation.

To stabilize the rectangular frame of Fig. 6.2, we add a fifth bar between points A and D as shown in Fig. 6.3. With the application of the horizontal force F, the joints will again tend to rotate. However, significant rotations of the structure require member AD to elongate by yielding. If member AD is designed with sufficient strength, the truss cannot deform and its stability is ensured.

Fig. 6.3 Converting the rectangular structure into a truss with triangular elements assures its stability.

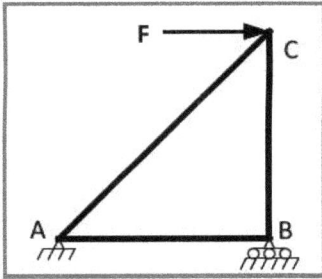

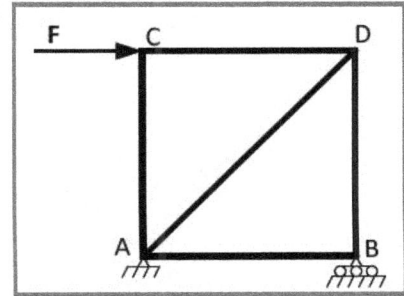

Fig. 6.4 Triangular structural elements are stable.

Note the addition of the member AD has converted the rectangular truss arrangement into one consisting of two triangles. The triangular element is the essence of stability of a structural arrangement. A simple three-member triangular truss is illustrated in Fig. 6.4 with a horizontal force F applied at point C. Even if the joints are made with frictionless pins (free to rotate), the structure remains stable. Stability is insured by the presence of member AC that must experience large deformations before rotation of member BC can occur.

Examination of the truss geometries presented in Fig. 1.6 shows the repeated use of the triangular elements to increase the size of the truss. With the addition of the new members, the number of joints used in fabricating the truss is also increased. For trusses fabricated from simple triangular elements, the number of joints and the number of members are related by:

$$n = 2k - 3 \qquad\qquad (6.1)$$

where n is the number of members and k is the number of joints.

## 6.1.2 The Truss Members

The members of a truss are long and thin. As such, they cannot support transverse (lateral) forces; consequently loads applied to the structure must be placed at the joints. The weight of most truss members is usually negligible in comparison to the applied forces. However, if the weight of a specific member is significant, because of an exceptionally high density or extremely long length, the weight of the member is divided by two and applied to its joints—thus accounting for the effect of the member's weight in the analysis.

The joints are usually constructed using a gusset plate to effect the connection of the various members that meet at a specified point. We illustrate a gusset plate used in fabricating a joint between four members in Fig. 6.5. The members are fastened to the gusset plate by riveting, bolting or welding. Even though the joint between two or more members may resist rotation and support moments, we assume that the joint is free to rotate. This assumption does not produce significant errors in the analysis if the individual truss members are long and flexible. Also this assumption permits us to treat each truss member as a **two-force member**—free of moments and transverse forces. For a two-force member, the force acting on the member must be transmitted through the axis of the member. Consequently, the direction of the force is established by the orientation of the member. The fact that the individual truss members are two-force members permits us to represent members, joints, and sections with the three different FBDs shown in Fig. 6.6.

Fig. 6.5 A joint formed with a gusset plate at the junction of four truss members.

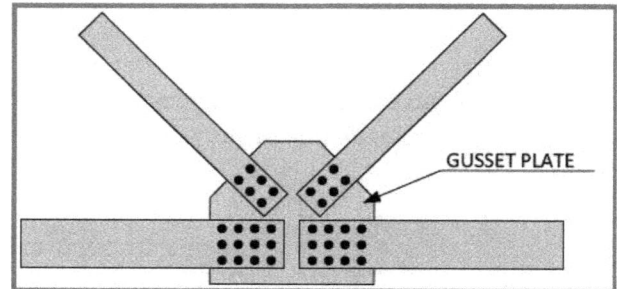

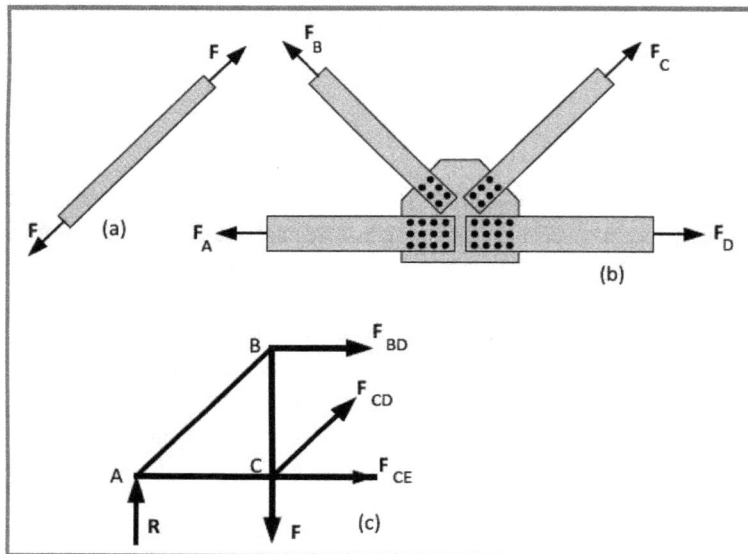

Fig. 6.6 FBDs for a truss: (a) an individual member from a truss, (b) a joint from a truss, and (c) a section of a truss.

## 6.2 METHOD OF JOINTS

In conducting a design analysis of a truss, we employ either the method of joints or the method of sections. Both methods employ a FBD of a portion of the truss as illustrated in Fig. 6.6. With the method of joints, we "remove" a specified joint from the truss and draw a free diagram of that joint including both internal and external forces that act on it. With the method of sections, we make a section cut to remove either the left or right side of the truss. We then prepare a FBD of one side or the other of the sectioned truss.

The method of joints is a simple extension of the material already covered in Chapters 3, 4 and 5. The analysis incorporates six steps:

1. Draw a FBD of the entire truss structure.
2. Apply the equations of equilibrium to solve for the reactions at the supports.
3. Select a joint and construct a FBD that includes both the internal and external forces.
4. Apply the equations of equilibrium to solve for the forces in the members that are connected at the subject joint. Since we have assumed that the joint acts like a particle, the joint is

represented as a point and equilibrium is satisfied if $\Sigma F = 0$. It is not necessary to consider $\Sigma M = 0$.

5. Determine the stresses in the individual members by using $\sigma = P/A$, and note if the stresses are tensile or compressive.

6. Compare the stress in each member with the strength of the material used to fabricate the truss and then establish the margin of safety for each member.

To illustrate the procedure for a design analysis of a simple truss, consider the following example of a Howe truss.

## EXAMPLE 6.1

Consider the Howe truss shown in Fig. E6.1. The total span of the truss is $S_p = 4s = 4 \times 12 = 48$ ft, and the height $h = 16$ ft. External forces $F_C$, $F_E$ and $F_G$ are applied at joints C, E and G. Determine the margin of safety for truss members AB and AC. The following parameters are necessary for a numerical solution for the margin of safety.

The forces are $F_C = 100,000$ lb, $F_E = 150,000$ lb, and $F_G = 100,000$ lb

The cross sectional areas of the bars employed in the construction of the truss are $A_{AC} = 4$ in². and $A_{AB} = 6$ in². These bars are fabricated from hot rolled mild steel with a yield strength of 38,000 psi.

Fig. E6.1

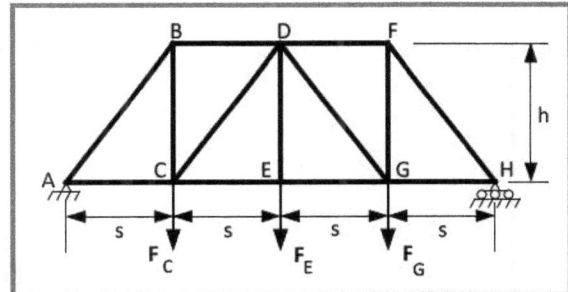

## Solution:

**Step 1:** Draw a FBD of the entire truss structure.

Since it is necessary to determine the reactions at the left and right hand supports of the truss, the supports are removed and replaced with reaction forces $R_L$, $R_R$ and $R_x$ in the FBD shown In Fig. E6.1a:

Fig. E6.1a

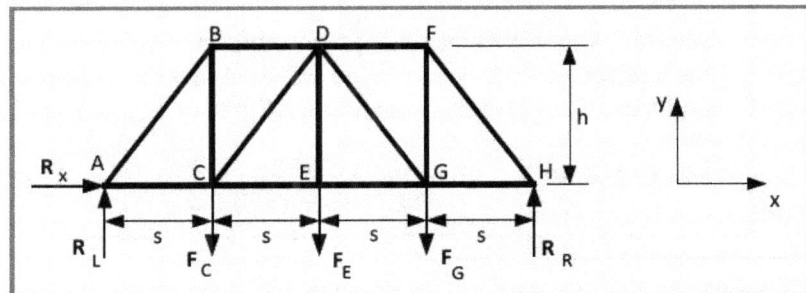

**Step 2:** Apply the equations of equilibrium to solve for the reactions at the supports. It is immediately evident from $\Sigma F_x = 0$ that the reaction force at the left support $R_x = 0$.

Consider the sum of the moments about point A and write:

$$\Sigma M_A = 4sR_R - 3sF_G - 2sF_E - sF_C = 0 \tag{a}$$

Solving Eq. (a) for $R_R$ yields:

$$R_R = (1/4)(3F_G + 2F_E + F_C) = (1/4)(300 + 300 + 100) = 175 \text{ kip} \tag{b}$$

From equilibrium we write:

$$\Sigma F_y = R_L + R_R - F_C - F_E - F_G = 0$$

$$R_L = 100 + 150 + 100 - 175 = 175 \text{ kip} \tag{c}$$

Both $R_L$ and $R_R$ are equal to 175 kip. The equality of the two reactions is expected because the geometry and the loading of the truss are both symmetrical.

**Step 3:** Select a joint and construct a FBD that includes both the internal and external forces acting at that joint. We choose the joint at point A because members AB and AC, the subject of the analysis, meet at this joint. The FBD is presented in Fig. E6.1b:

Fig. E6.1b

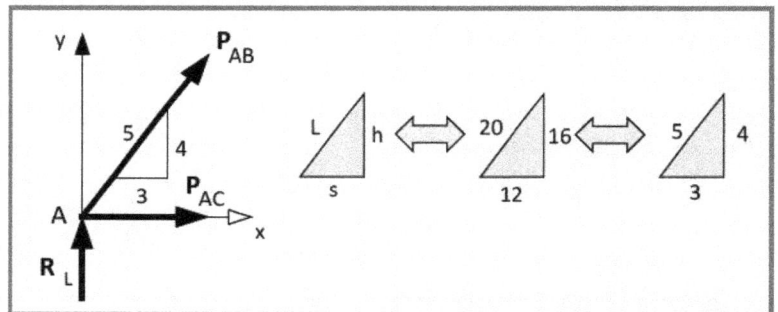

Note, the dimensions of the span and height of the primary triangular element in the truss provides the orientation of the internal force $P_{AB}$ in member AB. In this FBD we have used:

- The symbol P to represent internal forces.
- The symbol R to represent reaction (external) forces.
- Subscripts to identify the truss members.
- The internal forces $P_{AB}$ and $P_{AC}$ as both positive (tensile)[2].

**Step 4:** Apply the equations of equilibrium to solve for the forces in the members that connect at the subject joint. Since we have assumed that the joint acts like a pin, the joint is represented as a point and equilibrium is satisfied if $\Sigma F = 0$, or $\Sigma F_x = 0$ and $\Sigma F_y = 0$.

Let's begin with:

$$\Sigma F_y = R_L + (4/5)\, P_{AB} = 0$$

---

[2]It is usual practice to assume that the unknown forces in truss members are tensile.

Solve this relation for $P_{AB}$ by using the results of Eq. (c) to obtain:

$$P_{AB} = -(5/4)\ R_L = -(5/4)\ 175 = -218.8 \text{ kip} \tag{d}$$

The negative sign for $P_{AB}$ indicates that it is a compressive force. Our initial assumption indicating this force was tensile (+) was in error. This error does not cause a problem—we note the sign, declare that member AB is in compression, and treat $P_{AB}$ as a negative quantity in subsequent steps in the analysis.

Next, write the equilibrium relation for the force components in the x direction.

$$\Sigma F_x = (3/5)P_{AB} + P_{AC} = 0$$

$$P_{AC} = (-3/5)P_{AB} = -(3/5)(-218.8) = +131.3 \text{ kip} \tag{e}$$

The plus sign for $P_{AC}$ indicates that the internal force in truss member AC is tensile. Our assumption about the direction of the force $P_{AC}$ in constructing the FBD for joint A was correct.

**Step 5:** Determine the stresses in the individual members of the truss from $\sigma = P/A$, and note if the stresses are tensile or compressive.

For member AB and AC, we write:

$$\sigma_{AB} = P_{AB}\ /A_{AB} = -218.8 \text{ kip} /6 \text{ in}^2 = -36.46 \text{ ksi} \tag{f}$$

$$\sigma_{AC} = P_{AC}\ /A_{AC} = +131.3 \text{ kip} /4 \text{ in}^2 = +32.82 \text{ ksi} \tag{g}$$

The minus sign for $\sigma_{AB}$ indicates that the stress in member AB is compressive[3]. The stress in member AC is tensile as indicated by the positive sign.

**Step 6:** Compare the stress in each member to the strength of the material used to fabricate the truss and then establish the margin of safety for the structure.

From Eq. 4.12, we determine the safety factor for members AB and AC as:

$$SF_{AB} = S_y/\sigma_{AB} = 38.0/36.46 = 1.042 \tag{h}$$

$$SF_{AC} = S_y/\sigma_{AC} = 38.0/32.82 = 1.158 \tag{i}$$

Finally, let's determine the margin of safety MOS according to its definition:

$$MOS = SF - 1 \tag{6.2}$$

Then for members AB and AC, we determine:

$$MOS_{AB} = 1.042 - 1 = 0.042 = 4.2\% \tag{j}$$

---

[3]Since the stress in member AB is compressive, there is a possibility that the member may fail by buckling. Another analysis must be conducted to determine if the critical buckling load on this member has been exceeded. Methods for determining the critical buckling load on bars loaded in axial compression will be provided in a later course.

$$\text{MOS}_{AC} = 1.158 - 1 = 0.158 = 15.8\% \qquad (k)$$

Okay! We have completed our solution, and it is mandatory to interpret the results for the safety of both members AB and AC. **Clearly, these margins of safety are too small.** If the external loads are increased by only five percent, member AB will fail by yielding and begin to deform. This deformation will change the shape of the structure and redistribute the internal forces. The structure is in significant danger—collapse is pending.

        **When designing structures, it is common to incorporate safety factors of two or three which provide margins of safety of 100 to 200%. These safety factors are to accommodate for uncertainties in the construction and operation over the life of the structure.** Uncertainties include:

- A possible increase in loading, deliberate or accidental, any time during the operation of the structure.
- Variations in the strength of the materials used in construction.
- Effect of fatigue (cyclic loading) in generating flaws (cracks) in the structure.
- Effects of corrosion on reducing section size or on inducing crack initiation in one or more members of the structure.

In this example problem, it is necessary to increase either the area of the cross section of members AB and AC or to increase the yield strength of the material from which the members are fabricated.

## EXAMPLE 6.2

Determine the internal forces in members AB, AC, BC, BD, BE, DE, and DF of the truss defined in Fig. E6.2. Also determine the margin of safety for the highest stressed member. The external forces applied at the joints C, E, G, K, and I are all equal to 100 kip. The cross sectional area of all of the members used to construct the truss is equal to 10 in². The material used in fabricating the truss is hot rolled steel with a yield strength of 54 ksi. Note, s and h are 10 and 20 ft respectively.

Fig. E6.2

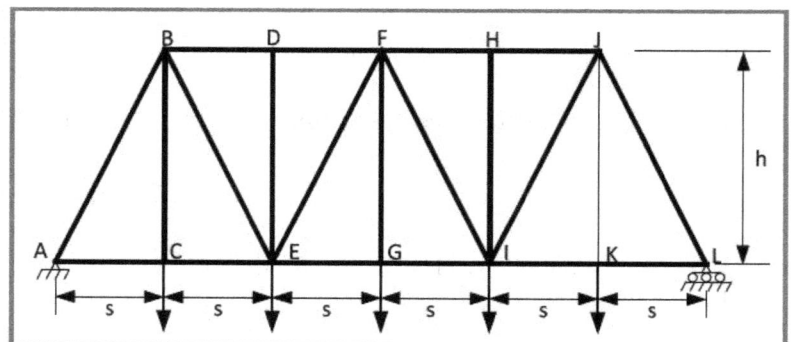

**Solution:**

**Step 1:** Construct a FBD of the entire truss:

Fig. E6.2a

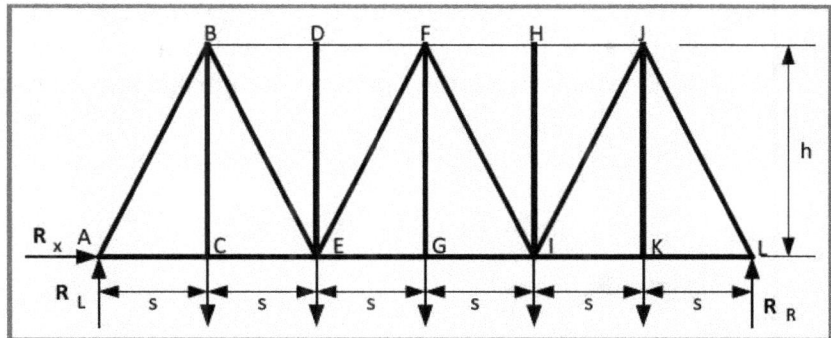

**Step 2:** Solve for the reactions at the supports.

Again it is clear from $\Sigma F_x = 0$ that $R_x = 0$.

Inspection of the structure shows that it is symmetric with respect to both geometry and loading. For this reason, it is evident that $R_L = R_R = R$.

$$\Sigma F_y = 2R - 5F = 0$$

$$R = (5/2)100 = 250 \text{ kip} \qquad\qquad (a)$$

**Step 3:** Select the joints necessary for solution of the unknown forces and prepare FBDs as shown in Fig. E6.2b.

Fig. E6.2b

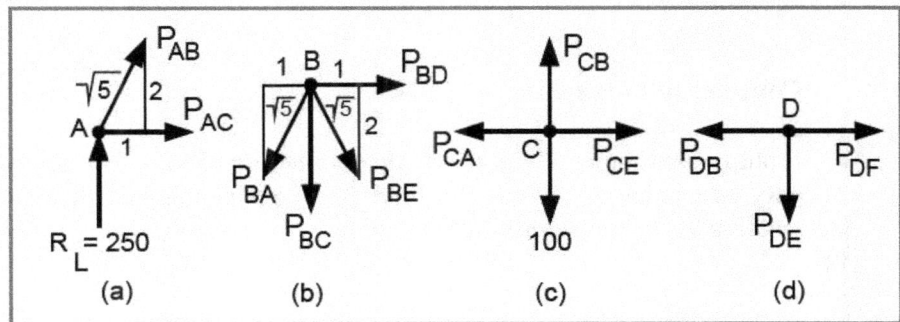

We have selected joints A, B, C and D because the forces occurring at these joints in the FBDs correspond to the unknown forces listed in the problem statement. Note, we have assumed all of the internal forces to be tension as is the usual practice. Also, the first subscript gives the letter of the joint described in the FBD. The second subscript gives the direction of the force. Since the truss members are two-force members in equilibrium, $P_{AB} = P_{BA}$, etc.

**Step 4:** Use equilibrium relations as necessary to solve for the unknown forces.

We have four FBDs—one for joint A, B, C and D. It is possible to write a total of eight relevant equilibrium relations if they are required. Let's write equilibrium equations beginning with the FBD of joint A, then B, C and D.

**For joint A:**

$$\Sigma F_y = R_L + (2/\sqrt{5})\, P_{AB} = 0$$

$$P_{AB} = -(\sqrt{5}/2)R_L = -(\sqrt{5}/2)(250) = -279.5 \text{ kip} \qquad (b)$$

Note, the minus sign for $P_{AB}$ indicates the internal force in member AB is compressive.

$$\Sigma F_x = P_{AC} + (1/\sqrt{5})\, P_{AB} = 0$$

$$P_{AC} = -(1/\sqrt{5})P_{AB} = -(1/\sqrt{5})(-279.5) = 125 \text{ kip} \qquad (c)$$

**For joint B:**

Examining the FBD for joint B shows that four forces act at this point. We know only one of these forces, namely $P_{BA}$. The remaining three forces are unknown quantities. Since only two meaningful equilibrium relations may be written for a concurrent force system, the analysis of joint B is statically indeterminate. The analysis of joint B is possible only if we reduce the number of unknown forces to two. With the reduction of forces at joint B as a motivation, let's move onto joint C.

**For joint C:**

$$\Sigma F_x = P_{CE} - P_{CA} = 0$$

$$P_{CE} = P_{CA} = P_{AC} = 125 \text{ kip} \qquad (d)$$

$$\Sigma F_y = P_{CB} - 100 = 0$$

$$P_{CB} = 100 \text{ kip} \qquad (e)$$

**Consider joint B again:**

From the analysis of Joint C, we determined that $P_{CB} = 100$ kip. Since $P_{CB}$ and $P_{AB}$ are known, only two unknown forces ($P_{BE}$ and $P_{BD}$ ) act at joint B; therefore, we may proceed with the solution by writing equilibrium relations:

$$\Sigma F_y = -P_{BC} - (2/\sqrt{5})(P_{BA} + P_{BE}) = 0$$

$$P_{BE} = -(\sqrt{5}/2)P_{BC} - P_{BA} = -(\sqrt{5}/2)100 - (-279.5) = +167.7 \text{ kip} \qquad (f)$$

$$\Sigma F_x = P_{BD} - (1/\sqrt{5})(P_{BA} - P_{BE}) = 0$$

$$P_{BD} = (1/\sqrt{5})(P_{BA} - P_{BE}) = (1/\sqrt{5})[-279.5 - 167.7] = -200.0 \text{ kip} \qquad (g)$$

**Finally, consider joint D:**

$$\Sigma F_y = P_{DE} = 0 \qquad (h)$$

$$\Sigma F_x = - P_{DB} + P_{DF} = 0$$

$$P_{DB} = P_{DF} = - 200.0 \text{ kip} \tag{i}$$

Since $P_{DE} = 0$, member DE of the truss structure is a zero-force member. We will discuss the arrangement of members at a joint in a truss that results in zero-force members later in this chapter.

Let's summarize the results of the analysis in Table 6.1.

**Table 6.1**
**Summary of forces and stresses in Example 6.2**

| MEMBER | FORCE kip | AREA in$^2$ | STRESS ksi | SAFETY FACTOR | MARGIN OF SAFETY |
|--------|-----------|-------------|------------|---------------|------------------|
| AB | 279.5 C | 10 | 27.95 C | 1.93 | 0.93 |
| AC | 125.0 T | 10 | 12.50 T | 4.32 | 3.32 |
| BC | 100.0 T | 10 | 10.00 T | 5.40 | 4.40 |
| BD | 200.0 C | 10 | 20.00 C | 2.70 | 1.70 |
| BE | 167.7 T | 10 | 16.77 T | 3.22 | 2.22 |
| DE | 0 | 10 | 0 | ∞ | ∞ |
| DF | 200.0 C | 10 | 20.00 C | 2.70 | 1.70 |

**Step 5:** Solve for the stresses.

Let's determine the stresses in member AB of the truss.
$$\sigma_{AB} = P_{AB}/A_{AB} = -279.5/10 = -27.95 \text{ ksi}$$

The results for the stresses in the other members in the truss are shown in Table 6.1.

**Step 6:** Determine the safety factor and the margin of safety for the most highly stressed member.

Inspection of Table 6.1 indicates that member AB is the most highly stressed member in the truss. We determine its safety factor as:

$$SF = S_y/\sigma$$

For member AB:
$$SF = S_y/\sigma = 54/27.95 = 1.93$$

$$MOS = SF - 1 = 1.93 - 1 = 0.93 = 93\%$$

Finally, let's interpret the results that are presented in Table 6.1 from a designer's viewpoint. First, we note a significant difference in the stresses from one member to another. It is clear that the choice of 10 in$^2$ for the cross sectional areas of all of the members was not appropriate. A more uniformly safe design would have been possible if the areas had been adjusted for the different forces imposed on the different members. Area adjustment is possible, but it is preferable to use structural members with standard section sizes. It is not usually economically feasible to specify sectional areas that differ from the areas available with standard size members.

Second, the safety factor and the margin of safety for members AB, BD and DF are probably too low. We normally seek a safety factor of 2 to 3 to account for contingencies that have been described previously in Example 6.1.

Third, two **zero-force members** have been incorporated in the design: members DE and IH carry no force. The designer may have added them to the truss structure for aesthetic reasons, to provide backup if certain diagonal members failed by yielding, or in anticipation of loads not considered in this example. However, if a good reason for using these members does not exist, they should be eliminated. They add no strength to the structure and increase its cost, weight, and construction time.

## EXAMPLE 6.3

A scissors truss illustrated in Fig. E6.3 is loaded with forces $F_1$, $F_2$ and $F_3$ at the joints located at points B, C and D, respectively. If the truss is fabricated from HR 1020 steel with a yield strength of 290 MPa, determine the size of members AB, AF, BC and BF. The safety factor is specified as 2.6. The forces and dimensions are:

$$F_1 = 10 \text{ kN}, F_2 = 12 \text{ kN}, F_3 = 15 \text{ kN} \quad \Rightarrow\Rightarrow \quad h = 5 \text{ m}, s = 8 \text{ m}$$

Fig. E6.3

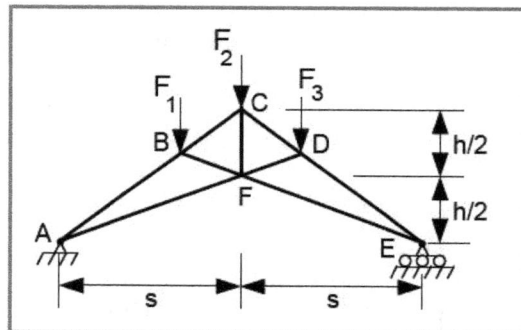

**Solution:**

**Step 1:** Construct a FBD of the entire truss as indicated in Fig. E6.3a. We remove the supports and replace them with the reactions for a pivot and roller. However, we encounter a problem in dimensioning the FBD. The dimensions locating points B and D (the points of application of forces $F_1$ and $F_3$) are not given explicitly in Fig. E6.3. It will be necessary to consider the geometry of the truss and compute the x coordinate of these points. Let's begin by forming the triangle ADE as shown in Fig. E6.3b. Note that the member EF has been removed to make the triangle more apparent.

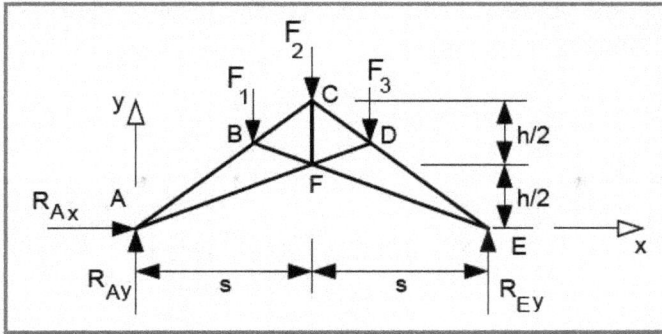

Fig. E6.3a

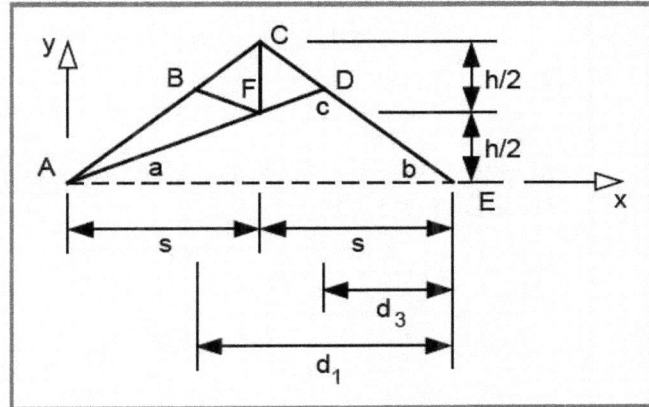

Fig. E6.3b

Solving for the angles a, b and c in the triangle ADE gives

$$a = \tan^{-1}(h/2s) = \tan^{-1}(5/16) = 17.35°$$
$$b = \tan^{-1}(h/s) = \tan^{-1}(5/8) = 32.00°$$
$$c = 180° - a - b = 180° - 17.35° - 32.00° = 130.65°$$

The law of sines is employed to write:

$$\sin(c)/2s = \sin(a)/DE$$

Solving this relation for DE gives:

$$DE = (2s)[\sin(a)/\sin(c)] = 16[\sin(17.35°)/\sin(130.65)] = 6.289 \text{ m}$$

The distances $d_3$ and $d_1$ locating the points of application of forces $F_3$ and $F_1$ relative to point E are given by:

$$d_3 = DE \cos(b) = 6.289 \cos(32.00°) = 5.333 \text{ m}$$

Because of geometric symmetry, we may write:

$$d_1 = 2s - d_3 = 16 - 5.333 = 10.67 \text{ m}$$

**Step 2:** Solve for the reactions at the supports.

Again it is clear from $\Sigma F_x = 0$ that $R_{Ax} = 0$.

Inspection of the structure shows that it is not symmetric with respect to the loading. For this reason, $R_{Ay} \neq R_{Ey}$. Let's consider moments about point E and write:

$$\Sigma M_E = -2sR_{Ay} + d_1 F_1 + s F_2 + d_3 F_3 = 0$$

Solving for $R_{Ay}$ gives:

$$R_{Ay} = (d_1 F_1 + s F_2 + d_3 F_3)/(2s) = [(10.67)(10) + (8)(12) + (5.333)(15)]/16 = 17.67 \text{ kN}$$

Let's solve for $R_{Ey}$ by writing:

$$\Sigma F_y = R_{Ay} + R_{Ey} - F_1 - F_2 - F_3 = 0$$

$$R_{Ey} = 10 + 12 + 15 - 17.67 = 19.33 \text{ kN}$$

**Step 3:** Select the joints necessary for solution of the unknown forces and prepare FBDs as shown in Fig. E6.3c.

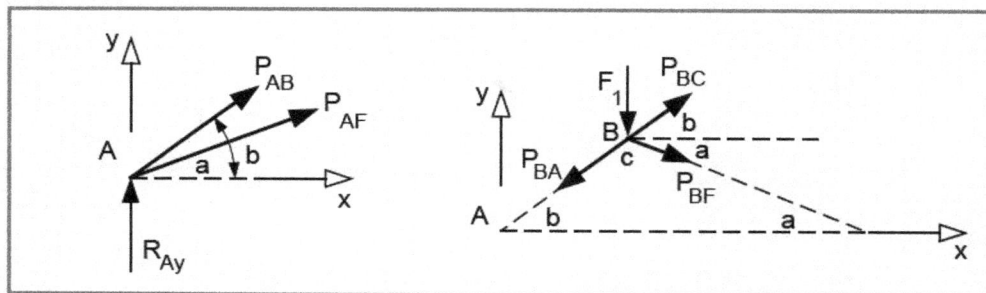

Fig. E6.3c

We have selected joints A, and B because the forces occurring at these joints in the FBDs correspond to the unknown forces listed in the problem statement. Note, we have assumed all of the internal forces to be tension as is the usual practice. Also, the first subscript gives the letter of the joint described in the FBD. The second subscript gives the direction of the force. Since the truss members are two-force members in equilibrium, $P_{AB} = P_{BA}$, etc.

**Step 4:** Use equilibrium relations as necessary to solve for the unknown forces.

We have two FBDs—one for joint A and another for joint B. It is possible to write a total of four relevant equilibrium relations if they are required. Let's begin to write equilibrium equations beginning with the FBD of joint A.

**For joint A:**
$$\Sigma F_y = R_{Ay} + P_{AB} \sin (b) + P_{AF} \sin (a) = 0$$

$$P_{AB} = -[R_{ay} + P_{AF} \sin (a)] / \sin (b) = -[17.67 + P_{AF} \sin (17.35°)]/\sin (32.00°)$$

$$P_{AB} = -33.34 \text{ kN} - 0.5627 P_{AF} \tag{a}$$

$$\Sigma F_x = P_{AB} \cos (b) + P_{AF} \cos (a) = 0$$

$$P_{AF} = -P_{AB} \cos (b)/ \cos (a) = -P_{AB} \cos (32.00°)/ \cos (17.35°) = -0.8884 P_{AB} \tag{b}$$

Combining the results from Eqs. (a) and (b) yields:

$$P_{AB} = -66.67 \text{ kN} \qquad \Rightarrow\Rightarrow \qquad P_{AF} = 59.23 \text{ kN} \qquad (c)$$

Note, the minus sign for $P_{AB}$ indicates the internal force in member AB is compressive.

**For joint B:**

Examining the FBD for joint B shows that four forces act at this point. We know two of these forces, namely $P_{BA}$ and $F_1$. The remaining two forces are unknown quantities that are determined from the two equilibrium relations that apply for a concurrent force system.

$$\Sigma F_x = P_{BC} \cos (b) + P_{BF} \cos (a) - P_{BA} \cos (b) = 0$$

$$P_{BC} \cos (32.00°) + P_{BF} \cos (17.35°) - P_{BA} \cos (32.00°) = 0$$

$$(0.8480) P_{BC} + (0.9545)P_{BF} - (-66.67)(0.8480) = 0$$

$$(0.8480) P_{BC} + (0.9545)P_{BF} + 56.54 = 0 \qquad (d)$$

$$\Sigma F_y = P_{BC} \sin (b) - P_{BF} \sin (a) - P_{BA} \sin (b) - F_1 = 0$$

$$P_{BC} \sin (32.00°) - P_{BF} \sin (17.35°) - P_{BA} \sin (32.00°) - 10.0 = 0$$

$$(0.5299) P_{BC} - (0.2982)P_{BF} - (-66.67)(0.5299) - 10.0 = 0 \qquad (e)$$

Solving Eqs. (d) and (e) for the two unknowns gives:

$$P_{BF} = -11.20 \text{ kN} \qquad \Rightarrow\Rightarrow \qquad P_{BC} = -54.10 \text{ kN} \qquad (f)$$

The minus signs in these results indicate that both members BC and BF are subjected to internal compression forces.

**Step 5:** Solve for the allowable design stresses in the members.

Let's determine the allowable design stresses from Eq. (4.12) as:

$$\sigma_{design} = S_y/ \text{ } SF_y = 290/2.6 = 111.5 \text{ MPa} \qquad (g)$$

**Step 6:** Determine the areas required to limit the design stresses to the allowable level determined in Eq. (g).

$$A = P/\sigma_{dessign} = P/111.5 \qquad (h)$$

The results from Eq. (h) are presented in Table 6.2.

**Table 6.2**
**Results showing required dimensions for scissors truss members**

| Member | P (kN) | A (mm) | $d_1$ (mm) | $d_2$ (mm) |
|--------|--------|--------|-----------|-----------|
| AB | −66.67 | 597.9 | 27.59 | 30 |
| AF | 59.23 | 531.2 | 26.01 | 30 |
| BC | −54.10 | 485.2 | 24.86 | 30 |
| BF | − 11.20 | 100.4 | 11.31 | 15 |

Inspection of Table 6.2 indicates the areas for each member. We have assumed that rods with a diameter d are to be employed in fabricating the scissors truss. In Table 6.2, the diameter $d_1$ is the computed value. Since rods of these sizes are not commercially available, we have increased the diameter to $d_2$ that corresponds to standard sizes that are available without special order.

## 6.3 ZERO-FORCE MEMBERS

Zero-force members are usually easy to identify in a structure if we examine all of the joints in the structure that do not carry externally applied loads. Let's consider two examples of joints with a zero-force member as shown in Fig. 6.7.

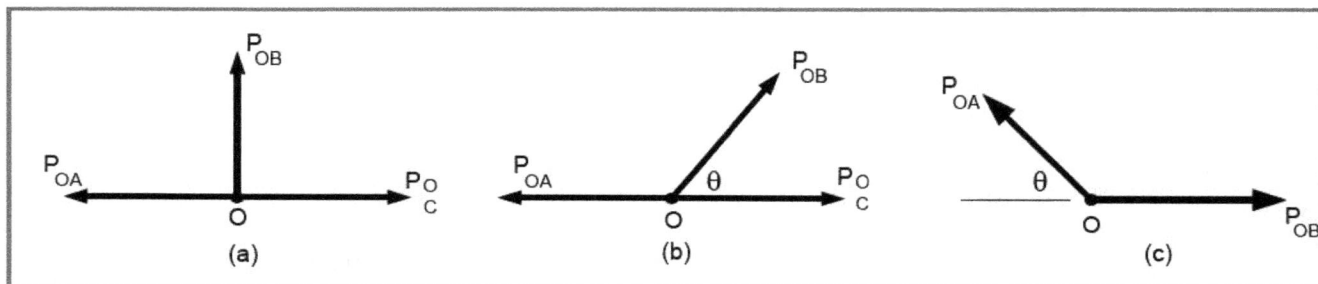

Fig. 6.7 Joint O with a zero-force member OB.

Let's begin by considering the equilibrium of joint O in Fig. 6.7a. Write $\Sigma F_y = 0$ and it is clear that:

$$\Sigma F_y = P_{OB} = 0$$

because the forces $P_{OA}$ and $P_{OC}$ do not have force components in the y direction. Also, from $\Sigma F_x = 0$, we may write that:

$$\Sigma F_x = - P_{OA} + P_{OC} = 0$$

$$P_{OA} = P_{OC}$$

Next, examine joint O in Fig. 6.7b. While force $P_{OB}$ is inclined at some angle $\theta$ with respect to the x-axis, its component in the y direction is zero because neither $P_{OA}$ nor $P_{OC}$ have components in the y direction. Since $P_{OB}$ is transmitted through a two-force member, the component of $P_{OB}$ in the x direction is also zero. Thus, OB is a zero-force member.

Finally, consider the joint in Fig. 6.7c. Summing forces in the y direction yields $P_{OA} = 0$. Then summing forces in the x direction gives $P_{OB} = 0$. In this case both forces at joint O are zero. The three cases illustrated in Fig. 6.7 provide the only three configurations where zero force members can be identified by inspection.

## EXAMPLE 6.4

For the truss defined in Fig. E6.4, determine the forces, stresses, and safety of members GI, HI, JI and HJ. The dimensions h = s = 5 m. The yield strength of the structural steel from which the truss is fabricated is 270 MPa. The following cross sectional areas have been specified for these members:

$$A_{GI} = 2000 \text{ mm}^2, A_{HI} = 2500 \text{ mm}^2, A_{JI} = 1200 \text{ mm}^2, \text{ and } A_{HJ} = 1200 \text{ mm}^2$$

Fig. E6.4

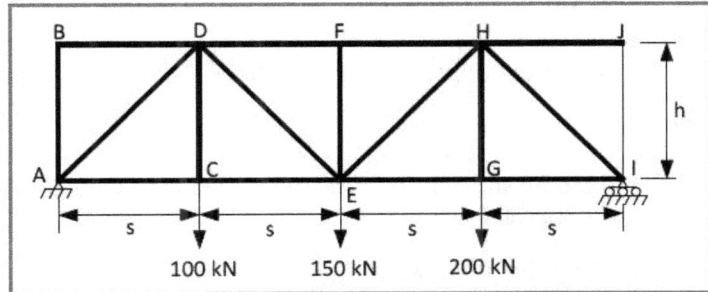

## Solution:

**Step 1:** Draw the FBD to permit solution for the unknown reaction forces as illustrated in Fig. E6.4a.

Fig. E6.4a

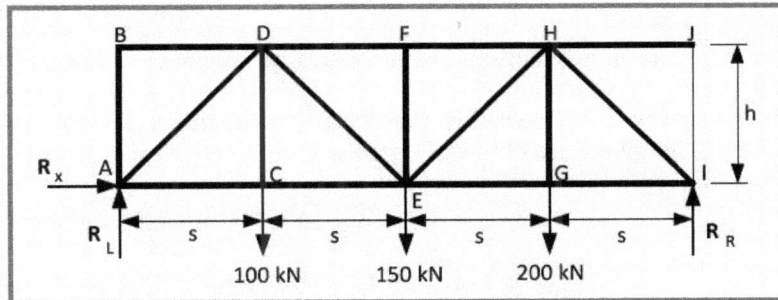

**Step 2:** Use the equilibrium relations to solve for the reaction forces $R_x$, $R_R$ and $R_L$.

$$\Sigma M_A = 4sR_R - 3s(200) - 2s(150) - s(100) = 0$$

$$R_R = \tfrac{1}{4}(600 + 300 + 100) = 250 \text{ kN} \qquad\qquad (a)$$

$$\Sigma F_y = R_L + R_R - 100 - 150 - 200 = 0$$

$$R_L = 450 - 250 = 200 \text{ kN} \qquad\qquad (b)$$

$$\Sigma F_x = R_x = 0$$

**Step 3:** Select joints J and I and construct FBDs of each showing the internal and external forces acting at these joints as shown in Fig. 6.4b.

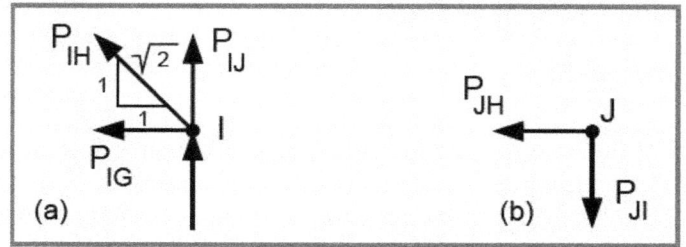

Fig. 6.4b

**Step 4:** Use equilibrium relations as necessary to solve for the unknown forces.

**Consider joint J.**

It is evident by inspection of the FBD for the joint at point J that:

$$P_{JH} = P_{JI} = 0 \qquad (c)$$

Both JI and JH are zero-force members that do not contribute to the strength or rigidity of the structure.

**For joint I,** we write:

$$\Sigma F_y = 250 + (1/\sqrt{2})P_{IH} = 0$$

$$P_{IH} = -(\sqrt{2})(250) = -353.6 \text{ kN} \qquad (d)$$

$$\Sigma F_x = -P_{IG} - (1/\sqrt{2})P_{IH} = 0$$

$$P_{IG} = -(1/\sqrt{2})P_{IH} = -(1/\sqrt{2})(-353.6) = 250 \text{ kN} \qquad (e)$$

The analysis of the forces in the four members defined in the problem statement is complete.

**Step 5:** Determine the stresses in member IH and IG. Note, the stresses in the zero-force members are obviously zero.

$$\sigma_{IH} = P_{IH}/A_{IH} = -353.6 \times 10^3/2500 = -141.4 \text{ N/mm}^2 = -141.4 \text{ MPa} \quad (f)$$

$$\sigma_{IG} = P_{IG}/A_{IG} = 250 \times 10^3/2000 = 125 \text{ N/mm}^2 = 125 \text{ MPa} \qquad (g)$$

**Step 6:** Compute the safety factor and the margin of safety for these two members.

For member IH:

$$SF = S_y/\sigma = 270/141.4 = 1.91$$

$$MOS = SF - 1 = 1.91 - 1 = 0.91 = 91\%$$

For member IG

$$SF = S_y/\sigma = 270/125 = 2.16$$

$$MOS = SF - 1 = 2.16 - 1 = 1.16 = 116\%$$

An examination of the results indicates that the zero-force members contribute nothing to the strength and rigidity of the structure. They should be removed to reduce the weight and cost of the structure unless there is another compelling reason for their inclusion in the design. The other two members of the truss (IG and IH) appear to be properly sized with safety factors of about two. One member IH is loaded in compression and its resistance to buckling must be ascertained in a subsequent analysis.

## 6.4 METHOD OF SECTIONS

The method of sections, like the method of joints is based on construction of a FBD and the subsequent application of the equations of equilibrium. The difference is in the selection of the portion of the truss used for the FBD. With the method of sections, we cut the truss and remove a complete section of it for the FBD. A typical example of a section cut through a truss and the two FBDs that are produced is shown in Fig. 6.8. The section cut passes through three members of the truss—BD, CD and CE. We have the option of examining either the left or right side of the truss. FBDs of both sides are shown in Fig. 6.8. At the location of the section cut, the internal forces in the members are displayed in the FBDs.

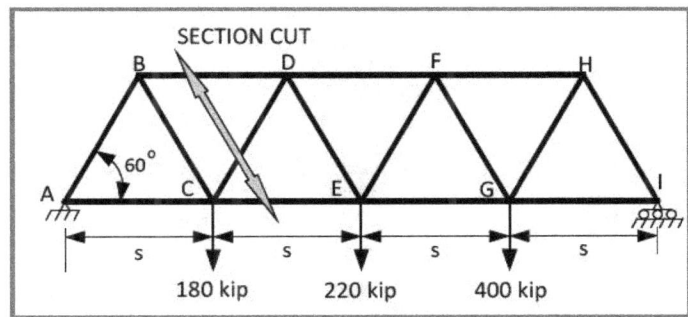

Fig. 6.8 FBDs resulting from the section cut of the truss.

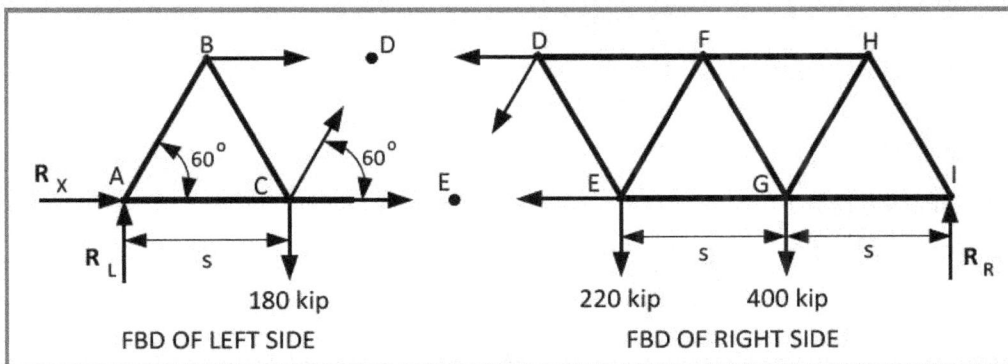

The FBDs provide a guide in writing the appropriate equilibrium relations. For the sections of the truss, either the right side or the left, we recognize the system of forces is planar but not concurrent. Accordingly, we write the following three equilibrium equations:

$$\Sigma F_x = 0; \qquad \Sigma F_y = 0; \qquad \text{and} \qquad \Sigma M_O = 0$$

and solve for three unknown internal forces in members of the truss that have been exposed by the section cut. Let's consider an example to demonstrate the procedure for the method of sections. You will note that we follow a similar six-step process as described for the method of joints.

## EXAMPLE 6.5

Consider the equilateral triangular section truss defined in Fig. 6.8, and determine the margin of safety for members BD, CD and CE. The span s of each triangular section is 20 ft and the yield strength of the structural steel used in fabrication is 42 ksi. The cross sectional areas of the three members are $A_{BD} = 20$ in$^2$, $A_{CD} = 12$ in$^2$ and $A_{CE} = 24$ in$^2$.

**Solution:**

**Step 1:** Construct a FBD of the entire truss structure as shown in Fig. E6.5.

Fig. E6.5

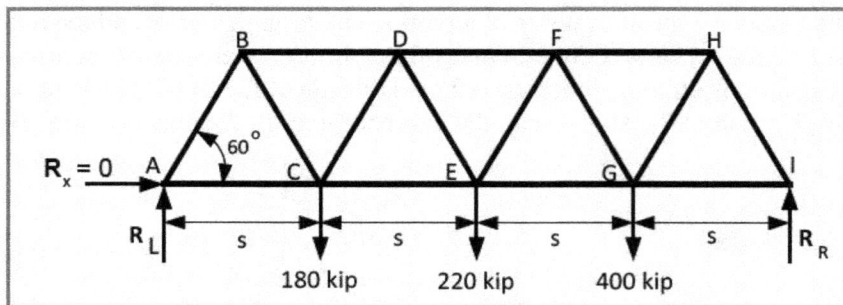

**Step 2:** Apply the equilibrium relations to determine the reactions at the supports.

$$\Sigma F_x = R_x = 0$$

$$\Sigma M_A = (4s)R_R - (3s)(400) - (2s)(220) - (s)(180) = 0$$

$$R_R = \tfrac{1}{4}(1200 + 440 + 180) = 455 \text{ kip} \qquad (a)$$

$$\Sigma F_y = R_L - 180 - 220 - 400 + 455 = 0$$

$$R_L = 345 \text{ kip} \qquad (b)$$

**Step 3:** Select a portion of the truss and make a section cut through the three members BD, CD and CE. Construct a FBD of that portion as shown in Fig. E6.5a. In this example, the left side is used for the FBD because it is easier to analyze than the right side.

Fig. E6.5a

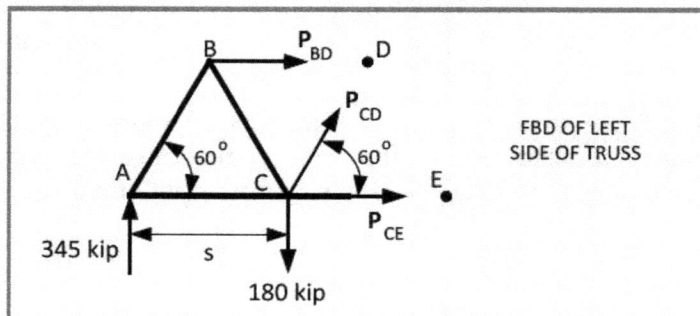

FBD OF LEFT SIDE OF TRUSS

**Step 4:** Apply the equilibrium relations to solve for the forces $P_{BD}$, $P_{CD}$ and $P_{CE}$.

$$\Sigma M_C = -345(s) - P_{BD}(s) \sin (60°) = 0$$

$$P_{BD} = -345/0.8660 = -398.4 \text{ kip} \qquad (c)$$

$$\Sigma F_y = 345 - 180 + P_{CD} \sin(60°) = 0$$

$$P_{CD} = (180 - 345)/0.8660 = -190.5 \text{ kip} \qquad (d)$$

$$\Sigma F_x = P_{BD} + P_{CD} \cos(60°) + P_{CE} = 0$$

$$P_{CE} = -(-398.4) - (-190.5)(0.5) = 493.6 \text{ kip} \qquad (e)$$

**Step 5:** Determine the stresses in the members.

$$\sigma_{BD} = P_{BD}/A_{BD} = -398.4/20 = -19.92 \text{ ksi} \qquad (f)$$

$$\sigma_{CD} = P_{CD}/A_{CD} = -190.5/12 = -15.88 \text{ ksi} \qquad (g)$$

$$\sigma_{CE} = P_{CE}/A_{CE} = 493.6/24 = 20.57 \text{ ksi} \qquad (h)$$

**Step 6:** Determine the safety factors and margins of safety for the three members.

For member BD:

$$SF = S_y/\sigma = 42/19.92 = 2.11; \qquad MOS = SF - 1 = 2.11 - 1 = 111\% \qquad (i)$$

For member CD:

$$SF = S_y/\sigma = 42/15.88 = 2.64; \qquad MOS = SF - 1 = 2.64 - 1 = 164\% \qquad (j)$$

For member CE:

$$SF = S_y/\sigma = 42/20.57 = 2.04; \qquad MOS = SF - 1 = 2.04 - 1 = 104\% \qquad (k)$$

From an examination of the results for these three truss members, it is apparent that the design is reasonable with safety factors exceeding two in all cases. The design is not perfectly balanced because the margin of safety for member CD is somewhat higher than the other two members. If standard section sizes are commercially available, cost and weight can be reduced by reducing the cross sectional area of this member from 12 in$^2$ to 9 or 10 in$^2$. Of course we note that members BD and CD are loaded in compression and their critical buckling load must be determined to complete the design analysis.

## EXAMPLE 6.6

For the truss shown in Fig. E6.6, determine the cross sectional area required for members DF, EF and EG if the safety factor for the design is to exceed 2.8. The span s for each section of the truss is 8 m. The yield strength of the hot rolled steel members used in fabricating the truss is 320 MPa.

Fig. E6.6

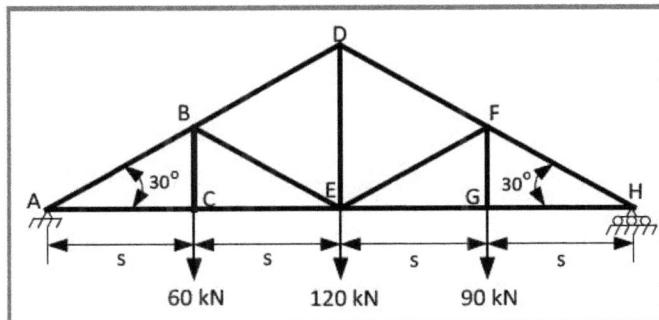

**Solution:**

**Step 1:** We construct a FBD of the entire structure as shown in Fig. E6.6a.

Fig. 6.6a

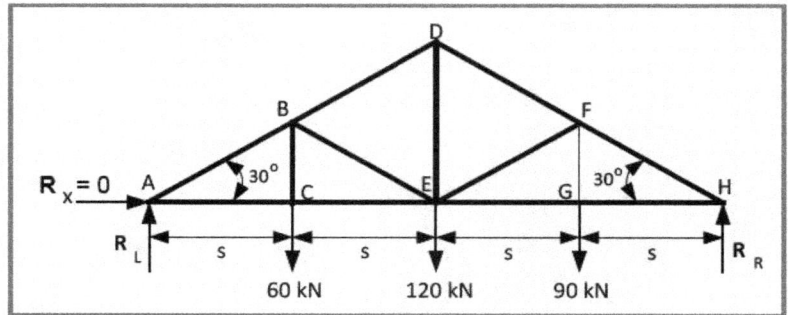

**Step 2:** Apply the equilibrium relations to solve for the reaction forces $R_L$ and $R_R$.

$$\Sigma F_x = R_x = 0$$

$$\Sigma M_A = -60(s) - (120)(2s) - (90)(3s) + (4s)R_R = 0$$

$$R_R = \tfrac{1}{4}(60 + 240 + 270) = 142.5 \text{ kN} \qquad (a)$$

$$\Sigma F_y = R_L - 60 - 120 - 90 + R_R = 0$$

$$R_L = -270 + 142.5 = 127.5 \text{ kN} \qquad (b)$$

**Step 3:** Make a section cut through the truss to expose the internal forces in members DF, EF and EG. Construct a FBD of the right hand side of the truss showing the unknown internal forces $P_{DF}$, $P_{EF}$, and $P_{EG}$ as presented in Fig. E6.6b.

Fig. E6.6b

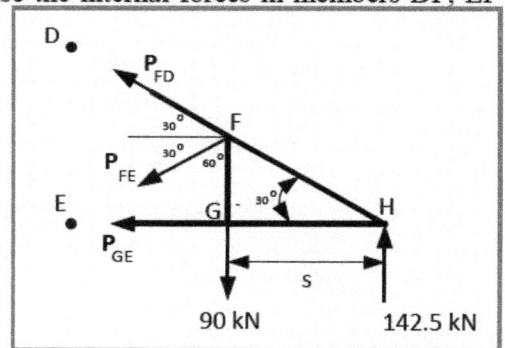

**Step 4:** Apply the equilibrium relations to solve for the unknown internal forces.

$$\Sigma M_F = (142.5)(s) - (P_{GE})(s)\tan(30°) = 0$$

$$P_{GE} = 142.5/\tan(30°) = 246.8 \text{ kN} \qquad (c)$$

$$\Sigma F_y = 142.5 - 90 - P_{FE} \sin(30°) + P_{FD} \sin(30°) = 0$$

$$P_{FE} = P_{FD} + 105 \qquad (d)$$

$$\Sigma F_x = -P_{GE} - P_{FD} \cos(30°) - P_{FE} \cos(30°) = 0$$

$$P_{FD} = -[P_{FE} + P_{GE}/\cos(30°)] \qquad (e)$$

Substituting Eqs. (c) and (d) into Eq. (e) yields:

$$P_{FD} = - [P_{FE} + 246.8/0.8660] = - [P_{FD} + 105 + 285.0]$$

$$P_{FD} = - 195.0 \text{ kN} \qquad\qquad (f)$$

Next, substitute Eq. (f) into Eq. (d) to obtain:

$$P_{FE} = P_{FD} + 105 = - 195 + 105 = - 90.0 \text{ kN} \qquad\qquad (g)$$

**Step 5:** Determine the stresses in the three members.

$$\sigma_{GE} = P_{GE}/A_{GE}$$

$$\sigma_{FD} = P_{FD}/A_{FD}$$

$$\sigma_{FE} = P_{FE}/A_{FE} \qquad\qquad (h)$$

It is impossible to solve for the stresses implicitly because the cross sectional areas of the three members are unknown. We will use Eqs. (h) after developing a relation for the stresses in terms of the safety factor. The result will give us a relationship between the allowable stress and the internal force on each member.

**Step 6:** Determine the allowable stresses and then the cross sectional areas by using the information regarding the safety factor and the yield strength required for the truss members.

$$SF = S_y/\sigma \qquad\qquad \sigma = S_y/SF \qquad\qquad (i)$$

Substitute Eq. (i) into Eqs. (h) and solve for the cross sectional area of each member.

$$\sigma_{GE} = P_{GE}/A_{GE} = S_y/SF = 320/2.8 = 114.3 \text{ MPa}$$

$$\sigma_{FD} = P_{FD}/A_{FD} = S_y/SF = 320/2.8 = 114.3 \text{ MPa}$$

$$\sigma_{FE} = P_{FE}/A_{FE} \quad = S_y/SF = 320/2.8 = 114.3 \text{ MPa} \qquad\qquad (j)$$

From Eqs. (j), (c), (f) and (g), it is evident that:

$$A_{GE} = P_{GE}/114.3 = 246.8 \times 10^3/114.3 = 2159 \text{ mm}^2$$

$$A_{FD} = P_{FD}/114.3 = 195 \times 10^3/114.3 = 1706 \text{ mm}^2$$

$$A_{FE} = P_{FE}/114.3 = 90 \times 10^3/114.3 = 787.4 \text{ mm}^2 \qquad\qquad (k)$$

The results of Eq. (k) give the cross sectional areas required to achieve the specified safety factors; however, sizes such as these are not normally used in design. Instead, we seek a standard size section slightly larger than that determined in the analysis. Hot rolled steel sections are available in a wide range of sizes and shapes from a large number of suppliers of metal products. We use standard sizes because it reduces costs and delivery time.

## EXAMPLE 6.7

Determine the forces and the stresses in members CE, DE and DF, of the bowstring truss shown in Fig. E6.7. Note that the numerical parameters that control the forces and stresses are $F_1 = 6.0$ kip, $F_2 = 5.0$ kip and $F_3 = 4.0$ kip, $s = 9$ ft, $A_{CE} = 1.2$ in.$^2$, $A_{DE} = 1.0$ in.$^2$, and $A_{DF} = 1.4$ in.$^2$

Fig. E6.7

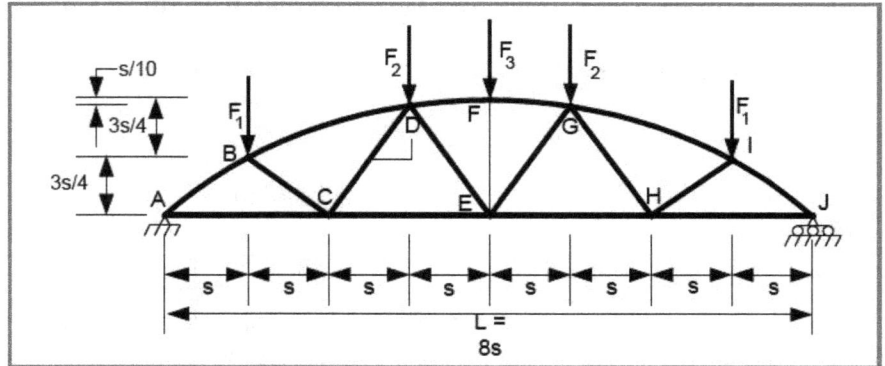

Step 1: We construct a FBD of the entire structure as shown in Fig. E6.7a.

Fig. E6.7a

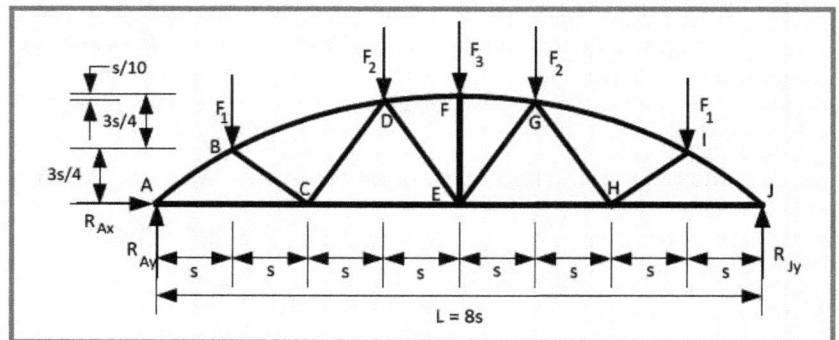

Step 2: Apply the equilibrium relations to solve for the reaction forces $R_{Ax}$, $R_{Ay}$ and $R_{Jy}$.

$$\Sigma F_x = R_{Ax} = 0 \qquad (a)$$

Since the loading and the geometry of the truss are symmetric, we may write:

$$R_{Ay} = R_{Jy} \qquad (b)$$

$$\Sigma F_y = R_{Ay} - 2F_1 - 2F_2 - F_3 + R_{Jy} = 0$$

$$R_{Ay} = R_{Jy} = (1/2)[2(6.0) + 2(5.0) + 4.0] = 13.0 \text{ kip} \qquad (c)$$

Step 3: Make a section cut through the truss to expose the internal forces in members CE, DE, and DF. Construct a FBD of the left hand side of the truss showing the unknown internal forces $P_{CE}$, $P_{DE}$, and $P_{DF}$ as presented in Fig. E6.7b.

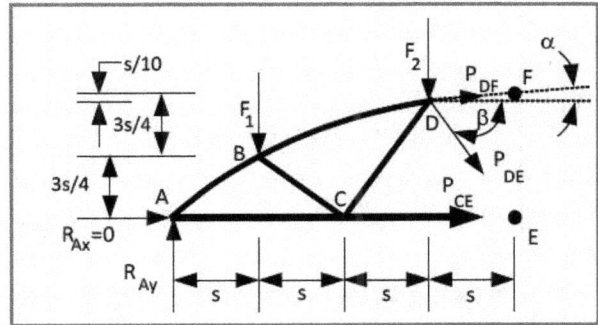

Fig. E7.7b

$\alpha = \tan^{-1}(1/10) = 5.710°$
$\beta = \tan^{-1}[(6/4) - (1/10)]$
$\beta = \tan^{-1}(1.4) = 54.46°$

**Step 4:** Apply the equilibrium relations to solve for the unknown internal forces. Select point E and write $\Sigma M_E = 0$ to obtain:

$$\Sigma M_E = (s)F_2 + (3s)F_1 - (4s)R_{Ay} - (P_{DF})[(s)\sin \alpha + (1.4s)\cos \alpha)] = 0 \qquad (d)$$

Substituting numerical parameters into Eq. (d) and solving for $P_{DF}$ yields:

$$P_{DF} = -19.43 \text{ kip} \qquad (e)$$

Consider equilibrium of the sectioned structure in the y direction:

$$\Sigma F_y = R_{Ay} - F_1 - F_2 + P_{DF} \sin \alpha - P_{DE} \sin \beta = 0 \qquad (f)$$

Substituting numerical parameters into Eq. (f) and solving for $P_{DE}$ yields:

$$P_{DE} = 0.08215 \text{ kip} \qquad (g)$$

Consider equilibrium of the sectioned structure in the x direction:

$$\Sigma F_x = R_{Ax} + P_{CE} + P_{DF} \cos \alpha + P_{DE} \cos \beta = 0 \qquad (h)$$

Substituting numerical parameters into Eq. (h) and solving for $P_{CE}$ yields:

$$P_{CE} = 19.29 \text{ kip} \qquad (i)$$

The minus sign for the numerical result for member DF indicates that it is subjected to a compressive force. However, the other two members of the bowstring truss are in tension.

**Step 5:** Determine the stresses in the three members. We may solve for the stresses in members DE and CE using Eq. (5.4) as:

$$\sigma_{DE} = P_{DE}/A_{DE} = 0.08215 \text{ kip}/1 \text{ in.}^2 = 0.08215 \text{ ksi}$$

$$\sigma_{CE} = P_{CE}/A_{CE} = 19.29 \text{ kip}/1.2 \text{ in.}^2 = 16.08 \text{ ksi}$$

It is not possible for us to determine the stresses in member DF because it is curved. The curvature of the member does not affect the determination of the force $P_{DF}$ since DF is a two-force member. For this reason, the internal forces acts along a line of action passing through points D and F on the bowstring truss as shown in Fig. E6.7c.

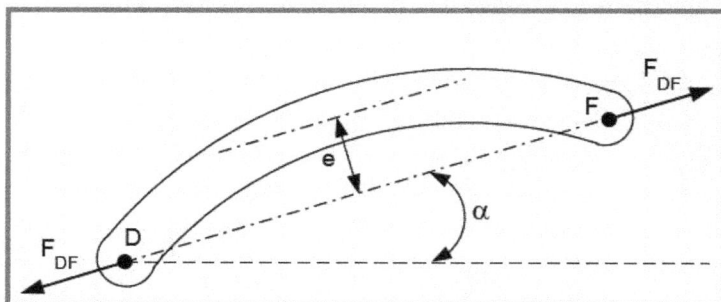

Fig. E6.7c

The force external $F_{DF}$ is equal to the internal force $P_{DF}$ both of which act along the line of action defined with respect to the x-axis by the angle $\alpha$. However, the centerline of the curved member does not coincide with the line of action of the two-forces. Indeed, an eccentricity e is evident in Fig. E6.7c. This eccentricity is important when determining the stresses in the curved member because it causes bending of the member. The relation for stresses given by Eq. (4.4) is not valid in this case because it does not account for either the curvature of the member or the eccentricity. Both of these topics are beyond the scope of this textbook and will be covered later in a more advanced course.

## 6.5 SUMMARY

Trusses are fabricated from bars and cables to provide a structure capable of spanning long distances. As such they are often used in the design of bridges and roofs to cover large buildings. Stability of the truss is achieved by constructing it from a series of triangular arrangements.

The bars or cables in a truss are two force members connected together by means of joints. We assume that the joints are capable of rotation and cannot support a moment. Since the bars and cables are long and flexible, it is necessary to restrict the point of load application to only the joints of the truss.

Two methods for analysis of trusses are described. The first, the method of joints, entails removing a specified joint from the truss and constructing a FBD showing both the internal and external forces acting on it. The joint is represented as a particle acted upon by a planar concurrent force system. As such two equilibrium relations apply:

$$\Sigma F_x = 0 \qquad \text{and} \qquad \Sigma F_y = 0$$

These relations are sufficient to solve for any two unknown forces that act on the joint.

The method of sections involves making a section cut through the truss to divide it into two or more parts. A FBD is prepared for one section of the truss showing both the internal and external forces. The forces acting on a particular section are planar but not concurrent; hence, the equilibrium relations that apply are:

$$\Sigma F_x = 0; \qquad \Sigma F_y = 0; \qquad \text{and} \qquad \Sigma M_O = 0$$

These relations are sufficient to solve for any three unknown forces that act on the portion of the truss under consideration. Examples illustrating the procedure for using the method of joints and the method of sections were provided.

A procedure was described for identifying zero-force members in a truss by examining the joints not subjected to externally applied loads. When a joint is acted upon by three internal forces and two of these forces are collinear, then the remaining force must be zero. Also, when two non-collinear members meet at a joint both are zero force members. Zero force members do not contribute to the strength or rigidity of a structure and should be eliminated unless there is a compelling reason for their inclusion in the design.

## PROBLEMS

6.1    Why is a truss used in constructing a bridge or a roof covering for a large structure?

6.2    What is the inherent geometric form that is used repeatedly in all truss designs?  Why?

6.3    If a truss is fabricated from 37 members, determine the number of joints that will be required in constructing the truss.

6.4    Why is it necessary to employ an odd number of members in constructing a truss?

6.5    Prepare a sketch of a joint using a gusset plate for connecting four uniaxial bars in a truss structure. Identify all the components included in the joint that you have designed.

6.6    Why is it important to restrict the application of external loads on a truss to only the joints?

6.7    Consider the four-unit Howe truss shown in the figure to the right.  If the external forces are as specified in this figure, use the method of joints to determine the forces in members AC, BC, CD and CE.

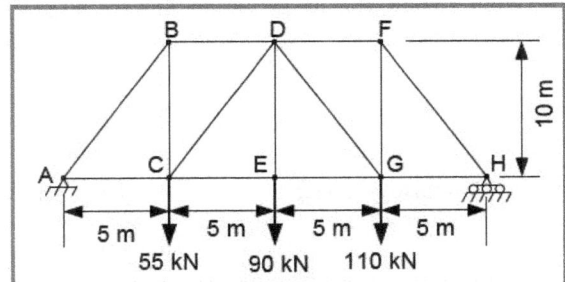

6.8    Consider the four-unit Howe truss shown in the figure to the left.  If the external forces are as specified in this figure, use the method of joints to determine the forces in members AC, BC, CD and CE.

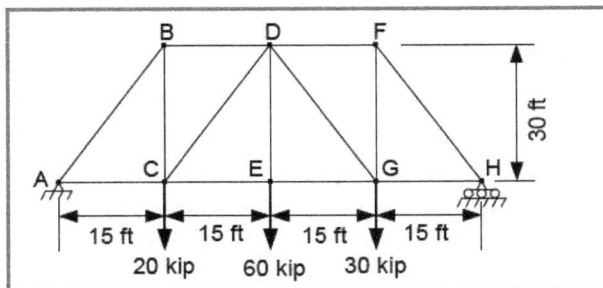

6.9    Consider the four-unit Howe truss shown in the figure to the right.  If the external forces are as specified in this figure, use the method of joints to determine the forces in members GE, GD, GF and GH.

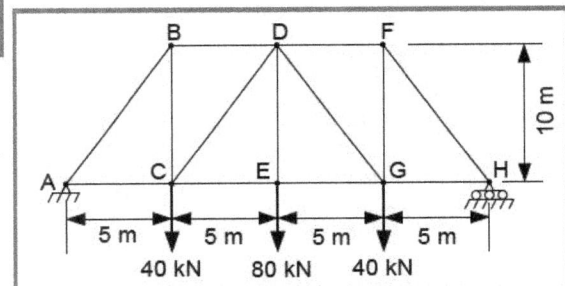

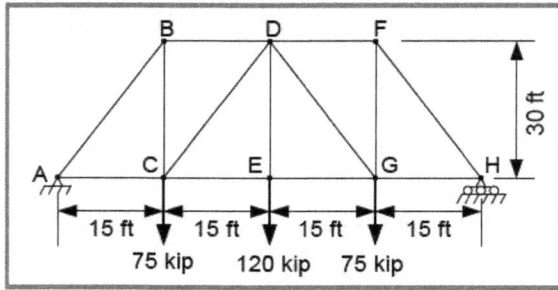

6.10 Consider the four-unit Howe truss shown in the figure to the left. If the external forces are as specified in this figure, use the method of joints to determine the forces in members AC, BC, CD and CE.

6.11 If the safety factor SF for each of the members in the truss of Problem 6.7 is specified as 3.0, determine the minimum required cross sectional area for members AC, BC, CD and CE. Note the yield strength $S_y$ of the hot rolled structural steel used in fabricating the truss is 200 MPa.

6.12 If the safety factor SF for each of the members in the truss of Problem 6.8 is specified as 3.2, determine the minimum required cross sectional area for members AB, AC, BD and BC. Note the yield strength $S_y$ of the hot rolled structural steel used in fabricating the truss is 30 ksi.

6.13 A scissors truss, illustrated in the figure to the right, is loaded with three forces at the joints B, C and D. If the truss is fabricated from 1018 A steel, determine the size of members AB, AF, BC and BF. The safety factor is specified as 3.0.

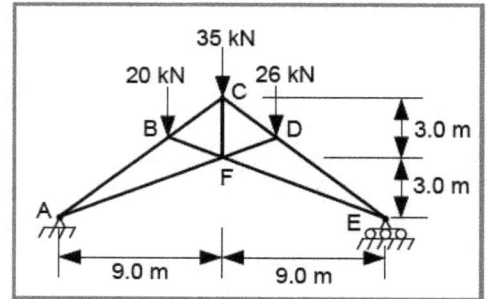

6.14 A scissors truss, illustrated in the figure to the left, is loaded with three forces at the joints B, C and D. If the truss is fabricated from 1020 HR steel, determine the size of members AB, AF, BC and BF. The safety factor is specified as 2.9.

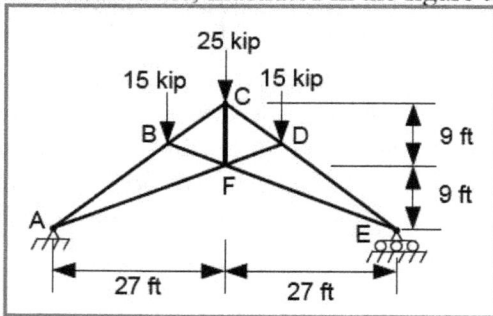

6.15 For the Howe truss, shown in the figure to the right, use the method of sections to determine the forces in the members BD, CD and CE.

6.16 For the Howe truss, shown in the figure to the right, use the method of sections to determine the forces in the members DF, DG and EG.

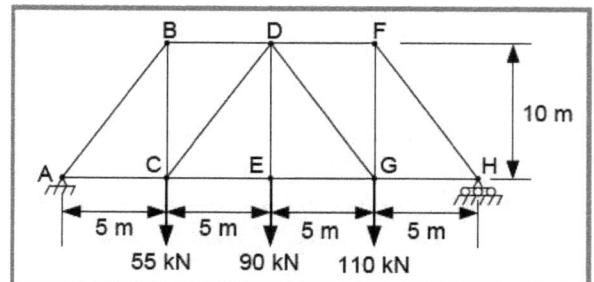

6.17 For the truss, defined in the figure to the right, use the method of sections to determine the forces in members DF, DB and DE. The external forces applied at joints C, E, G, I, and K are each equal to 15 kip.

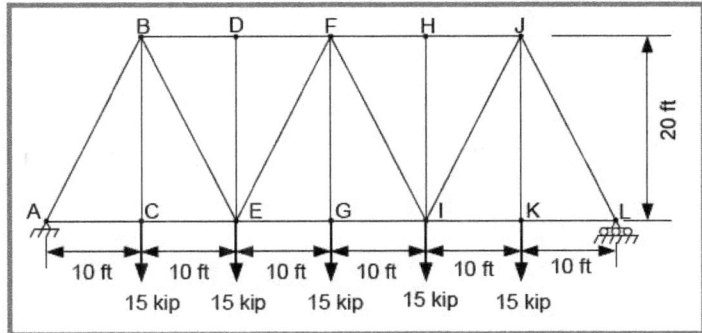

6.18 If the cross sectional areas of all of the members in the truss defined in Problem 6.17 are equal to A = 2.3 in.$^2$, determine the safety factor for structural members DF, DB and DE. The yield strength of the steel used in the truss members is 35 ksi.

6.19 By inspection identify the zero-force members in the truss defined in Problem 6.17.

6.20 For the truss, defined in the figure to the right, use the method of joints to determine the forces in members DC, DE and DF.

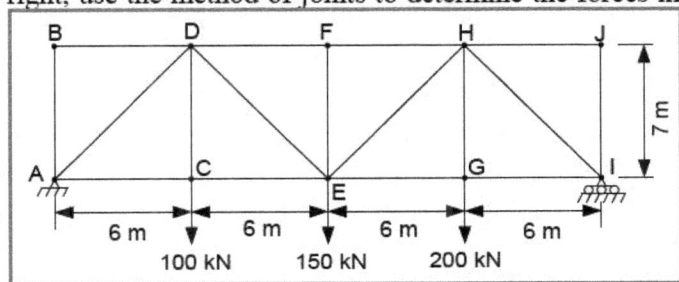

6.21 If the cross sectional areas of all of the members in the truss defined in Problem 6.20 are equal to 3,000 mm$^2$, determine the safety factor for the members DC, DE and DF. Note the yield strength of the steel used in the truss is 310 MPa.

6.22 Use the method of sections with the truss, shown in the figure for Problem 6.20, to determine the forces in members FH, EH, and EG.

6.23 Let s/h be a variable in the truss structure defined in the figure to the left. Determine the forces in the members AC and AD as a function of the s/h ratio. We suggest you use a spreadsheet to perform the calculations. Consider the ratio s/h over the range from 0.5 to 2.0 varying in steps of 0.1. Prepare a graph of the results for the forces in members AC and AD as a function of s/h.

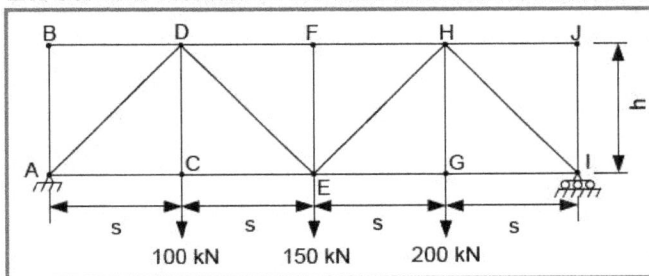

6.24 Determine the forces in members CE, DE and DF of the bowstring truss shown in the figure to the right. The span s = 12 ft.

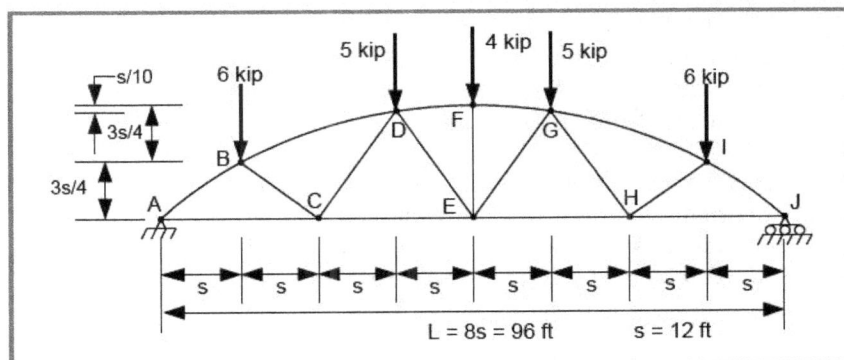

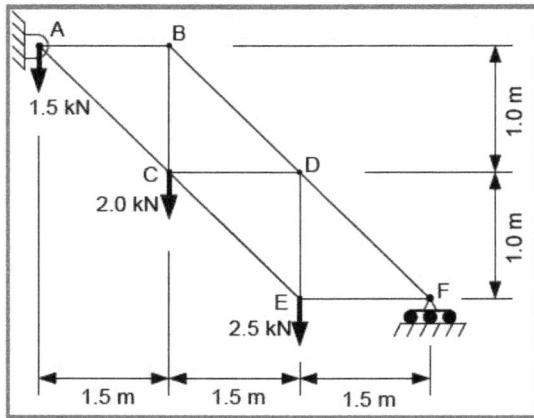

6.25 Determine the forces in members AC, CD, CE and DF of the inclined truss shown in the figure to the left.

6.26 Determine the forces in members CD, CJ, and KJ of the truss shown in the figure below and to the right.

6.27 Determine the forces in members DE, EJ, and JI of the truss shown in the figure to the right.

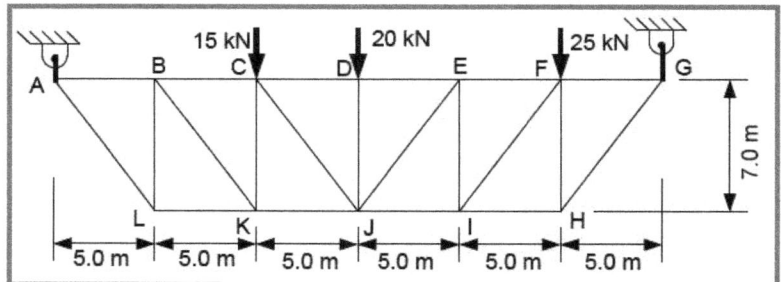

6.28 Determine the forces in members EF, IF, and IH, of the truss shown in the figure below and to the right.

6.29 Determine the forces in members BC, BK, and LK, of the truss shown in the figure to the right.

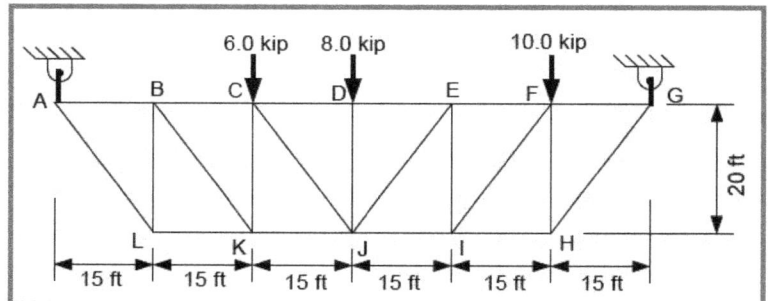

6.30 For the cantilever truss, shown in the figure to the right, determine the stress in member BC if its cross sectional area is 2.5 in². Also determine the safety factor for this truss member if it is fabricated from 1020 HR steel.

6.31 For the cantilever truss, shown in the figure to the right, determine the stress in member AB if its cross sectional area is 3.5 in². Also determine the safety factor for this truss member if it is fabricated from 1020 HR steel.

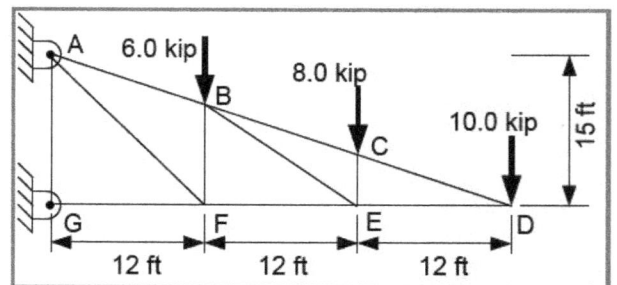

6.32 For the cantilever truss, shown in the figure to the right, determine the force in member GF. Specify the minimum cross sectional area if the member is to exhibit a safety factor of 3.0 relative to the yield strength of its material. This truss member is fabricated from 1020 HR steel.

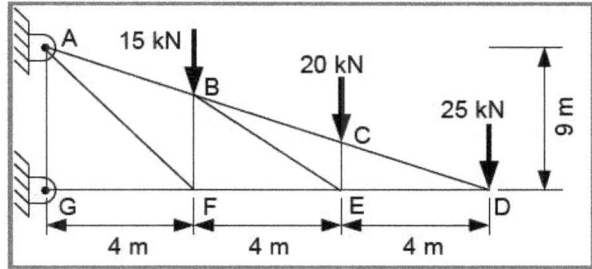

6.33 For the cantilever truss, shown in the figure to the right, determine the stress in member CD if its cross sectional area is 900 mm$^2$. Also determine the safety factor for this truss member if it is fabricated from 1020 HR steel.

6.34 For the cantilever truss shown in Problem 6.33, determine the force in member ED. Specify the minimum cross sectional area if this member is to exhibit a safety factor of 3.0 relative to the yield strength of its material. This truss member is fabricated from 1212 HR steel.

6.35 For the truss, presented in the figure below, determine the cross sectional area required for members AB, BD, DE, EF and CF if they are fabricated from 1020 HR steel and a safety factor of 3.2 is specified.

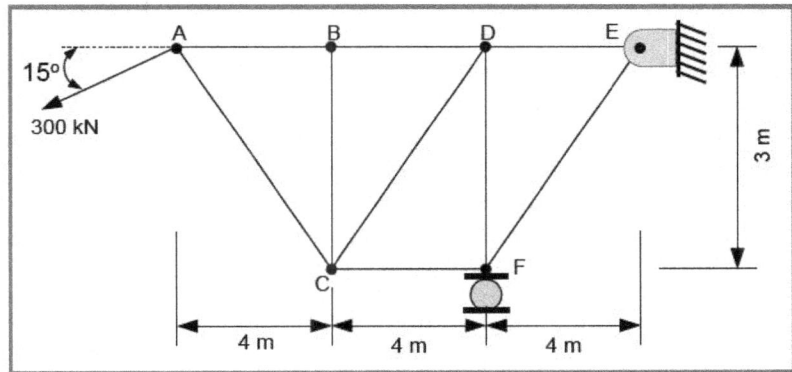

# CHAPTER 7

# SPACE STRUCTURES AND 3-D EQUILIBRIUM

## 7.1 INTRODUCTION

A space structure is a three-dimensional system, fabricated from uniaxial members, that spans in two-directions. The space structure incorporates members that extend into space with force components in the x-y, x-z and y-z planes. Space structures may be fabricated with rigid joints, which support moments, or pinned joints, which do not. In this chapter, we will first consider space structures with pinned joints that cannot support moments. Then we will study more complex three-dimensional structures. Trusses, described in Chapter 6, are two-dimensional structures because all of the members involved in their construction lie in the x-y plane. Trusses span in only one direction. We can perform a design analysis of a two-dimensional truss by using only two or possibly three of the six equilibrium relations. However, the three-dimensional nature of space structures is more complex requiring the repeated application of the six equations of equilibrium to determine the forces in its uniaxial members.

The geometry of space structures is quite diverse with domes, tetrahedrons and half-octahedrons. An example of a space structure constructed with the half octahedrons is presented in Fig. 7.1. Space structures are often employed to support horizontal, flat roofs that cover large buildings. However, they may be employed in other applications including walls and sloped roofs. Because space structures are efficient, they cost less than a system of trusses spanning the primary direction with beams (purlins) spanning in the opposite direction.

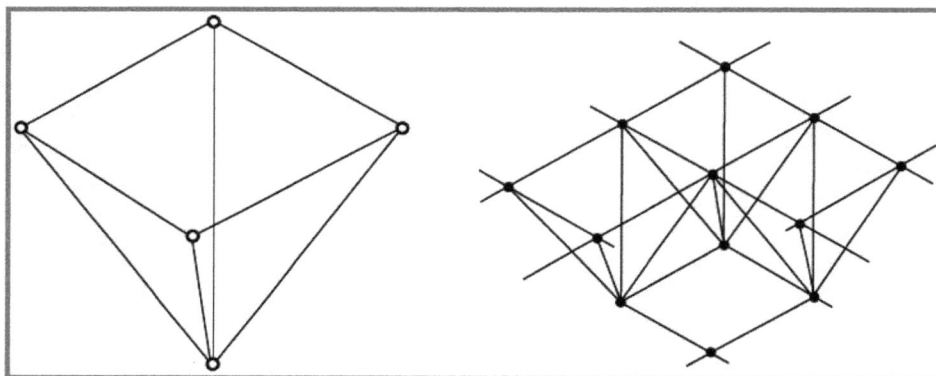

Fig. 7.1 Half-octahedrons arranged to form a space structure.

Another illustration of a space structure is a communication tower that is an extremely tall structure usually built at a high elevation to facilitate the distance of unobstructed propagation of microwave, radio or television signals. The structure has a very high aspect ratio and is inherently unstable. Stability is achieved with cables mounted near the top of the transmission tower that extend downward and outward. These cables are anchored into the ground to provide a large footprint for the tower. Since stability is achieved with cables, the base of the tower is usually pinned. In three dimensions the equivalent for a pinned joint is a ball and socket joint. A ball and socket joint provides reaction forces $R_x$, $R_y$ and $R_z$ but moments are not supported. We show a schematic illustration of a communication tower with four cables anchored to the ground plane in Fig. 7.2.

The forces acting on a communications tower are due to:

- The dead weight of the tower and cables.
- The pretension, if any, applied to the cables
- The weight of any communication equipment installed on the tower.
- Snow and/or ice accumulations.
- Transverse loads due to winds
- In some parts of the country earthquake loading should be considered.

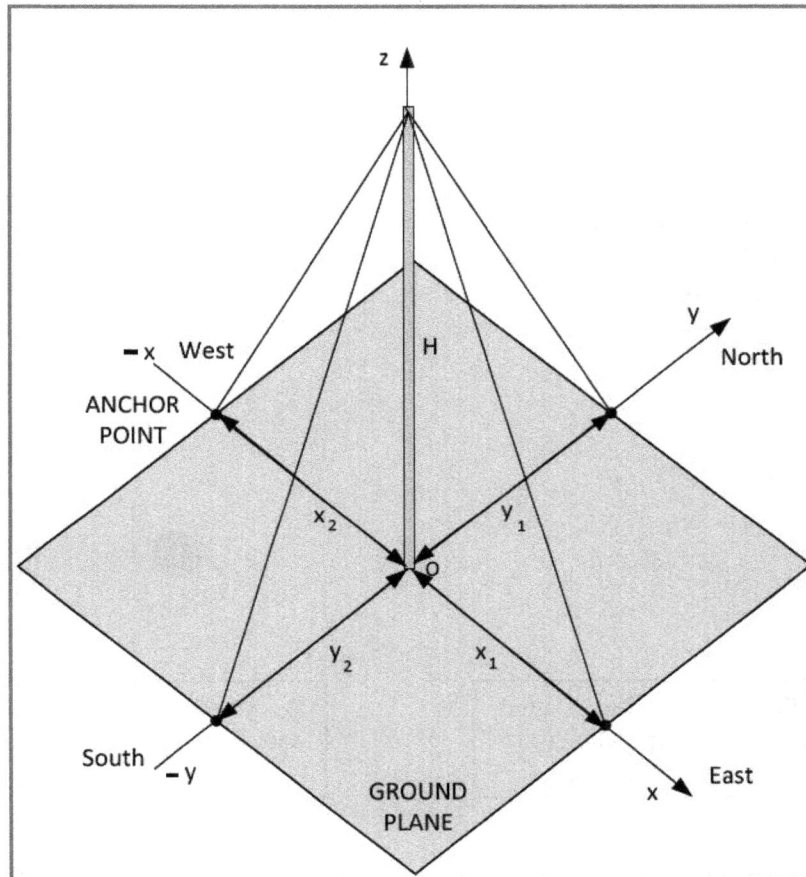

Fig. 7.2 Schematic illustration of a cable stabilized communications tower.

To perform a design analysis of three-dimensional structures, such as a communications tower, we follow the same six-step procedure as outlined previously for trusses. The difference in the analysis involves the added complexity due to the inclusion of the third dimension. The additional dimension requires additional FBDs, and the application of all six of the equilibrium relations.

The structure is always constrained by supports at several locations. These supports are modeled in FBDs with some combination of concentrated forces and moments that effectively provides the constraint the supports provide. The forces and the moments produced by common supports are listed in Table 7.1.

## Table 7.1
## Different types of supports or connections and their reactive forces and moments.

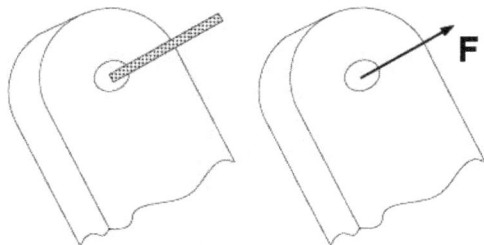

A. A cable connection is modeled with a single force acting in the direction of the cable and away from the structure.

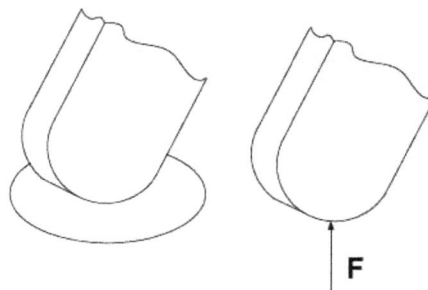

B. A rocker on a frictionless surface is modeled with a concentrated force perpendicular to the surface at the point of contact.

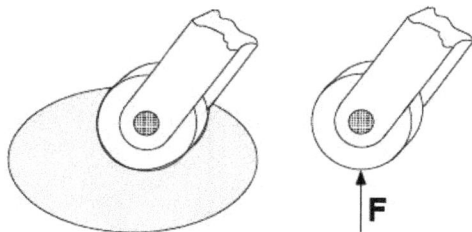

C. A roller on a flat surface is represented with a concentrated force perpendicular to the surface at the point of contact.

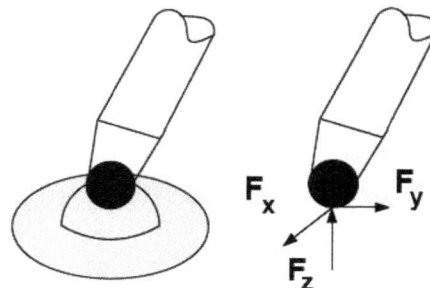

D. A ball and socket joint is modeled with three Cartesian force components.

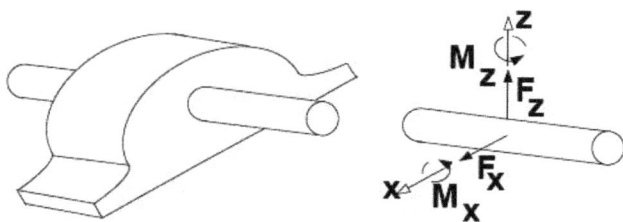

E. A journal bearing is represented with two forces perpendicular to the shaft and two moments about axes perpendicular to the shaft.

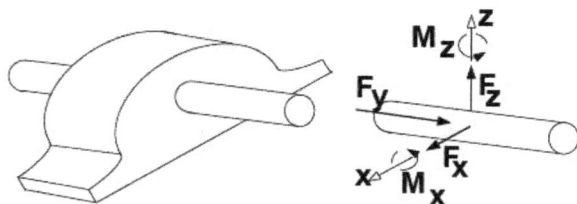

F. A thrust bearing is represented with three Cartesian forces and two moments about axes perpendicular to the shaft.

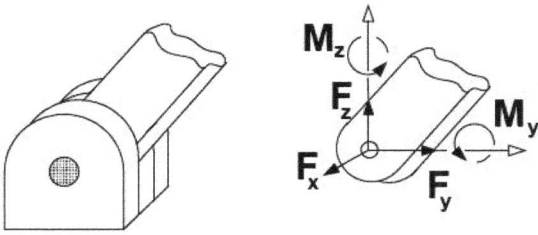

G. A pin and clevis connection is represented with three Cartesian forces and two moments about axes perpendicular to the pin.

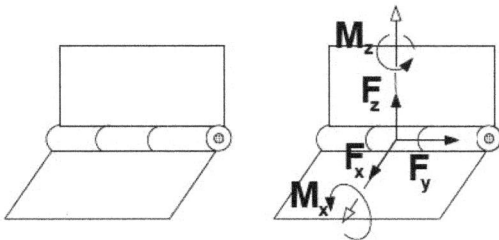

H. A hinge connection is represented with three Cartesian forces and two moments about axes perpendicular to the hinge pin.

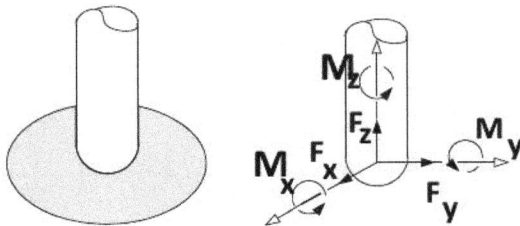

I. A fixed support is modeled with six possible reactions (three Cartesian forces and three Cartesian moments).

## 7.2 SOLUTIONS USING TRIGONOMETRIC ANALYSIS

In the discussion of three-dimensional structures, we will present two different mathematical approaches. The first approach, consistent with previous treatments, employs trigonometric methods for determining force and moment components in each of the three directions. The second approach employs vector analysis. The first approach is intuitively obvious, while the second is more systematic and mathematically more elegant.

Let's begin with the trigonometric approach to determine the force and moment components by performing a design analysis of the communications tower.

### EXAMPLE 7.1

For the communications tower described in Fig. 7.2, determine the safety factors[1] for the cable and the tower structure if:

- The dead weight of the tower is 20,000 lb.
- The receivers and transmitters mounted on the tower weigh 2,000 lb.

---

[1] We will neglect bending in this analysis; however the tower is subjected to bending stresses due to the transverse loads imposed by the wind.

- The tower height H is 500 ft and the material used in fabrication has a yield strength of 35,000 psi.
- Each of the four cables is 5/8 in. in diameter with a rated breaking strength of 41,200 lb. The length of the cables is adjusted so that they are snug but they are not pretensioned.
- The anchor points for the tower are $x_1 = x_2 = y_1 = y_2 = 200$ ft.
- The maximum transverse load anticipated from a southwest wind is 1600 lb. We assume that this load is applied at a distance $z = 2H/3$ from the ground plane.
- The anchor points are oriented so that the x-axis is due east and the y-axis is due north.
- The cross sectional area at the base of the tower is 11 in.$^2$.
- A ball and socket joint supports the tower at point O.

**Solution:**

**Step 1:** Draw a FBD of the structure as shown in Fig. E7.1.

Fig. E7.1

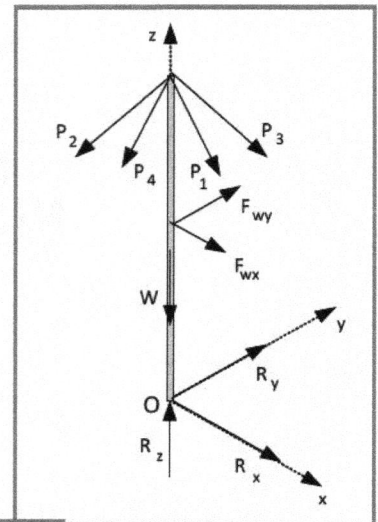

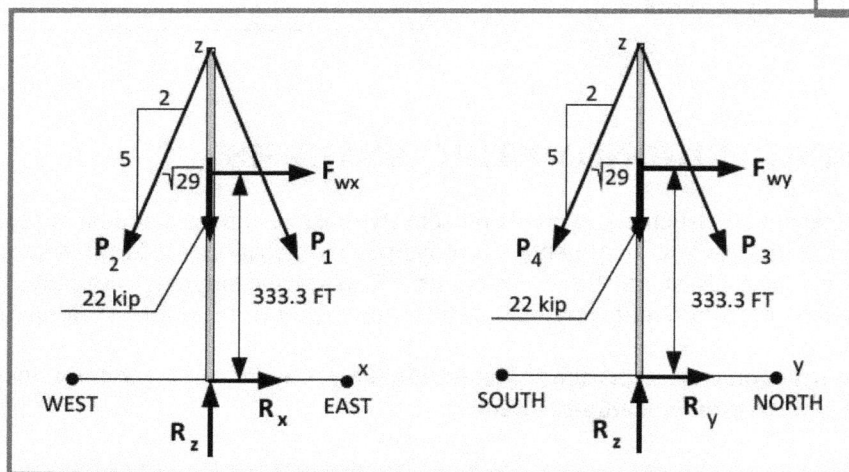

Fig. E7.1a

The view on the left is observed by standing to the south of the tower and looking north. The view on the right is observed by standing in the east and looking west. By drawing two different views of the FBDs, we are able to show the components of the wind loading and the forces in all four of the cables. The FBDs are not complete because we have not shown the magnitude of the wind forces $F_{wx}$ and $F_{wy}$. Since the wind is from the southwest, we represent the wind force with the vector diagram shown in Fig. E7.1b, and solve for the force components in the x and y directions.

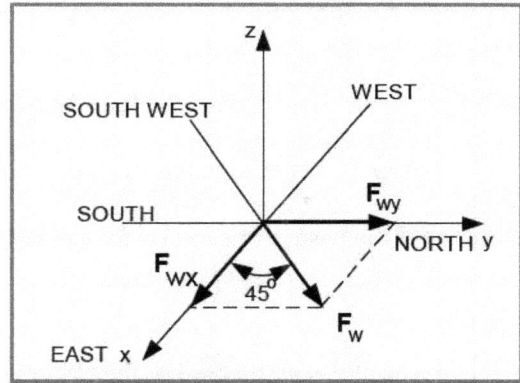

Fig. E7.1b

It is evident that:

$$F_{wy} = F_w \sin(45°) = (1600)(0.7071) = 1131 \text{ lb}$$

$$F_{wx} = F_w \cos(45°) = 1131 \text{ lb} \tag{a}$$

**Step 2:** Consider equilibrium of the structure. From the FBDs in Fig. E7.0a, we will write several moment equations.

First, consider moments about the y-axis and write:

$$\Sigma M_y = (2/\sqrt{29})P_2 (500) - F_{wx}(333.3) = 0$$

$$P_2 = [(1131)(333.3)(\sqrt{29})]/(1000) = 2030 \text{ lb} \tag{b}$$

Since the cables supporting the tower are not pretensioned, $P_1 = P_3 = 0$ when the wind has components from the west and south. Next, consider the moments about the x-axis and write:

$$\Sigma M_x = (2/\sqrt{29})P_4(500) - F_{wy}(333.3) = 0$$

$$P_4 = [(1131)(333.3)(\sqrt{29})]/(1000) = 2030 \text{ lb} \tag{c}$$

Finally, note that the relation $\Sigma M_z = 0$ is satisfied regardless of the values of the applied forces. There are no forces that tend to twist the structure, because all the forces intersect the z-axis.

Consider equilibrium of forces in the three directions and write:

$$\Sigma F_x = -(2/\sqrt{29})P_2 + F_{wx} + R_x = 0$$

$$R_x = (2/\sqrt{29})P_2 - F_{wx} = (2/\sqrt{29})(2030) - 1131 = -377 \text{ lb} \tag{d}$$

$$\Sigma F_y = -(2/\sqrt{29})P_4 + F_{wy} + R_y = 0$$

$$R_y = (2/\sqrt{29})P_4 - F_{wy} = (2/\sqrt{29})(2030) - 1131 = -377 \text{ lb} \tag{e}$$

The negative signs for $R_x$ and $R_y$ indicate that the directions for these reactions should be reversed in the FBDs presented in Fig. E7.1.

$$\Sigma F_z = - 22{,}000 - (5/\sqrt{29})P_2 - (5/\sqrt{29})P_4 + R_z = 0$$

$$R_z = 22{,}000 + (10/\sqrt{29})(2030) = 25{,}770 \text{ lb} \qquad\qquad (f)$$

**Step 3:** Determine the stresses in the structure.

At the base of the tower, the normal stress $\sigma_z$ is given by:

$$\sigma_z = R_z/A = - 25{,}770/11 = - 2343 \text{ psi}$$

The stress at the base of the tower is compressive.

It is not necessary to compute the stress in the cables because the breaking strength is given in terms of the force applied to the cable, and we have already made this determination.

**Step 4:** Determine the safety factor and the margin of safety for the tower and the cables.

For the tower:
$$SF = S/\sigma = 35{,}000/2343 = 14.9 \qquad\qquad MOS = SF - 1 = 13.9$$

For the cables:
$$F = P_b/P = 41{,}200/2030 = 20.3 \qquad\qquad MOS = SF - 1 = 19.3$$

Interpretation: The results indicate that the safety factors are very large. Develop arguments supporting the use of these large margins of safety in the design of a communication tower. Also, develop arguments for redesign with smaller diameter cables, smaller footprint and smaller cross sectional area for the tower structure.

## EXAMPLE 7.2

A small hoist is constructed with two cables (AE and CE) and a strut (DE) as shown in Fig. E7.2. If the applied force F = 3 tons, determine the size of the cables and strut if a safety factor of 3.5 is specified. The yield strength of the cables is specified by the supplier as 150,000 psi, and the yield strength of the low carbon steel used for the strut is 40,000 psi. The dimensions for the crane assembly are AB = BC = 6 ft; BD = 8 ft; BE = 12 ft.

Fig. E7.2

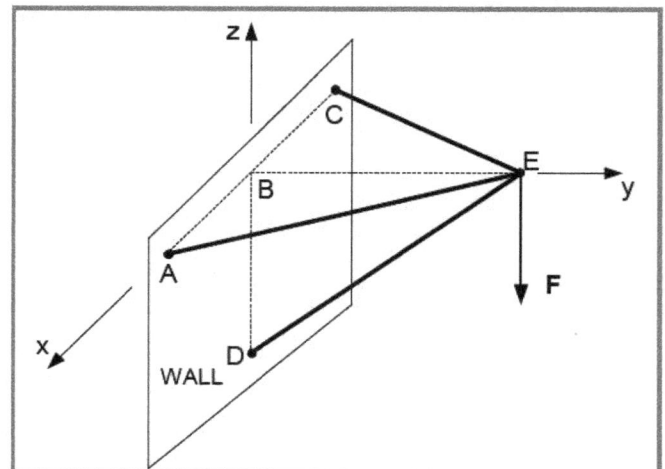

**Solution:**

**Step 1:** Draw a FBD of the structure as shown in Fig. E7.2a.

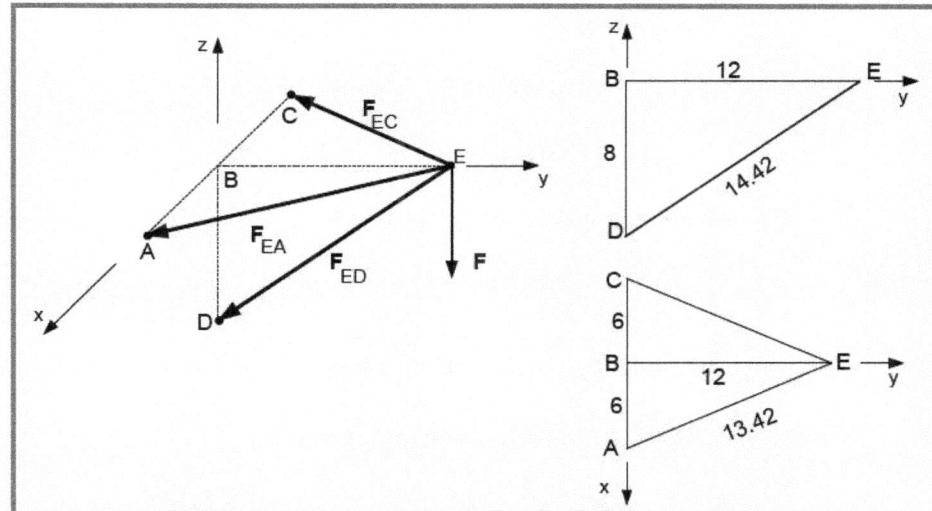

Fig. E7.2a

We have removed the wall and have drawn force vectors for the cables and strut. We have also presented two drawings—one for the y-z plane and the other for the x-y plane. These drawings, showing the dimensions of the arrangement, are useful in determining the components of the forces.

**Step 2:** Consider equilibrium of the structure. From the FBDs, it is clear that the system is non-coplanar and concurrent. Since the system is concurrent it is not necessary to write moment equations. The forces in the cables and strut are determined from equilibrium of the forces at point E as:

$$\Sigma F_x = F_{EA}\,(6/13.42) - F_{EC}\,(6/13.42) = 0 \tag{a}$$

$$\Sigma F_y = -\,F_{EA}\,(12/13.42) - F_{EC}\,(12/13.42) - F_{ED}\,(12/14.42) = 0 \tag{b}$$

$$\Sigma F_z = -\,F_{ED}\,(8/14.42) - F = 0 \tag{c}$$

From Eq. (c), it is clear that $F_{ED}$ is given by:

$$F_{ED} = -\,(14.42/8.00)\,F = -\,(1.8025)(3) = (5.408 \text{ ton})(2000 \text{ lb/ton}) = -\,10{,}815\,\text{lb} \tag{d}$$

From Eq. (a), it is evident that:

$$F_{EA} = F_{EC} \tag{e}$$

Substituting Eqs. (d) and (e) into Eq. (b) yields:

$$-\,2\,F_{EA}\,(12/13.42) - (-\,10{,}815)\,(12/14.42) = 0$$

$$F_{EA} = F_{EC} = 5{,}032\,\text{lb} \tag{f}$$

**Step 3:** Determine the stresses in the cables and the strut.

The stresses are determined from the safety factor and the strength of the materials used in fabricating the components.

$$\sigma = S_y/SF \tag{g}$$

For the cables, the allowable design stress is:

$$\sigma_{design} = S_y/SF = 150{,}000/3.5 = 42{,}860 \text{ psi} \tag{h}$$

For the strut, the allowable design stress is:

$$\sigma_{design} = S_y/SF = 40{,}000/3.5 = 11{,}430 \text{ psi} \tag{i}$$

**Step 4:** Determine the cross sectional area required for the cables and the strut.

For the cables, the cross sectional area required is:

$$A = P/\sigma = F_{EA}/\sigma = 5{,}032/42{,}860 = 0.1174 \text{ in}^2 \tag{j}$$

The diameter of the cable specified would be d = 7/16 in. providing an area A = 0.1503 in². The cross sectional area required for the strut is:

$$A = P/\sigma = F_{EA}/\sigma = 10{,}815/11{,}430 = 0.9462 \text{ in}^2 \tag{k}$$

## EXAMPLE 7.3

The Patcenter in Princeton, NJ is a large open structure with a cable-stayed roof. In Fig. E7.3, we show one of the nine tubular steel masts that are uniformly spaced at 9 m intervals along the depth of the building to support the roof structure. In this design, the roof hangs from cables[2] since the vertical side members are used only to prevent wind uplift. The tubular mast, 15 m high is supported with a 9 m wide by 6 m high rectangular steel frame. Determine the force in the primary rod stay and the tubular steel masts. Assume the roof is pinned to the steel mast at point C.

Fig. E7.3

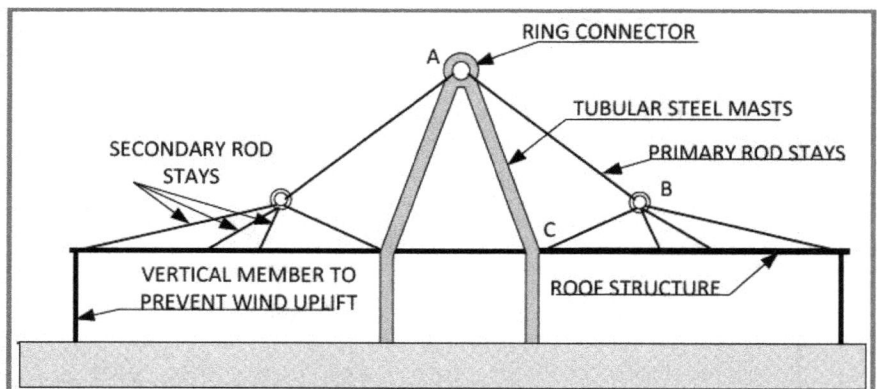

---

[2] Thin steel tie rods were actually employed instead of cables.

**Solution:**

**Step 1.** Prepare FBDs for the structural members. Add dimensions, reactions and applied forces to the FBDs. The dimensions shown in Fig. 7.3a were obtained by scaling a drawing of the front view of the Patcenter [1].

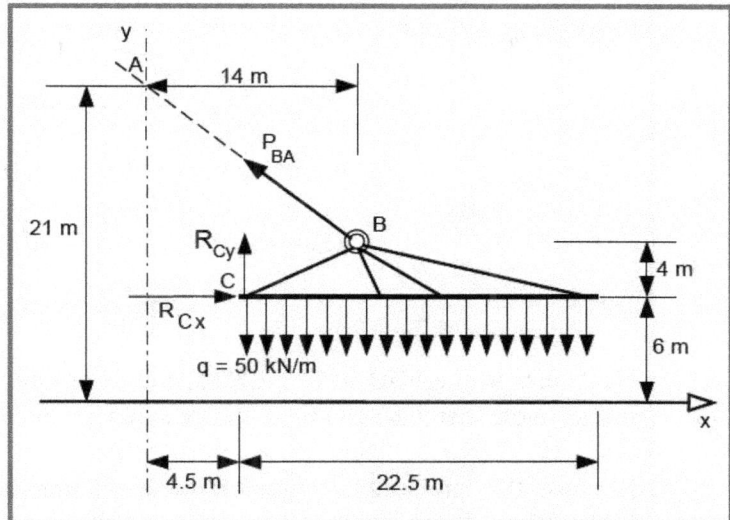

Fig. 7.3a

The uniformly distributed load of 50 kN/m was specified to account for the dead weight of the roof structure and snow/water accumulation[3]. To obtain this distributed load, we multiplied the roof area (9 m by 22.5 m) by a pressure of approximately 5.56 kN/m$^2$, and then divided by the length (22.5 m) over which the force was distributed.

The second FBD is a model of the tubular steel frame as shown in Fig. E7.3b. We have recognized the symmetry of the structure and have equated internal forces on the left and right sides of the structure. We have also considered that the internal forces in the tubular members act along their axes. The angles $\alpha$ and $\beta$ in Fig. E7.3b are given by:

$$\alpha = \tan^{-1}(11/14) = 38.16° \qquad\qquad \beta = \tan^{-1}(4.5/15) = 16.70° \qquad\qquad (a)$$

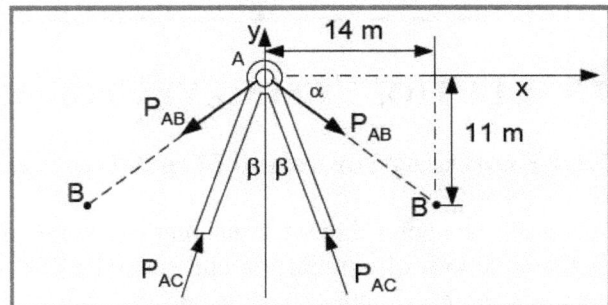

Fig. E7.3b

**Step 2:** Consider equilibrium of the structure. From the FBD in Fig. E7.3a, we write the moment equations about point C as:

$$\Sigma M_C = -(50)(22.5)(22.5/2) + P_{BA}\sin\alpha\,(14 - 4.5) + P_{BA}\cos\alpha\,(4) = 0 \qquad (b)$$

---

[3] The snow loading assumes a water-saturated layer of snow 0.3 m deep.

Solving Eq. (b) for $P_{BA}$ yields:

$$P_{BA} = 1404 \text{ kN} \tag{c}$$

Consider equilibrium in the x direction and write:

$$\Sigma F_x = R_{Cx} - P_{BA} \cos \alpha = 0 \tag{d}$$

$$R_{Cx} = 1404 \cos (38.16°) = 1104 \text{ kN} \tag{e}$$

$$\Sigma F_y = R_{Cy} + P_{BA} \sin \alpha - (50)(22.5) = 0 \tag{f}$$

$$R_{Cy} = 257.5 \text{ kN} \tag{g}$$

Next refer to the FBD in Fig. E7.3b, and determine the internal forces acting on the tubular A frame. Note that this FBD represents a coplanar concurrent force system.

Because the structure is symmetric the relation $\Sigma F_x = 0$ does not provide any additional information. However, considering the equilibrium equation in the y direction yields:

$$\Sigma F_y = 2P_{AC} \cos \beta - 2 P_{AB} \sin \alpha = 0 \tag{h}$$

Solving Eq. (h) for $P_{AC}$ gives:

$$P_{AC} = [(1404) \sin (38.16°)] /\cos (16.70°) = 905.7 \text{ kN} \tag{1}$$

The Patcenter is a modern structure with large open spaces that provide flexibility in the arrangement of laboratories and offices. Except for the vertical members supporting the A frame and the vertical tie bar to stabilize the roof against wind loading, it is free of columns. Modeling with FBDs and applying the simple equations of equilibrium enabled the analysis of the internal forces in selected members of this structure.

# 7.3 SOLUTIONS USING VECTOR MECHANICS

## 7.3.1 Expressing Forces and Moments as Vectors

When we encounter three-dimensional structures, the analysis becomes complex—often involving all six of the equations of equilibrium and several FBDs. Visualization of the structure and force components becomes more difficult. To alleviate the complexity and reduce visualization difficulties, we often use vector mechanics. When we express each force **F** or moment **M** in a complete vector representation, the equilibrium relations reduce to two vector equations:

$$\Sigma \mathbf{F} = 0 \qquad \text{and} \qquad \Sigma \mathbf{M} = 0 \tag{7.1}$$

Moments are a product of a force times a distance. In most of our previous solutions to example problems, we were careful to define the distance as the perpendicular distance from the line of action of the force to the point about which the moments were determined. The definition of the moment does not change, but with the vector mechanics approach described previously[4] in Chapter 2, we define the moment in terms of a vector cross product.

$$\mathbf{M} = \mathbf{r} \times \mathbf{F} \tag{7.2}$$

where r is a position vector given by:

$$\mathbf{r} = r_x\,\mathbf{i} + r_y\,\mathbf{j} + r_z\,\mathbf{k} \tag{7.3}$$

The force vector is written as:

$$\mathbf{F} = F_x\,\mathbf{i} + F_y\,\mathbf{j} + F_z\,\mathbf{k} \tag{7.4}$$

Recall, the properties of the cross vector product of the unit vectors:

$$\mathbf{i} \times \mathbf{i} = 0, \qquad \mathbf{j} \times \mathbf{j} = 0, \qquad \text{and} \quad \mathbf{k} \times \mathbf{k} = 0$$

$$\mathbf{i} \times \mathbf{j} = -\,\mathbf{j} \times \mathbf{i} = \mathbf{k}, \qquad \mathbf{j} \times \mathbf{k} = -\,\mathbf{k} \times \mathbf{j} = \mathbf{i}, \qquad \mathbf{k} \times \mathbf{i} = -\,\mathbf{i} \times \mathbf{k} = \mathbf{j} \tag{7.5}$$

Combining Eqs. (7.2) to (7.5), we obtain:

$$\mathbf{M} = \mathbf{r} \times \mathbf{F} = (r_y\,F_z - r_z\,F_y)\mathbf{i} + (r_z\,F_x - r_x\,F_z)\mathbf{j} + (r_x\,F_y - r_y\,F_x)\mathbf{k} \tag{7.6}$$

or

$$\mathbf{M} = \mathbf{r} \times \mathbf{F} = M_x\,\mathbf{i} + M_y\,\mathbf{j} + M_z\,\mathbf{k} \tag{7.7}$$

Comparing the results of Eqs. (7.6) with those of Eq. (7.7) yields:

$$M_x = (r_y\,F_z - r_z\,F_y)$$

$$M_y = (r_z\,F_x - r_x\,F_z)$$

$$M_z = (r_x\,F_y - r_y\,F_x) \tag{7.8}$$

When employing the vector cross product **r × F** to determine the moment **M,** we simultaneously write the equations for the moments $M_x$, $M_y$ and $M_z$ about the three Cartesian axes. In writing Eq. (7.6), we often employ the determinant given by:

$$\mathbf{M} = \begin{vmatrix} \mathbf{i} & \mathbf{j} & \mathbf{k} \\ r_x & r_y & r_z \\ F_x & F_y & F_z \end{vmatrix} \tag{7.9}$$

Let's apply these results to a previous example and solve for the forces acting on the communications tower. As we repeat the solution to this example problem, a step-by-step procedure to follow in a design analysis of a three-dimensional structure is outlined.

---

[4] For the convenience of the reader several of the more important equations from Chapter 2 are summarized in this section.

## EXAMPLE 7.4

Consider again the communications tower of Example 7.1. However, use the vector mechanics approach in the solution for the unknown forces acting on the tower.

**Solution:**

**Step 1:** Construct a FBD of the communications tower. We always begin a solution with a FBD regardless of the mathematical approach used in the solution of the problem. We will use the FBD shown previously in Example 7.1 to assist us in writing the necessary equations for the vectors.

**Step 2:** Count the forces acting on the structure and write the vector equation for each force. We have five forces acting on the tower including:

- Two-forces applied by the cables (two of the four cables are slack).
- The wind force.
- The reaction force at the base (origin) of the tower.
- The weight of the tower and transmission equipment.

The vector representing the force $P_2$ on the cable anchored on the x-axis (west) is written as:

$$\mathbf{P_2} = P_2[(-2/\sqrt{29})\mathbf{i} + (-5/\sqrt{29})\mathbf{k}] \tag{a}$$

Similarly, the force $P_4$ on the cable anchored on the y-axis (south) is:

$$\mathbf{P_4} = P_4[(-2/\sqrt{29})\mathbf{j} + (-5/\sqrt{29})\mathbf{k}] \tag{b}$$

Recall that $P_1 = P_3 = 0$ because these cables are slack under the specified wind conditions and do not support the tower. However, when the wind direction changes these cables become functional.

The wind force $F_3$ is written as:

$$\mathbf{F_3} = F_3[\cos(45°)\,\mathbf{i} + \sin(45°)\,\mathbf{j}] = (\sqrt{2}/2)F_3[\mathbf{i} + \mathbf{j}] \tag{c}$$

Note, the vector diagram shown in Fig. E7.1b is helpful in writing Eq. (c).

The reaction force $R_4$ is written as:

$$\mathbf{R_4} = R_{4x}\,\mathbf{i} + R_{4y}\,\mathbf{j} + R_{4z}\,\mathbf{k} \tag{d}$$

The dead weight of the tower structure and the equipment is expressed as:

$$\mathbf{F_5} = -22{,}000\,\mathbf{k} \tag{e}$$

Okay! We have described the five active forces acting on the communications tower in vector format.

**Step 3:** Write the equations for the position vectors for the five active forces. Let's consider the forces in order (i.e. 1, 2, …..5).

For the two active cables that are attached at the top of the tower, we write the relations for the position vectors as:

$$\mathbf{r}_1 = H\,\mathbf{k} \tag{f}$$

$$\mathbf{r}_2 = H\,\mathbf{k} \tag{g}$$

For the wind force, the position vector is given by:

$$\mathbf{r}_3 = (2/3)H\,\mathbf{k} \tag{h}$$

Finally, for both the reaction and dead weight forces, we note that their lines of action pass through the origin and they do not produce a moment about the origin. Hence:

$$\mathbf{r}_4 = \mathbf{r}_5 = 0 \tag{i}$$

**Step 4:** Employ the equilibrium relations to solve for the unknown forces.

First, consider the moments:

$$\Sigma\mathbf{M} = \Sigma\mathbf{r} \times \mathbf{F} = \mathbf{r}_1 \times \mathbf{P}_2 + \mathbf{r}_2 \times \mathbf{P}_4 + \mathbf{r}_3 \times \mathbf{F}_3 \tag{j}$$

Note, the forces due to the reaction and the dead weight forces do not appear in Eq. (j) because $\mathbf{r}_4 = \mathbf{r}_5 = 0$.

Substituting the results for the forces and the position vectors in Eq. (7.9) yields three determinants:

$$\mathbf{M} = P_2 \begin{vmatrix} \mathbf{i} & \mathbf{j} & \mathbf{k} \\ 0 & 0 & H \\ \dfrac{-2}{\sqrt{29}} & 0 & \dfrac{-5}{\sqrt{29}} \end{vmatrix} + P_4 \begin{vmatrix} \mathbf{i} & \mathbf{j} & \mathbf{k} \\ 0 & 0 & H \\ 0 & \dfrac{-2}{\sqrt{29}} & \dfrac{-5}{\sqrt{29}} \end{vmatrix} + \dfrac{\sqrt{2}}{2}F_3 \begin{vmatrix} \mathbf{i} & \mathbf{j} & \mathbf{k} \\ 0 & 0 & \dfrac{2H}{3} \\ 1 & 1 & 0 \end{vmatrix} = 0 \tag{k}$$

Expanding these determinants gives:

$$\mathbf{M} = [(2/\sqrt{29})HP_4 - (\sqrt{2}/2)F_3\,(2H/3)]\,\mathbf{i} + [-(2/\sqrt{29})HP_2 + (\sqrt{2}/2)F_3\,(2H/3)]\,\mathbf{j} = 0 \tag{l}$$

It is evident from Eq. (l) that:

$$M_x = [(2/\sqrt{29})HP_4 - (\sqrt{2}/2)F_3\,(2H/3)] = 0$$

$$M_y = [-(2/\sqrt{29})HP_2 + (\sqrt{2}/2)F_3\,(2H/3)] = 0 \tag{m}$$

Recall that $F_3 = 1600$ lb and solve Eqs. (m) to obtain:

$$P_2 = P_4 = 2030\ \text{lb} \tag{n}$$

The results for $P_2$ and $P_4$ as expected are identical with the solutions determined previously in Example 7.1.

Next, consider the forces:

$$\Sigma F = P_2 + P_4 + F_3 + R_4 + F_5 = 0 \tag{o}$$

Substituting Eqs. (a) through (e) into Eq. (o) yields:

$$\Sigma F = P_2[(-2/\sqrt{29})i + (-5/\sqrt{29})k] + P_4[(-2/\sqrt{29})j + (-5/\sqrt{29})k]$$

$$+ F_3[\cos (45°) i + \sin (45°) j] + R_{4x} i + R_{4y} j + R_{4z} k + (-22,000) k = 0 \tag{p}$$

We collect the coefficients of the unit vectors $i$, $j$, and $k$ and set each of these to zero. It is clear that the components of the sum of the forces must be zero in each of the three directions.

The coefficient of the unit vector $i$ is:

$$\Sigma F_x = P_2(-2/\sqrt{29}) + F_3 \cos (45°) + R_{4x} = 0 \tag{q}$$

The coefficient of the unit vector $j$ is:

$$\Sigma F_y = P_4(-2/\sqrt{29}) + F_3 \sin (45°) + R_{4y} = 0 \tag{r}$$

The coefficient of the unit vector $k$ is:

$$\Sigma F_z = P_2(-5/\sqrt{29}) + P_4(-5/\sqrt{29}) + R_{4z} - 22,000 = 0 \tag{s}$$

From Eqs (q) and (r), we show that:

$$R_{4x} = R_{4y} = -377.5 \text{ lb} \tag{t}$$

The negative sign indicates that the direction of the forces $R_{4x}$ and $R_{4y}$ shown in the FBD should be reversed.

And from Eq. (s), we solve for $R_{4z}$ as:

$$R_{4z} = (5/\sqrt{29})(P_2 + P_4) + 22,000$$

$$R_{4z} = (10/\sqrt{29})(2030) + 22,000 = 25,770 \text{ lb} \tag{u}$$

Comparisons of the results for the forces acting on the communication tower are identical with those determined in Example 7.1. The approach in these two examples was the same—we employed FBDs and the equations of equilibrium in both cases. However, in Example 7.1, we employed the six equations of equilibrium involving the components of the forces and moments. In Example 7.4 we used the two vector equations of equilibrium, a complete vector representation for the moments and forces, and determined the moments by using the vector cross product.

## EXAMPLE 7.5

For the derrick shown in Fig. E7.5, determine the margin of safety for the three supporting cables if the force F = 32 kN. The anchor points P and Q for two of the cables lay in the x-y plane with coordinates P(4, − 3, 0) and Q(− 4, −3, 0). The derrick pole is 5 m high and the boom is 6 m long. The derrick pole and boom are separate components. The boom is attached to the pole with a sleeve type bearing at point R that enables the boom to rotate about the pole. However, the boom is constrained from sliding up or down the pole. The cables are all 20 mm in diameter with a specified strength of 700 MPa. The pole is supported by the floor at point O with a ball and socket joint.

Fig. E7.5

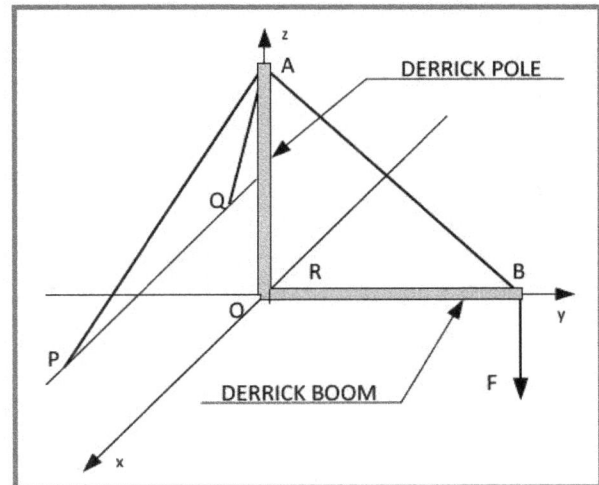

Fig. E7.5

### Solution:

**Step 1:** Let's consider the boom and pole of the derrick separately. First, prepare a FBD of the boom showing all of the unknown forces as shown in Fig. E7.5a. Examination of this figure indicates that we have one cable force $F_{AB}$, the force F due to the load, and the reaction forces between the boom and the derrick pole. Note that no reactive moments are present at point R, only reactive forces.

Fig. E7.5a

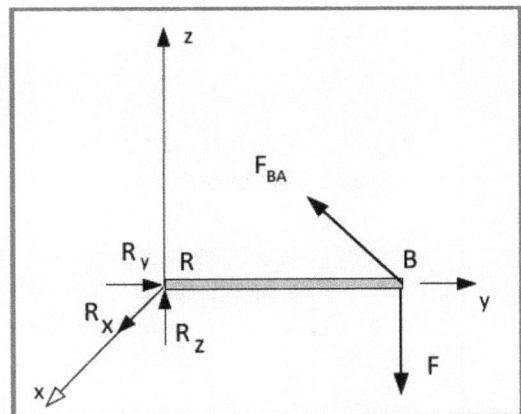

Fig. E7.5a

**Step 2:** Let's write the relations for the forces in vector format.

$$\mathbf{F_{BA}} = F_{BA} \, \mathbf{u_{BA}} \qquad \mathbf{R} = R_x\mathbf{i} + R_y\mathbf{j} + R_z\mathbf{k} \qquad \mathbf{F} = -F\,\mathbf{k} = -32\,\mathbf{k} \qquad \text{(a)}$$

**Step 3:** Next write the equation for the unit vector that provides the direction of the cable force $F_{BA}$ and the position vector defining the location of point B relative to the origin:

$$\mathbf{u_{BA}} = (-6\,\mathbf{j} + 5\,\mathbf{k})/(36 + 25)^{1/2} = -0.7682\,\mathbf{j} + 0.6402\,\mathbf{k} \qquad \text{(b)}$$

$$\mathbf{r_B} = 6\,\mathbf{j} \tag{c}$$

**Step 4:** Apply the equilibrium equations to the derrick boom.

$$\sum \mathbf{M_O} = \mathbf{r_B} \times \mathbf{F_{BA}} + \mathbf{r_B} \times \mathbf{F} = 0 \tag{d}$$

$$\sum \mathbf{M_O} = \begin{vmatrix} \mathbf{i} & \mathbf{j} & \mathbf{k} \\ 0 & 6 & 0 \\ 0 & -0.7682 & 0.6402 \end{vmatrix} F_{BA} + \begin{vmatrix} \mathbf{i} & \mathbf{j} & \mathbf{k} \\ 0 & 6 & 0 \\ 0 & 0 & -32 \end{vmatrix} = 0$$

$$\sum \mathbf{M_O} = (6)(0.6402)F_{BA} - (6)(32)\,\mathbf{i} = 0$$

$$F_{BA} = 49.98 \text{ kN} \tag{e}$$

$$\sum \mathbf{F} = \mathbf{F_{BA}} + \mathbf{R} + \mathbf{F} = 0 \tag{f}$$

$$\sum \mathbf{F} = (R_x)\,\mathbf{i} + (-0.7682\,F_{BA} + R_y)\,\mathbf{j} + (0.6402\,F_{BA} + R_z - 32)\,\mathbf{k} = 0$$

Considering the coefficient of the $\mathbf{i}$, $\mathbf{j}$ and $\mathbf{k}$ terms yields:

$$R_x = 0$$

$$R_z = 32 - 0.6402\,F_{BA} \qquad R_z = 32 - (0.6402)(49.98) = 0 \tag{g}$$

$$R_y = 0.7682\,F_{BA} \qquad R_y = (0.7682)(49.98) = 38.4 \text{ kN} \tag{h}$$

We observe that $R_y$ acts along the axis of the boom tending to compress it. Having completed the analysis of the boom, we continue by considering the pole.

**Step 5:** Let's prepare a FBD of the pole showing all of the unknown forces as shown in Fig. E7.5b. Inspection of this FBD shows five unknown forces ($F_{AP}$, $F_{AQ}$, $R_{xp}$, $R_{yp}$ and $R_{zp}$) and two known forces ($F_{AB}$ and $R_y$). The reaction $R_y$ is the boom pushing on the pole; whereas, the reactions $R_{xp}$, $R_{yp}$ and $R_{zp}$ are due to the interaction of the pole with the deck supporting the derrick.

**Step 6:** Let's write the vector equations for the forces acting on the pole:

$$\mathbf{F_{AB}} = 49.98\,\mathbf{u_{AB}} \qquad\qquad \mathbf{F_{AP}} = F_{AP}\,\mathbf{u_{AP}}$$

$$\mathbf{F_{AQ}} = F_{AQ}\,\mathbf{u_{AQ}} \qquad\qquad \mathbf{R_p} = R_{xp}\,\mathbf{i} + R_{yp}\,\mathbf{j} + R_{zp}\,\mathbf{k} \tag{i}$$

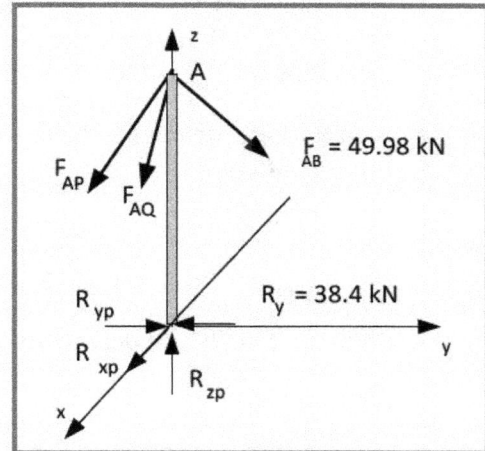

Fig. E7.5b

**Step 7:** Next write the equations for the unit vectors and position vector that provide the directions of the cable forces and their location relative to the origin:

$$\mathbf{u}_{AB} = (6\,\mathbf{j} - 5\,\mathbf{k})/(36 + 25)^{1/2} = 0.7682\,\mathbf{j} - 0.6402\,\mathbf{k}$$

$$\mathbf{u}_{AP} = (4\,\mathbf{i} - 3\,\mathbf{j} - 5\,\mathbf{k})/(16 + 9 + 25)^{1/2} = 0.5657\,\mathbf{i} - 0.4243\,\mathbf{j} - 0.7071\,\mathbf{k} \tag{j}$$

$$\mathbf{u}_{AQ} = (-\,4\,\mathbf{i} - 3\,\mathbf{j} - 5\,\mathbf{k})/(16 + 9 + 25)^{1/2} = -\,0.5657\,\mathbf{i} - 0.4243\,\mathbf{j} - 0.7071\,\mathbf{k}$$

$$\mathbf{r}_A = 5\,\mathbf{k}$$

**Step 8:** Apply the equilibrium equations to the derrick pole.

$$\sum \mathbf{M_O} = \mathbf{r}_A \times \mathbf{F}_{AB} + \mathbf{r}_A \times \mathbf{F}_{AP} + \mathbf{r}_A \times \mathbf{F}_{AQ} = 0 \tag{k}$$

$$\sum \mathbf{M_O} = \begin{vmatrix} \mathbf{i} & \mathbf{j} & \mathbf{k} \\ 0 & 0 & 5 \\ 0 & 0.7682 & -0.6402 \end{vmatrix} 49.98 + \begin{vmatrix} \mathbf{i} & \mathbf{j} & \mathbf{k} \\ 0 & 0 & 5 \\ 0.5657 & -0.4243 & -0.7071 \end{vmatrix} F_{AP} + \begin{vmatrix} \mathbf{i} & \mathbf{j} & \mathbf{k} \\ 0 & 0 & 5 \\ -0.5657 & -0.4243 & -0.7071 \end{vmatrix} F_{AQ} = 0$$

Evaluation of this relation gives the coefficients of the unit vectors. For **i**, we obtain:

$$(-5)(0.7682)(49.98) + (5)(0.4243)\,F_{AP} + (5)(0.4243)F_{AQ} = 0 \tag{l}$$

For **j**, we obtain:

$$(5)(0.5657)\,F_{AP} - (5)(0.5657)\,F_{AQ} = 0 \tag{m}$$

Note that the coefficient of the unit vector **k** is zero. From Eq. (m), it is clear that:

$$F_{AP} = F_{AQ} \tag{n}$$

Substituting Eq. (n) into Eq. (l) gives:

$$F_{AP} = F_{AQ} = [(5)(0.7682)(49.98)]/4.243 = 45.24 \text{ kN} \tag{o}$$

This completes the solution for the forces in the cables. In a more complete analysis, we would employ $\Sigma \mathbf{F} = \mathbf{0}$ to determine the reactions at the base of the pole. It is suggested that you verify the results for $R_{px} = 0$, $R_{py} = 38.4$ kN and $R_{pz} = 95.99$ kN.

**Step 9:** Let's determine the stresses in each of the cables using the results for the forces given in Eqs. (e) and (o).

$$\sigma_{AB} = F_{AB}/A = (49.98 \times 10^3) / [\pi (10)^2] = 159 \text{ MPa}$$

$$\sigma_{AP} = \sigma_{AQ} = F_{AP}/A = (45.24 \times 10^3) / [\pi (10)^2] = 144 \text{ MPa}$$

**Step 10:** Finally, we determine the margin of safety for each of the cables.

For cable AB:

$$MOS = SF - 1 = S/\sigma - 1 = 700/159 - 1 = 3.40$$

For cables AP and AQ:

$$MOS = SF - 1 = S/\sigma - 1 = 700/144 - 1 = 3.86$$

**Step 11:** To complete the analysis, the results should be interpreted. We find, in this example, margins of safety of about 3.5 for all of the cables. The design is reasonably well balanced with stresses on all three cables nearly equal. The margin of safety is comfortable and certainly not excessive for a derrick application where large loads are lifted in the presence of workers.

## EXAMPLE 7.6

A space truss shown in Fig. E7.6, is supported at point A with a ball and socket joint and with cables that are anchored at points B and D as shown in the illustration. A force **F** is applied to joint E. **F** is represented in vector format as:

$$\mathbf{F} = [10\,\mathbf{i} + 8\,\mathbf{j} - 15\,\mathbf{k}] \text{ kips}$$

Determine the internal forces in the three members that intersect at joint E. Also specify the diameter of the rods, if the truss is fabricated from 1020 HR steel. A safety factor of 3.4 based on yield strength is specified by the building code.

Fig. E 7.6

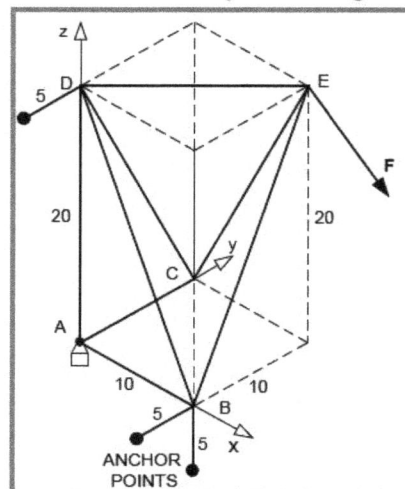

**Solution:**

**Step 1:** Let's consider point E of the space truss and prepare a FBD showing all of the unknown forces as shown in Fig. E7.6a. Examination of this figure indicates that we have a concurrent three-dimensional force system.

**Step 2:** Let's write the relations for the forces in vector format.

$$\mathbf{P_{EB}} = P_{EB} \, \mathbf{u_{EB}} \qquad \mathbf{P_{EC}} = P_{EC} \, \mathbf{u_{EC}} \qquad \mathbf{P_{ED}} = P_{ED} \, \mathbf{u_{ED}}$$

$$\mathbf{F} = 10 \, \mathbf{i} + 8 \, \mathbf{j} - 15 \, \mathbf{k} \tag{a}$$

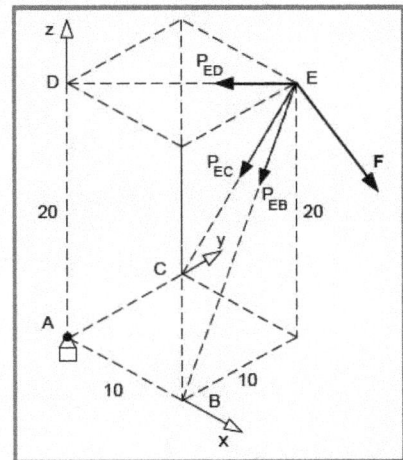

Fig. E7.6a

**Step 3:** Next write the equation for the unit vectors that gives the direction of the forces $F_{EB}$, $F_{EC}$, and $F_{ED}$.

$$\mathbf{u_{EB}} = (- 10 \, \mathbf{j} - 20 \, \mathbf{k})/(100 + 400)^{1/2} = - 0.4472 \, \mathbf{j} - 0.8944 \, \mathbf{k}$$

$$\mathbf{u_{EC}} = (- 10 \, \mathbf{i} - 20 \, \mathbf{k})/(100 + 400)^{1/2} = - 0.4472 \, \mathbf{i} - 0.8944 \, \mathbf{k}$$

$$\mathbf{u_{ED}} = (- 10 \, \mathbf{i} - 10 \, \mathbf{j})/(100 + 100)^{1/2} = - 0.7071 \, \mathbf{i} - 0.7071 \, \mathbf{j} \tag{b}$$

**Step 4:** Apply the equilibrium equation to the force system acting at point E.

$$\Sigma \, \mathbf{F} = 0 \tag{c}$$

Substituting Eqs. (a) and (b) into Eq. (c) yields:

$$\Sigma \, \mathbf{F} = \mathbf{i} \, (-0.4472 P_{EC} - 0.7071 \, P_{ED} + 10) + \mathbf{j} \, (-0.4472 \, P_{EB} - 0.7071 \, P_{ED} + 8)$$

$$+ \, \mathbf{k} \, (-0.8944 \, P_{EB} - 0.8944 \, P_{EC} - 15) = 0 \tag{d}$$

Setting each of the coefficients of the unit vectors to zero yields the three simultaneous equations given by:

$$-0.4472 P_{EC} - 0.7071 \, P_{ED} + 10 = 0$$

$$-0.4472\,P_{EB} - 0.7071\,P_{ED} + 8 = 0 \tag{e}$$

$$-0.8944 P_{EB} - 0.8944\,P_{EC} - 15 = 0$$

Solving Eq. (e) for the forces gives:

$$P_{EC} = -\,6.149 \text{ kip} \qquad P_{EB} = -\,10.62 \text{ kip} \qquad P_{ED} = 18.03 \text{ kip} \tag{f}$$

**Step 5:** Determine the allowable design stress from the yield strength of the 1020 HR steel and the specified safety factor.

$$\sigma_{design} = S_y\,/\mathbf{SF_y} = 42/3.4 = 12.35 \text{ ksi} \tag{g}$$

**Step 6:** Solve for the cross sectional areas and diameters of the three members:

$$A = P/\sigma_{design} \qquad\qquad d = (4A/\pi)^{1/2}$$

$$A_{EC} = 6.149/12.35 = 0.4979 \text{ in.}^2 \qquad d = 0.7962 \text{ in.} \quad \Rightarrow d = 7/8 \text{ in.}$$

$$A_{EB} = 10.62/12.35 = 0.8599 \text{ in.}^2 \qquad d = 1.046 \text{ in.} \quad \Rightarrow d = 1\ 1/4 \text{ in.}$$

$$A_{ED} = 18.03/12.35 = 1.460 \text{ in.}^2 \qquad d = 1.363 \text{ in.} \quad \Rightarrow d = 1\ 1/2 \text{ in.} \tag{h}$$

Our steel supplier lists the standard sizes for 1020 HR steel rods from 1 to 2 in. in diameter as 1, 1 1/4, 1 1/2, 1 3/4 and 2 in.

## EXAMPLE 7.7

A tetrahedral space truss, shown in Fig. E7.7, supports a massive scoreboard and several sets of remotely controlled spotlights in an amphitheater. The base ABC of the space truss lies in the x-y plane (horizontal). The base triangle is connected to anchors in the roof beams by long cables. Determine the size of the solid round rods required to fabricate members AB, AC and AD of the space truss. The safety factor for public arenas is 10, based on yield strength, and the material employed for all of the members is 1018A steel. The scoreboard and spotlights weigh 250 kN.

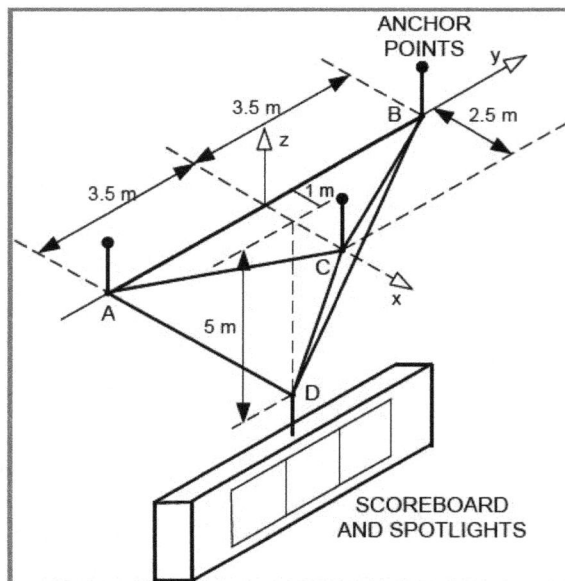

Fig. E7.7

**Solution:**

**Step 1:** Let's consider the space truss in its entirety and prepare a FBD showing the three unknown forces applied by the anchor cables as shown in Fig. E7.7a.

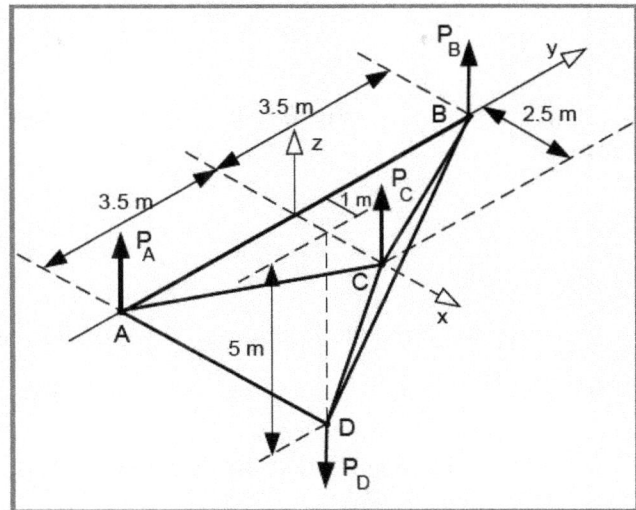

Fig. E7.7a

**Step 2:** Let's write the equations for the moments about point C in vector format and use the results to solve for $P_A$.

$$\Sigma M_C = r_{CA} \times P_A + r_{CB} \times P_B + r_{CD} \times P_D = 0 \qquad \text{(a)}$$

The force vectors and position vectors are written as:

$$P_A = P_A\, k \qquad P_B = P_B\, k \qquad P_C = P_C\, k \qquad P_D = -W\, k \qquad \text{(b)}$$

$$r_{CA} = -2.5\, i - 3.5\, j \qquad r_{CB} = -2.5\, i + 3.5\, j \qquad r_{CD} = -1.5\, i - 5\, k \qquad \text{(c)}$$

Substituting the results of Eqs. (b) and (c) into Eq. (a) yields:

$$M_C = \begin{vmatrix} i & j & k \\ -2.5 & -3.5 & 0 \\ 0 & 0 & P_A \end{vmatrix} + \begin{vmatrix} i & j & k \\ -2.5 & +3.5 & 0 \\ 0 & 0 & P_B \end{vmatrix} + \begin{vmatrix} i & j & k \\ -1.5 & 0 & -5 \\ 0 & 0 & -W \end{vmatrix} = 0 \qquad \text{(d)}$$

Expanding the determinants gives two equations:

$$i(-3.5\, P_A + 3.5\, P_B) = 0 \qquad\qquad j(2.5\, P_A + 2.5\, P_B - 1.5\, W) = 0 \qquad \text{(e)}$$

Recall W = 250 kN and solve Eq. (e) to obtain:

$$P_A = P_B \qquad\qquad\qquad\qquad P_A = 75 \text{ kN} \qquad\qquad \text{(f)}$$

**Step 3:** Let's consider joint A from the space truss and prepare a FBD showing the three unknown forces $P_{AB}$, $P_{AC}$, and $P_{AD}$ as shown in Fig. E7.7b.

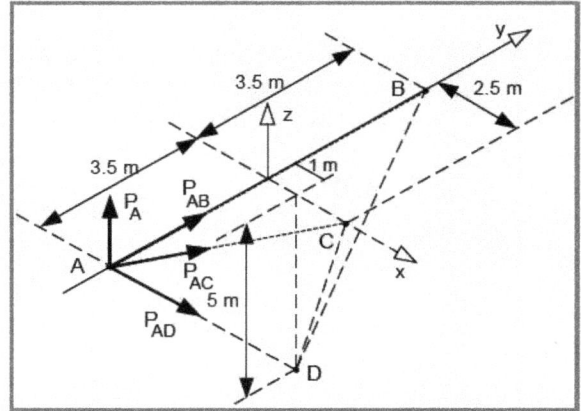

Fig. E7.7b

**Step 4:** Write the equilibrium equation for the concurrent force system acting on joint A as:

$$\Sigma F = P_{AB} + P_{AC} + P_{AD} + P_A = 0 \tag{g}$$

To employ this relation we write the forces in vector format as:

$$P_{AB} = P_{AB}\, u_{AB} \qquad\qquad P_{AC} = P_{AC}\, u_{AC} \qquad\qquad P_{AD} = P_{AD}\, u_{AD} \tag{h}$$

The unit vectors are given by:

$$u_{AB} = r_{AB}/r_{AB} = 7j/7 = j$$

$$u_{AC} = r_{AC}/r_{AC} = (2.5\, i + 3.5 j)/[(2.5)^2 + (3.5)^2]^{1/2} = 0.5812\, i + 0.8137\, j$$

$$u_{AD} = r_{AD}/r_{AD} = (1.0\, i + 3.5 j - 5k)/[(1.0)^2 + (3.5)^2 + (5)^2]^{1/2}$$

$$u_{AD} = 0.1617\, i + 0.5659\, j - 0.8084\, k \tag{i}$$

Substituting the results of Eqs. (h) and (i) into Eq. (g) yields:

$$i\,(0.5812\, P_{AC} + 0.1617\, P_{AD}) + j\,(0.8137\, P_{AC} + 0.5659\, P_{AD} + P_{AB}) + k\,(75 - 0.8084\, P_{AD}) = 0$$

Solving this equation gives:

$$P_{AD} = 92.77 \text{ kN} \qquad P_{AC} = -25.81 \text{ kN} \qquad P_{AB} = -31.50 \text{ kN} \tag{j}$$

**Step 5:** Determine the allowable design stress from the yield strength of the 1018A steel and the specified safety factor.

$$\sigma_{design} = S_y/SF_y = 221/10 = 22.1 \text{ MPa} \tag{k}$$

**Step 6:** Solve for the cross sectional areas and diameters of the three members:

$$A = P/\sigma_{design} \qquad\qquad d = (4A/\pi)^{1/2} \tag{l}$$

$$A_{AD} = 92.77 \times 10^3/22.1 = 4197 \text{ mm}^2 \qquad d = 73.10 \text{ mm} \qquad \Rightarrow d = 75 \text{ mm}$$

$A_{AC} = 25.81 \times 10^3/22.1 = 1168 \text{ mm}^2 \qquad d = 38.56 \text{ mm} \qquad \Rightarrow d = 40 \text{ mm}$

$A_{AB} = 31.50 \times 10^3/22.1 = 1425 \text{ mm}^2 \qquad d = 42.60 \text{ mm} \qquad \Rightarrow d = 45 \text{ mm} \qquad (m)$

The diameters of the bars have been converted to the standard metric sizes available. A second analysis will be necessary to insure that the long thin rods used for members AC and AB, that are subjected to compressive forces, do not buckle.

## EXAMPLE 7.8

The long horizontal boom of a construction crane is fabricated from a lattice of steel bars. A four-cell segment of the boom, presented in Fig. E7.8, shows that the lattice is periodic from points A to E with identical cells at 8 ft intervals. The rectangle AEOK lies in the x-y plane and the equilateral triangle AKF lies in the x-z plane. The lengths of the members AF = FK = AK = 6 ft. The crane is lifting a load F = 12 kip at point P. Determine the internal forces in members CD, MN, HI, HN, HD, and MD using the method of sections.

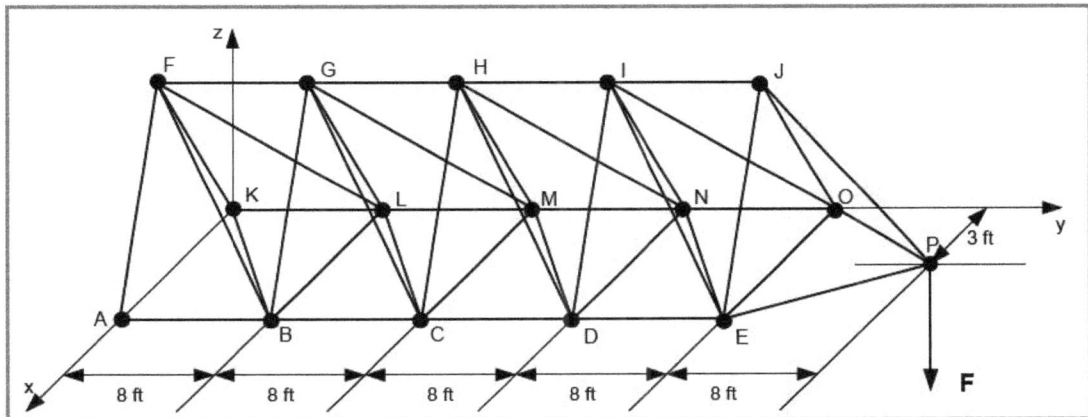
Fig. E7.8

## Solution:

**Step 1:** Let's make a section cut perpendicular to the y-axis between points C and D and prepare a FBD of the right side of the boom. This FBD, presented in Fig. E7.8a, shows six unknown internal forces acting on the members CD, MN, HI, HN, DH, and DM.

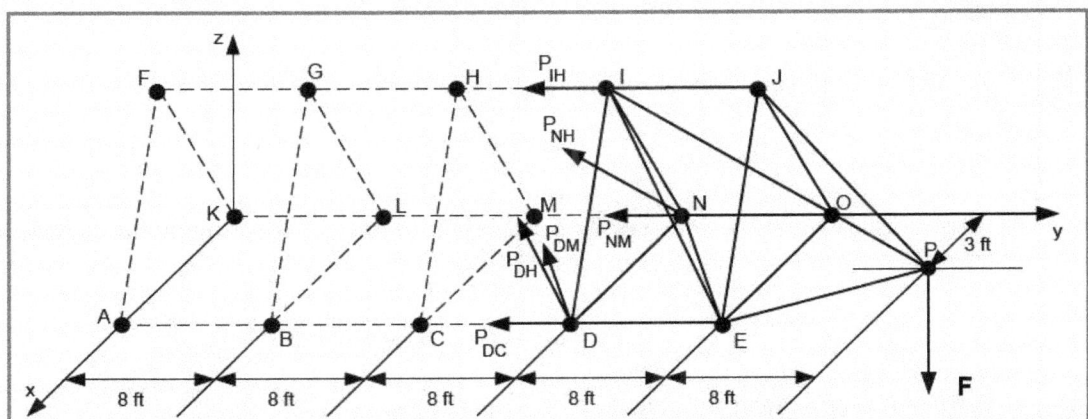
Fig. E7.8a

**Step 2:** Let's write the equations for the moments about point D in vector format and use the results to obtain three equations in terms of the unknown internal forces $P_{NM}$, $P_{NH}$ and $P_{IH}$.

$$\Sigma M_D = r_{DN} \times P_{NM} + r_{DN} \times P_{NH} + r_{DI} \times P_{IH} + r_{DP} \times F = 0 \tag{a}$$

The force vectors, unit vectors and position vectors in Eq. (a) are written as:

$$P_{NM} = P_{NM} \, u_{NM} \qquad P_{NH} = P_{NH} \, u_{NH} \qquad P_{IH} = P_{IH} \, u_{IH} \qquad F = -F \, k \tag{b}$$

$$u_{NM} = r_{NM}/ \, r_{NM} = -8 \, j/8 = -j$$

$$u_{NH} = r_{NH}/ \, r_{NH} = (+3i - 8j + 5.196 \, k)/[(3)^2 + (8)^2 + (5.196)^2]^{1/2} = 0.3 \, i - 0.8 \, j + 0.5196 \, k \tag{c}$$

$$u_{IH} = r_{IH}/ \, r_{IH} = -8 \, j/8 = -j$$

Substituting Eq. (c) into Eq. (b) gives the internal forces in vector notation as:

$$P_{NM} = -P_{NM} \, j \qquad P_{NH} = P_{NH} \, (0.3 \, i - 0.8 \, j + 0.5196 \, k) \qquad P_{IH} = -P_{IH} \, j \qquad F = -F \, k \tag{d}$$

The position vectors are written as:

$$r_{DN} = -6 \, i \qquad\qquad r_{DI} = -3 \, i + 5.196 \, k \qquad\qquad r_{DP} = -3 \, i + 16 \, j \tag{e}$$

Substituting the results of Eqs. (d) and (e) into Eq. (a) yields:

$$M_D = P_{NM} \begin{vmatrix} i & j & k \\ -6 & 0 & 0 \\ 0 & -1 & 0 \end{vmatrix} + P_{NH} \begin{vmatrix} i & j & k \\ -6 & 0 & 0 \\ 0.3 & -0.8 & 0.5196 \end{vmatrix} + P_{IH} \begin{vmatrix} i & j & k \\ -3 & 0 & 5.196 \\ 0 & -1 & 0 \end{vmatrix} + \begin{vmatrix} i & j & k \\ -3 & 16 & 0 \\ 0 & 0 & -F \end{vmatrix} = 0 \tag{f}$$

Expanding the determinants gives three equations:

$$i(5.196 \, P_{IH} - 16 \, F) = 0$$

$$j(3.118 \, P_{NH} - 3F) = 0 \tag{g}$$

$$k(6 \, P_{NM} + 4.8 \, P_{NH} + 3 \, P_{IH}) = 0$$

Recall F = 12 kip and solve Eqs. (g) to obtain:

$$P_{IH} = 36.95 \text{ kip} \qquad P_{NH} = 11.55 \text{ kip} \qquad P_{NM} = -27.71 \text{ kip} \tag{h}$$

**Step 3:** Let's write the equation for equilibrium of the forces acting on the right side of the crane boom that is freed by the section cut:

$$\Sigma F = P_{DC} + P_{DH} + P_{DM} + P_{NM} + P_{NH} + P_{IH} + F = 0 \tag{i}$$

Let's write each of these forces in vector notation as:

$$\mathbf{P_{DC}} = -P_{DC}\,\mathbf{j}$$

$$\mathbf{P_{DH}} = P_{DH}\,(-0.3\,\mathbf{i} - 0.8\,\mathbf{j} + 0.5196\,\mathbf{k})$$
$$\mathbf{P_{DM}} = P_{DM}\,(-0.6\,\mathbf{i} - 0.8\,\mathbf{j})$$

$$\mathbf{P_{NM}} = 27.71\,\mathbf{j} \tag{j}$$

$$\mathbf{P_{NH}} = 3.465\,\mathbf{i} - 9.240\,\mathbf{j} + 6.000\,\mathbf{k}$$

$$\mathbf{P_{IH}} = -36.95\,\mathbf{j}$$

$$\mathbf{F} = -12.0\,\mathbf{k}$$

Substitute Eqs. (j) into Eq. (i) and collect all of the coefficients of **i, j** and **k** to obtain three equations containing the three unknown internal forces:

$$-0.3\,P_{DH} - 0.6\,P_{DM} + 3.465 = 0$$

$$-P_{DC} - 0.8\,P_{DH} - 0.8\,P_{DM} + 27.71 - 9.240 - 36.95 = 0 \tag{k}$$

$$0.5196\,P_{DH} + 6.000 - 12.00 = 0$$

Solving Eqs. (k) yields:

$$P_{DH} = 11.55 \text{ kip} \qquad P_{DM} = 0 \qquad P_{DC} = -27.71 \text{ kip} \tag{l}$$

We note that the forces $P_{DC} = P_{NM} = -27.71$ kip are compressive along the stringers in the x-y plane. The force $P_{IH} = 36.95$ kip along the top stringer is tensile as expected. Because the internal force in the diagonal member DM = 0, the space truss exhibits symmetric behavior with $P_{DC} = P_{NM}$ and $P_{DH} = P_{NH}$.

## 7.4  THREE DIMENSIONAL EQUILIBRIUM

The emphasis in the preceding sections of this chapter has been on the application of the equilibrium equations to solve for unknown forces in space structures. However, applications of the equilibrium relations to other three-dimensional bodies such as machine components are often required in engineering studies. While the geometry of the bodies differs between structures and machine components, the approach is the same. We prepare FBDs, write appropriate equilibrium equations for each FBD and solve the equations for the unknown forces. To demonstrate these non-structural applications, let's consider two examples.

## EXAMPLE 7.9

A local firm has constructed a small crane consisting of a boom supported by two steel wires BC and DE as shown in Fig. E7.9. The boom is fixed to the supporting wall with a ball and socket joint at point A. The wires are anchored into the wall at points C and E. Each wire has a diameter of 0.0625 in. and an ultimate tensile strength of 200 ksi. Determine the maximum weight that can be supported by the boom, the support reactions at point A and the forces in both support wires.

Fig. E7.9

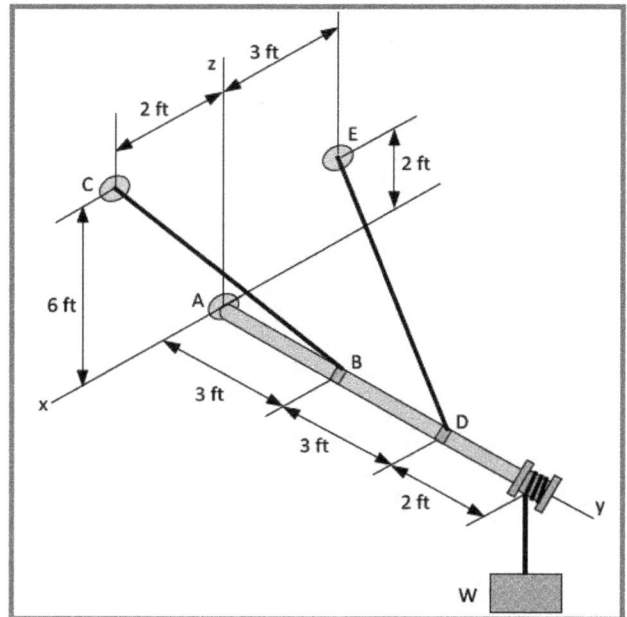

**Solution:**

**Step 1:** Let's consider the crane in its entirety and prepare a FBD showing the three unknown forces at the ball and socket joint and the forces in the wires as shown in Fig. E7.9a.

Fig. E7.9a

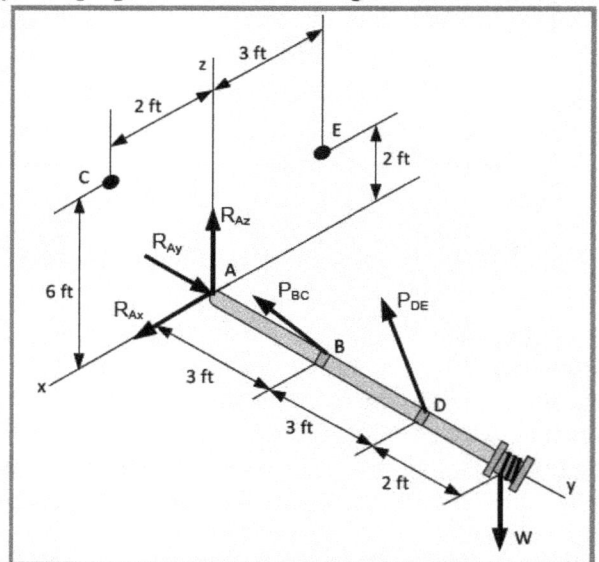

**Step 2:** Write the equations for the forces in the wires using a vector format.

$$\mathbf{P}_{BC} = P_{BC} \, \mathbf{u}_{BC} = P_{BC} \, (2\,\mathbf{i} - 3\,\mathbf{j} + 6\,\mathbf{k})/(7) \qquad (a)$$

$$\mathbf{P_{DE}} = P_{DE}\ \mathbf{u_{DE}} = P_{DE}\ (-3\ \mathbf{i} - 6\ \mathbf{j} + 2\ \mathbf{k})/(7) \qquad \text{(b)}$$

**Step 3:** Consider equilibrium of the structure. From the FBD in Fig. E7.9a, we write the six equations of equilibrium as:

$$\Sigma F_x = R_{Ax} + (2/7)P_{BC} - (3/7)\ P_{DE} = 0 \qquad \text{(c)}$$

$$\Sigma F_y = R_{Ay} - (3/7)P_{BC} - (6/7)\ P_{DE} = 0 \qquad \text{(c)}$$

$$\Sigma F_z = R_{Az} + (6/7)P_{BC} + (2/7)\ P_{DE} - W = 0 \qquad \text{(d)}$$

$$\Sigma M_x = (3)(6/7)P_{BC} + (6)(2/7)P_{DE} - (8)W = 0 \qquad \text{(e)}$$

$$\Sigma M_y = 0 \qquad \text{(f)}$$

$$\Sigma M_z = -(3)(2/7)P_{BC} + (6)(3/7)P_{DE} = 0 \qquad \text{(g)}$$

**Step 4:** Solve Eq. (g) and determine the maximum tension allowable in either wire to obtain:

$$P_{BC} = 3P_{DE} \qquad \text{(h)}$$

$$P_{Max} = SA = (200 \times 10^3)[(\pi/4)(0.0625)^2] = 613.6\ \text{lb} \qquad \text{(i)}$$

**Step 5:** Use Eqs. (h) and (i) to solve for the forces in the wires. Because $P_{BC} > P_{DE}$ it is evident that:

$$P_{BC} = P_{Max} = 613.6\ \text{lb} \quad \text{and} \quad P_{DE} = 204.5\ \text{lb} \qquad \text{(j)}$$

**Step 6:** Solve for the maximum weight that can be lifted by substituting Eq. (j) into Eq. (e):

$$W_{Max} = (1/8)[(18/7)(613.6) + (12/7)(204.5)] = 241.1\ \text{lb} \qquad \text{(k)}$$

**Step 7:** Solve for the reaction forces at point A by substituting Eqs. (j) and (k) into Eqs. (c), (d) and (e) to obtain:

$$R_{Ax} = -87.67\ \text{lb} \qquad R_{Ay} = 438.3\ \text{lb} \qquad R_{Az} = -343.3\ \text{lb} \qquad \text{(l)}$$

## EXAMPLE 7.10

A hand operated lifting mechanism called a windless utilizes a crank to rotate a drum as shown in Fig. E7.10. The shaft of the mechanism is supported by a wall mounted ball and socket joint at point A and a smooth journal bearing at point B. The handle of the crank in the position shown is in the y-z plane. For this position of the crank handle determine the force P required to lift a weight W = 150 lb. Also determine the reactions at the ball and socket joint and the journal bearing.

**Solution:**

**Step 1:** Let's consider the crane in its entirety and prepare a FBD showing the three unknown forces at the ball and socket joint, the forces at the journal bearing and the applied forces as shown in Fig. E7.10a.

**Step 2:** Let's employ vector mechanics in this solution by writing the equation for the moment vector as:

$$\Sigma M = r_W \times F_W + r_{AB} \times F_B + r_F \times F = 0 \tag{a}$$

The forces are given by:

$$W = -150\,k \qquad F_B = F_{Bx}\,i + F_{Bz}\,k \qquad F = F\,i \tag{b}$$

The position vectors are given by:

$$r_W = 0.5\,i + 2\,j \qquad r_{AB} = 4\,j \qquad r_F = 6\,j - 1\,k \tag{c}$$

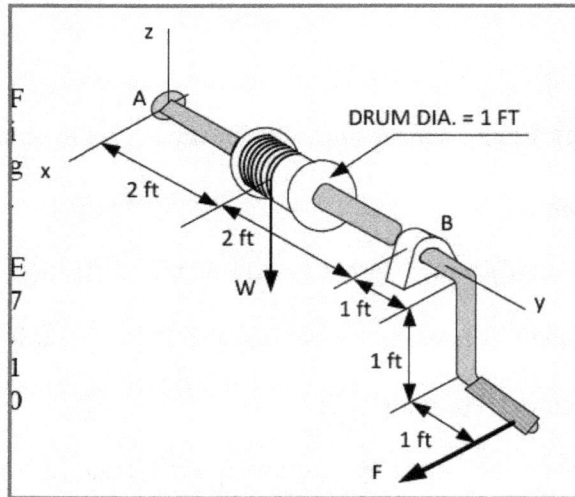

Fig. E7.10a

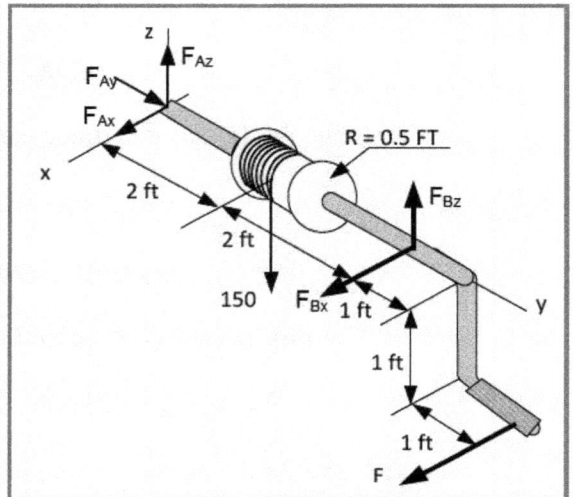

Fig. E7.10b

**Step 3:** Substitute the results from Eqs. (a) and (b) into Eq. (7.9) and solve the determinates to obtain:

$$\Sigma M = \begin{vmatrix} i & j & k \\ 0.5 & 2 & 0 \\ 0 & 0 & 0 \end{vmatrix} + \begin{vmatrix} i & j & k \\ 0 & 4 & 0 \\ F_{Bx} & F_{By} & F_{Bz} \end{vmatrix} + \begin{vmatrix} i & j & k \\ 0 & 6 & -1 \\ F & 0 & 0 \end{vmatrix} = 0 \tag{d}$$

$$(300\,i + 75\,j) + (4F_{Bz}\,i - 4\,F_{Bx}\,k) + F(-\,j - 6\,k) = 0 \tag{e}$$

Rearrange the terms in Eq. (e) and write:

$$(-300 + 4F_{Bz})i + (75 - F)\,j + (-4F_{Bx} - 6F)\,k = 0 \tag{f}$$

Equate each coefficient in Eq. (f) to zero to obtain:

$$(-300 + 4F_{Bz}) = 0 \qquad (75 - F) = 0 \qquad (-4F_{Bx} - 6F) = 0 \tag{g}$$

Solve Eqs. (g) for the unknowns F, $F_{Bx}$ and $F_{Bz}$ to obtain:

$$F = 75 \text{ lb} \qquad F_{Bz} = 75 \text{ lb} \qquad F_{Bx} = -112.5 \text{ lb} \qquad\qquad \text{(h)}$$

**Step 4:** To determine the forces at the ball and socket joint, let's use $\Sigma \mathbf{F} = 0$ and write:

$$\Sigma F_x = F_{Ax} + F_{Bx} + F = 0 \qquad \Sigma F_y = F_{Ay} = 0 \qquad \Sigma F_z = F_{Az} + F_{Bz} - 150 = 0 \qquad \text{(i)}$$

Substituting the results from Eqs. (h) into Eqs (i) yields:

$$F_{Ax} = 37.5 \text{ lb} \qquad\qquad F_{Ay} = 0 \qquad\qquad F_{Az} = 75 \text{ lb} \qquad\qquad \text{(j)}$$

## 7.5 SUMMARY

In the discussion of three-dimensional structures, we presented two different mathematical approaches. The first, consistent with most of the previous discussion, employed trigonometric methods for determining force and moment components in the three directions. The second approach employs vector analysis. The first approach is intuitively obvious and the second is more mathematically elegant. In both approaches, we begin with FBDs of the structure. We suggest the use of two or three-view FBDs to facilitate the visualization process.

With the trigonometric approach for determining the force and moment components, we employ six equilibrium relations in the solution of the unknown internal and external forces:

$$\Sigma F_x = 0; \qquad\qquad \Sigma F_y = 0; \qquad \text{and} \qquad \Sigma F_z = 0$$

$$\Sigma M_x = 0; \qquad\qquad \Sigma M_y = 0; \qquad \text{and} \qquad \Sigma M_z = 0$$

With the vector analysis approach, we write the vector relations defining all of the forces acting on the structure.

$$\mathbf{F_1} = F_{1x}\,\mathbf{i} + F_{1y}\,\mathbf{j} + F_{1z}\,\mathbf{k}$$
$$\cdots\cdots\cdots\cdots\cdots\cdots\cdots$$
$$\cdots\cdots\cdots\cdots\cdots\cdots\cdots$$
$$\mathbf{F_n} = F_{nx}\,\mathbf{i} + F_{ny}\,\mathbf{j} + F_{nz}\,\mathbf{k}$$

Then the position vectors that define the point of application of the forces relative to the point about which the moments act are written.

$$\mathbf{r_1} = r_{1x}\,\mathbf{i} + r_{1y}\,\mathbf{j} + r_{1z}\,\mathbf{k}$$
$$\cdots\cdots\cdots\cdots\cdots\cdots\cdots$$
$$\cdots\cdots\cdots\cdots\cdots\cdots\cdots$$
$$\mathbf{r_n} = r_{nx}\,\mathbf{i} + r_{ny}\,\mathbf{j} + r_{nz}\,\mathbf{k}$$

The moments are determined from the force vectors and the position vectors by employing the vector cross product.

$$\mathbf{M} = \mathbf{r} \times \mathbf{F}$$

Finally, the vector forms of the equilibrium relations are used to solve for the unknown internal and external forces acting on the three-dimensional structure.

$$\Sigma F = 0 \quad \text{and} \quad \Sigma M = 0$$

Three-dimensional structures are difficult to analyze in comparison to two-dimensional structures. In both cases, the approach is the same; construct the appropriate FBDs and apply the equilibrium relations. The difficulty in analyzing three-dimensional structures arises for two reasons. First, visualization is often a problem. Second, many forces are usually involved and the analysis becomes long and tedious. The need for visualization is mitigated to some degree by employing a vector mechanics approach where the process leads one to understand the geometry and dimensions of the body under consideration. Dividing the solution into different steps and completing each step as an individual analysis addresses the length of the analysis. This stepwise approach was demonstrated in several examples.

## REFERENCES

1. Zalewski, W. and E. Allen, Shaping Structures: Statics, John Wiley & Sons, New York, NY, 1998, p 167-169.

## PROBLEMS

7.1 The Patcenter in Princeton, NJ is a large open structure with a cable-stayed roof. In the figure presented below, one of the nine tubular steel masts that are uniformly spaced at 9 m intervals to support the roof structure is shown. In this design, the roof hangs from cables because the vertical side members are used only to prevent wind uplift. The tubular mast, 15 m high, is supported with a 9 m wide by 6 m high rectangular steel frame. Determine the force in the primary rod stay and the tubular steel masts if the uniformly distributed load on the roof is 70 kN/m. Assume the roof is pinned to the steel mast at point C. Hint: Refer to the Example 7.3 in the text for additional dimensions.

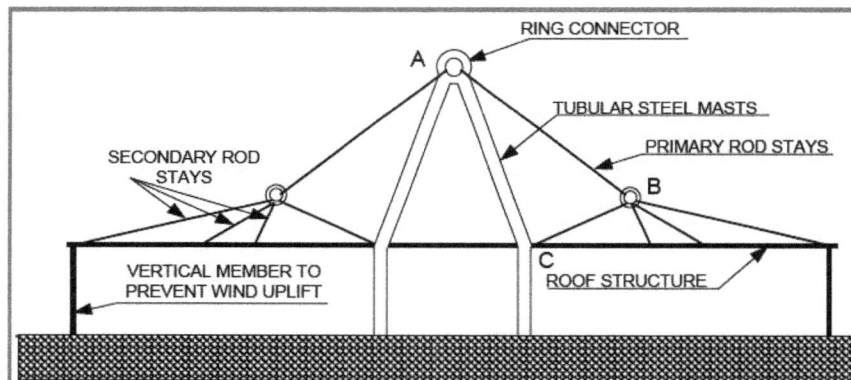

7.2 A new building similar to the Patcenter is under consideration by an architectural firm. They propose increasing the height of the mast from 15 m to 24 m to make the appearance of the structure more dramatic. If all of the other parameters are the same as given in Problem 7.1, prepare an analysis indicating the effect of this change.

7.3 A new building similar to the Patcenter is under consideration by an architectural firm. They propose a structure with the dimensions presented in the figure to the right. Determine the force in the primary rod stay AB and the tubular steel masts AC. Also find the reaction forces at point C.

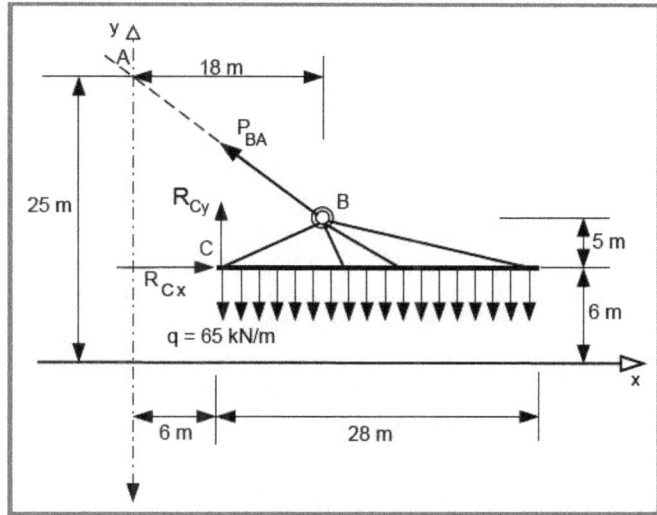

7.4 A space truss, shown in the figure to the left, is supported at point A with a ball and socket joint and with cables anchored at points B and D. A force F is applied to joint E. If F is represented in vector format as shown in the figure, determine the internal forces in the three members that intersect at joint E. Also specify the diameter of the rods, if the truss is fabricated from a 1018 A steel alloy. A safety factor of SF = 3.0 based on yield strength is specified by the building code. The truss dimensions are given in ft.

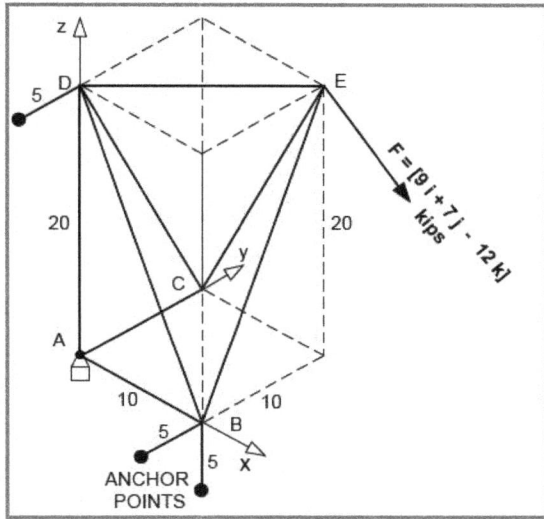

7.5 Determine the internal forces in members AB, DB and EB of the truss described in Problem 7.4.

7.6 Determine the internal forces in members AC, DC and EC of the truss described in Problem 7.4.

7.7 For the communications tower illustrated in the figure on the next page, a wind force of 4,500 lb from the West is applied at a position $z = 300$ ft. The tower has a height $H = 400$ ft and a dead weight $W = 50$ kips. The anchor points on the ground plane are defined by: $x_1 = 100$ft, $x_2 = 115$ ft, $y_1 = 130$ ft and $y_2 = 145$ ft. The cables have a breaking strength of 26,600 lb and a ball and socket joint is used to support the tower at point O. Determine the loads in the four cables and the reaction forces at the base of the tower. Also determine the minimum safety factor.

7.8 The anchor points on the ground plane for the communications tower shown in the figure on the next page are changed to: $x_1 = 60$ m, $x_2 = 70$ m, $y_1 = 80$ m, $y_2 = 90$ m. The tower has a height of 120 m, a dead weight of 200kN and is subjected to a wind force of 12.6 kN from the East, that is applied at a position $z = 80$ m. The cables have a breaking strength of 130 kN, and the tower is supported by a ball and socket joint at point O. Determine the loads on the four cables and the reaction forces at the ball and socket joint. Also find the minimum safety factor.

7.9 If the steel used in fabricating the tower described in Problem 7.7 has a yield strength of 50 ksi, determine the cross sectional area required at the base of the tower. The safety factor for the tower is specified as 4.0. Does the wind loading condition affect the result?

7.10 If the steel used in fabricating the tower in Problem 7.8 has a yield strength of 360 MPa, determine the cross sectional area required at the base of the tower. The safety factor for the tower is specified as 5.0. Does the wind loading condition affect the result?

7.11 The safety factors for the communications tower described in Problems 7.7 and 7.8 are very large. Develop arguments supporting the use of relatively large safety factors in the design of communications towers. Also develop arguments for redesign with smaller diameter cable, smaller footprint and smaller cross sectional area at the base of the tower structure.

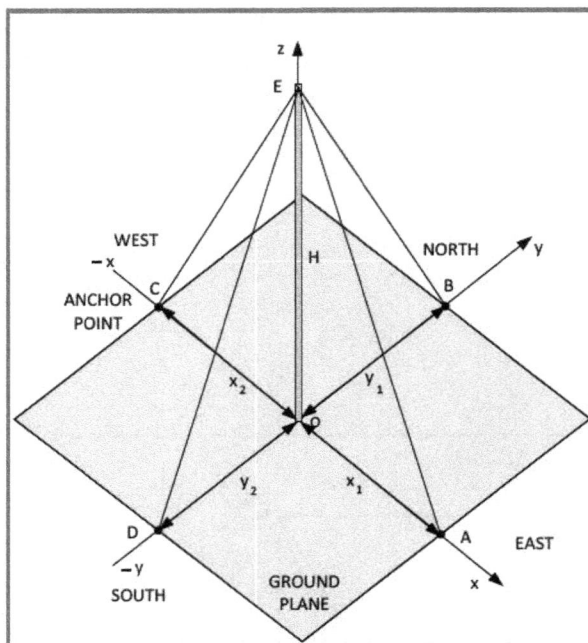

7.12 A severe ice storm strikes the communications tower coating all of the members with a thick layer of ice. The dead weight of the tower is increased from 200 kN to 320 kN. Determine the decrease in the safety factor for the conditions of Problem 7.8.

7.13 For the derrick shown in the figure below, determine the margin of safety for the three cables if the force F acting on the boom is 4.0 ton. The cable diameter is 0.5 in. for all three cables. The anchor points P and Q for two of the cables lie in the x-y plane with coordinates P= (12, −9, 0) ft and Q = (−12, − 9, 0) ft. The derrick pole is 15 ft high and the boom is 20 ft long. The derrick pole and the boom are separate components. The boom is attached to the pole at point R with a sleeve type bearing that enables the boom to rotate about the pole. However, the boom is constrained from sliding up or down the pole. The derrick pole is supported by a ball and socket joint at point O. Details of the bearing arrangement are also shown in the figure below. The cables have a specified breaking strength of 180 ksi.

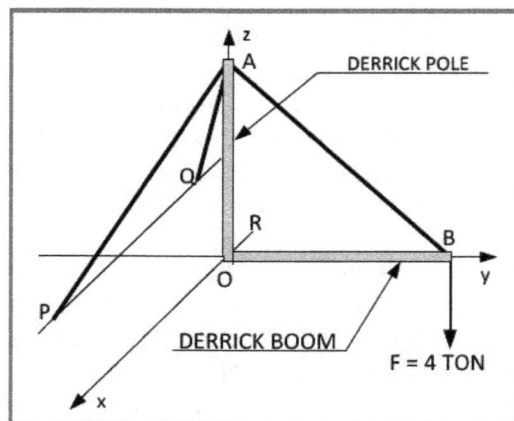

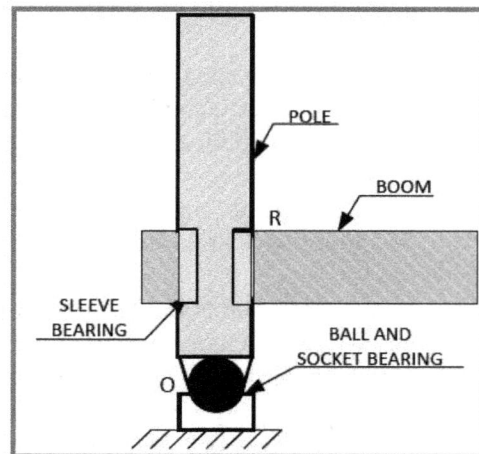

7.14  A tetrahedral space truss, shown in the figure to the right, supports a massive scoreboard and several sets of remotely controlled spotlights in an amphitheater. The base ABC of the space truss lies in the x-y plane (horizontal). The base triangle is connected to anchors in the roof beams by long cables. Determine the size of the solid round rods that are required to fabricate members AB, AC and AD of the space truss. The safety factor based on yield strength is specified as 6.0. 1020 HR steel is employed for all of the members and the weight of the scoreboard and spotlights is 20.0 kN.

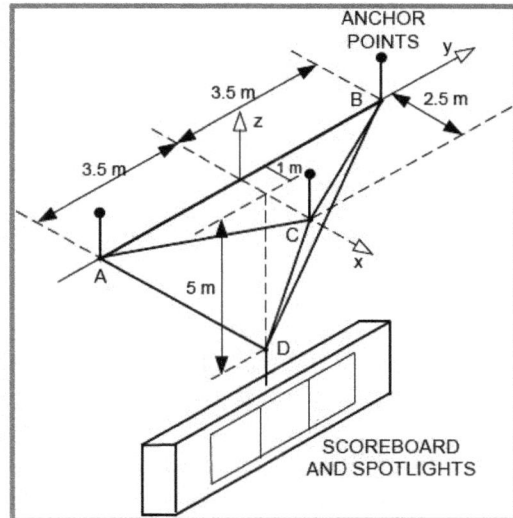

7.15  For the long horizontal boom of a construction crane, illustrated in the figure shown below, determine the internal forces in the members AB, BF, BK, FG, KL and FL. The load applied to the boom is F = 10 kip. Hint: See Example 7.8 in the text for additional details.

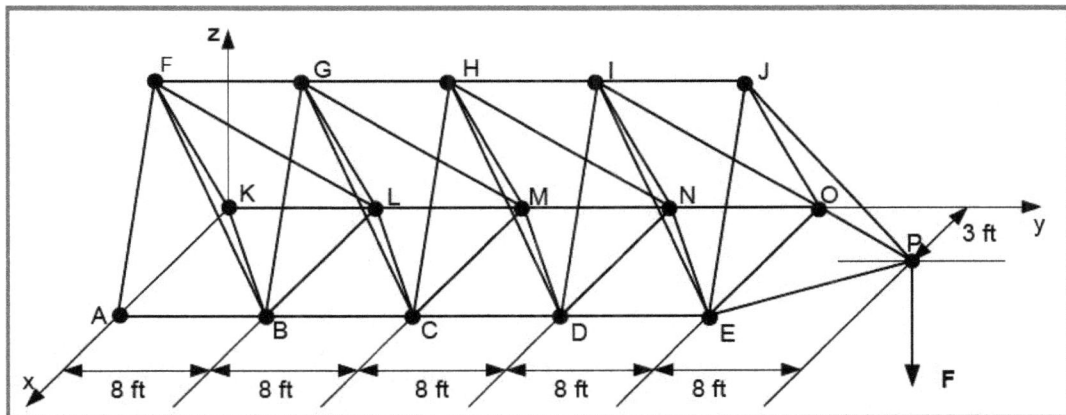

7.16  Repeat Problem 7.15 if the loading on the long horizontal boom is increased from 10 kip to 14 kip.

7.17  A hanging light assembly is positioned near the corner of a gymnasium as shown in the figure on the next page. Determine the gage of the stainless steel wire required for the support of a light assembly weighing 1,100 N if a safety factor of 5.0 is specified. The wire is fabricated from 302A stainless steel. Points B and C are anchors that lie in the x-y plane and point D is an anchor that lies along the z-axis. Point A is not anchored; however it lies in the x-y plane.

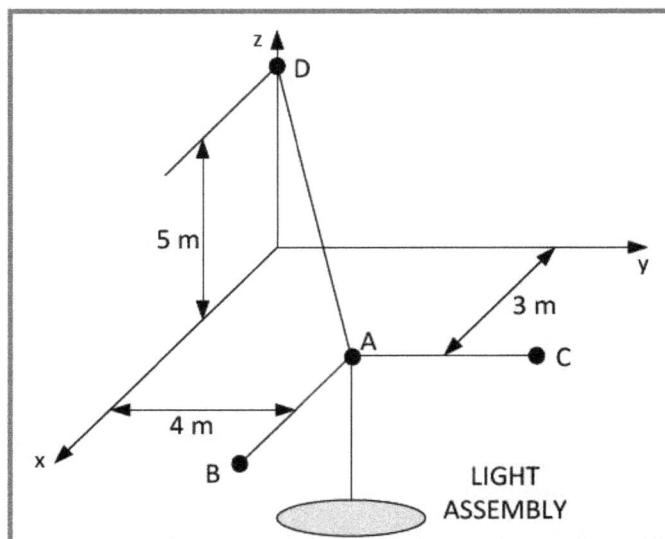

7.18    A hot air balloon, shown in the figure to the right, is moored to the ground with three cables that are anchored at points A, B and C. The coordinates (x, y, z) of the anchor points on the ground and on the basket are given in the figure. Determine the force exerted by each of the cables if the upward lift of the balloon is 950 lb. Assume that wind forces are negligible.

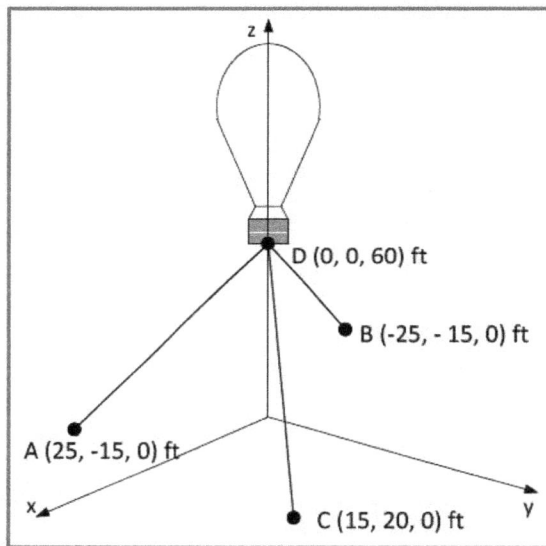

7.19    A circus cage, displayed in a large high ceiling auditorium, is supported above ground level by the three wires illustrated in the figure to the left. Determine the gage of the stainless steel wire required to support the cage weighing 2,500 lb if a safety factor of 3.5 is specified. The wire is fabricated from 4340 HR steel. The geometric parameters defining the assembly are listed in the table below. Points A, B and D locate the anchors for the cables.

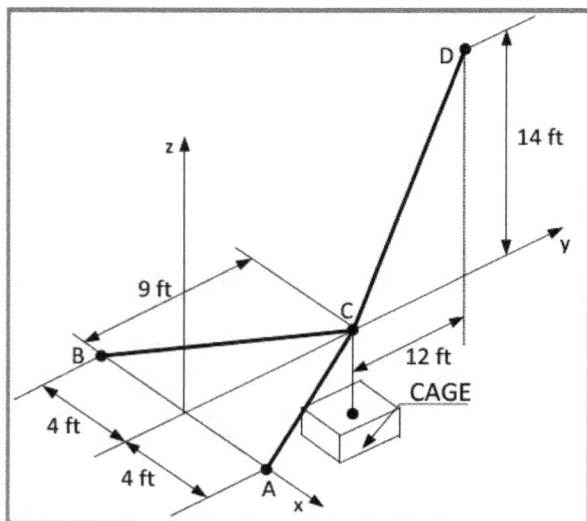

7.20 A local firm has constructed a small crane consisting of a boom supported by two steel wires BC and DE as shown in the figure to the right. The boom is fixed to the supporting wall with a ball and socket joint at point A. The wires are anchored into the wall at points C and E. Each wire has a diameter of 0.125 in. and an ultimate tensile strength of 150 ksi. Determine the maximum weight that can be supported by the boom, the support reactions at point A and the forces in wires BC and DE.

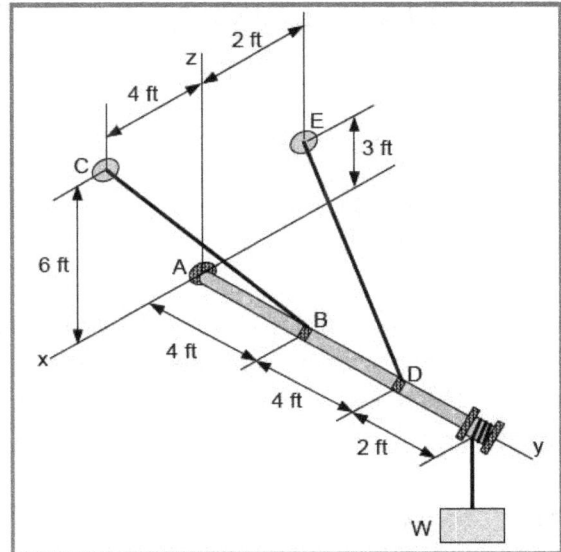

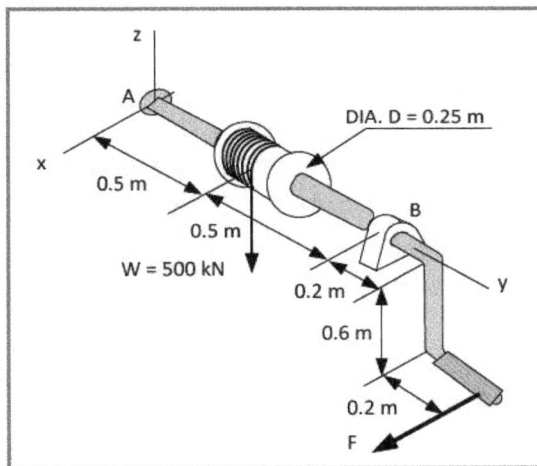

7.21 A hand operated lifting mechanism called a windless utilizes a crank to rotate a drum as shown in the figure to the left. The shaft of the mechanism is supported by a wall mounted ball and socket joint at point A and a smooth journal bearing at point B. The arm and handle of the crank in the position shown is in the y-z plane. For this position of the crank handle, determine the force F required to hold a weight of 500 kN in equilibrium. Also determine the reactions at the ball and socket joint and the journal bearing.

7.22 Repeat Problem 7.21 if the crank is rotated clockwise 90 degrees so that the arm and handle of the crank is in the x-y plane and the force F is acting in the positive z direction. For this position of the crank handle, determine the force F required to hold the weight of 500 kN in equilibrium. Also determine the reactions at the ball and socket joint and the journal bearing.

7.23 Repeat Problem 7.21 if the crank is rotated clockwise 180 degrees so that the handle of the crank is in the y-z plane and the force F is acting in the negative x direction. For this position of the crank handle determine, the force F required to hold the weight of 500 kN in equilibrium. Also determine the reactions at the ball and socket joint and the journal bearing.

# CHAPTER 8

# FRAMES AND MACHINES

## 8.1 INTRODUCTION

**Frames** are similar to trusses in that they both are fabricated from long thin members. However, there are two important differences.

1.  The joints in a truss are pinned and free to rotate; whereas, one or more joints in a frame may be rigid. Because rotation is constrained, the rigid joints are capable of providing reaction moments.
2.  The forces applied to a truss act only at the joints; whereas, forces may be applied at any location on a frame. Thus, one or more members in a frame may be subjected to more than two forces.

All of the members in a truss are subjected to only two external forces; hence, the internal force must coincide with the line of action of these two equal and opposite forces. In a frame one or more members is subjected to more than two external forces. When we make a section cut across a multiforce member[1], it is necessary to assume the existence of an internal axial force P, an internal shear force V, and an internal moment M. Because of this fact (introducing three unknowns for each member cut), the method of joints and the method of sections are usually not effective when analyzing multiforce members.

To analyze a frame, we follow the same general approach as with trusses while carefully avoiding section cuts on any multiforce member:

1.  Prepare a FBD of the entire frame.
2.  Apply the appropriate equilibrium relations to solve for the reactions.
3.  Dismember the frame and draw FBDs of each member.
4.  Use Newton's law of action and reaction in preparing these FBDs.
5.  Identify the two-force members, and show the two equal and opposite forces acting at the pinned joints.
6.  Apply the appropriate equilibrium relations to solve for the internal forces and moments.

**Machines** are similar to frames in that they contain one or more multiforce members. However, machines differ from both frames and trusses because motion occurs as they act to convert an input force (or moment) to an output force (or moment). Machines may act to amplify forces, to attenuate forces or change the direction of forces. Simple machines include the lever, screw, inclined plane, and pulley. While machines may amplify or attenuate forces when they are transmitted from the input to the output,

---

[1] A multiforce member is subjected to three or more external forces, or alternatively two or more external forces and one or more external moments.

energy is not gained in the process. Energy is conserved in the process except when heat is generated due to frictional effects and this heat is dissipated into the atmosphere.

The approach for solving problems involving frames and machines will be demonstrated in the following examples. The importance of drawing complete and accurate FBDs is stressed in these examples.

## 8.2 FRAMES—EXAMPLES AND SOLUTIONS

### EXAMPLE 8.1

To demonstrate the method for analyzing frames, consider the simple hoist presented in Fig. E8.1. First, identify the multiforce and two-force members. Then solve for the reactions at points A and E. Determine all of the external forces acting on member CFG. Determine the internal forces and the internal moment acting on this member. The load F on the hoist is 480 N. Point E is a pinned support and point A is a roller support.

Fig. E8.1

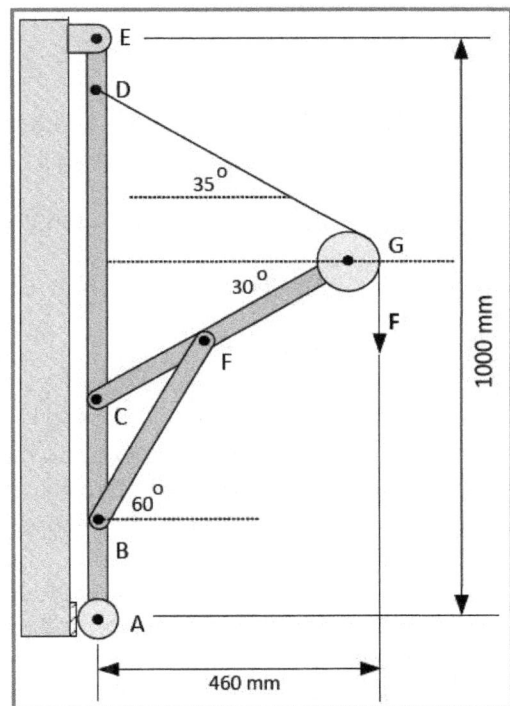

---

**Solution:**

**Step 1:** Identify the multiforce members and two force members.

The vertical bar ABCDE and the inclined bar CFG are multiforce members. The cable DG and the strut BF are two-force members.

**Step 2:** Draw a FBD of the entire frame as shown in Fig E8.1a.

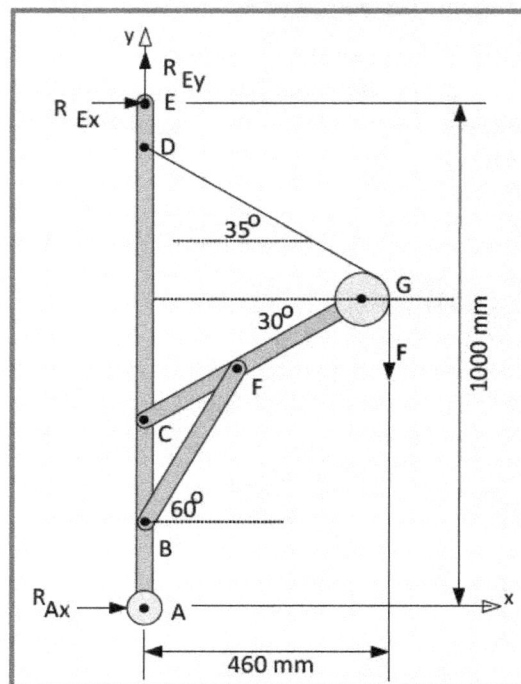

Fig. E8.1

**Step 3:** Let's employ the equilibrium relations to determine the reactions at points A and E.

$$\Sigma M_E = 1000\, R_{Ax} - 460\, F = 0 \quad \Rightarrow R_{Ax} = 0.46\,(480) = 220.8 \text{ N} \qquad (a)$$

$$\Sigma F_x = R_{Ax} + R_{Ex} = 0 \quad \Rightarrow R_{Ex} = -R_{Ax} = -220.8 \text{ N} \qquad (b)$$

$$\Sigma F_y = R_{Ey} - F = 0 \quad \Rightarrow R_{Ey} = F = 480 \text{ N} \qquad (c)$$

**Step 4:** Next, prepare the FBDs of the individual members as shown below. Four FBDs are presented in this illustration to show the forces acting on members ABCDE, BF, CFG and the

pulley at G. Note that we have used the law of action and reaction in preparing these FBDs. Let's examine member CFG to ascertain if it is possible to determine the forces applied to it. We have five forces acting on this member—$R_{Cx}$, $R_{Cy}$, $R_B$, $R_{Gx}$ and $R_{Gy}$. The force system is coplanar and non-concurrent; hence, only three equilibrium equations are applicable. Since we have five unknowns and only three equations, it is necessary to write two additional equations in terms of two of the unknown forces. Clearly, we can write these additional equations by considering the equilibrium of the pulley at point G.

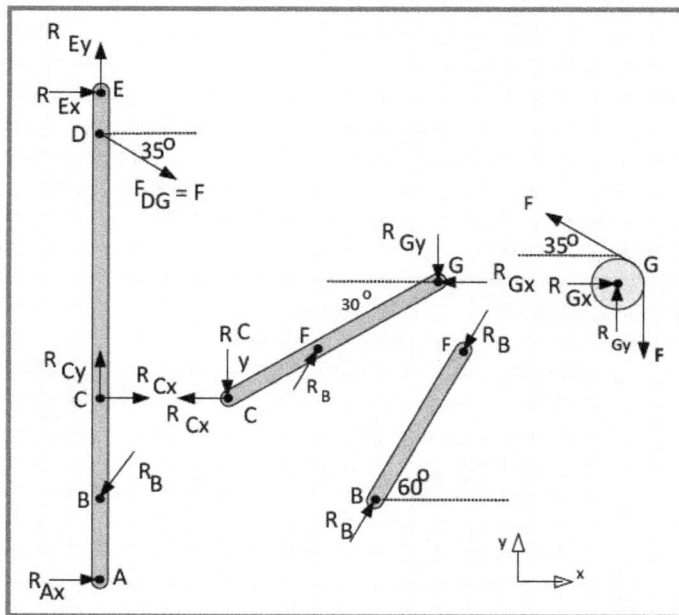

**Step 5:** Writing the equilibrium equations for the pulley gives:

$$\Sigma\, F_x = R_{Gx} - F \cos (35°) = 0 \qquad \Rightarrow R_{Gx} = 480 \cos (35°) = 393.2 \text{ N} \qquad \text{(d)}$$

$$\Sigma\, F_y = R_{Gy} + F \sin (35°) - F = 0 \Rightarrow R_{Gy} = F\,[1 - \sin (35°)] = 204.7 \text{ N} \qquad \text{(e)}$$

**Step 6:** Now it is possible to solve for the forces acting on member CFG from the equilibrium relations. Dimensions were not included in the FBDs presented in Fig E8.16b to avoid confusing the already crowded drawing. Note for member CFG that distances CF = 200 mm, FG = 320 mm and CG = 520 mm. Therefore, the pulley G has a radius of 10 mm.

$$\Sigma\, M_C = [R_B \sin (30°)](200) + R_{Gx}\,[(520)\sin (30°)] - R_{Gy}\,[(520)\cos (30°)] = 0 \qquad \text{(f)}$$

Substituting the results from Eqs. (d) and (e) into Eq. (f) yields:

$$R_B = -\,100.5 \text{ N}$$

The negative sign indicates that the direction of the force $R_B$ in the FBDs is not correct. We showed $R_B$ as a compressive force in the FBD of members BF and CFG. In reality it is a tensile force.

Continuing with the equilibrium analysis leads to:

$$\Sigma\, F_x = R_B \cos (60°) - R_{Cx} - R_{Gx} = 0 \qquad \Rightarrow \qquad R_{Cx} = -\,443.5 \text{ N}$$

$$\Sigma\, F_y = R_B \sin (60°) - R_{Cy} - R_{Gy} = 0 \qquad \Rightarrow \qquad R_{Cy} = -\,291.7 \text{ N}$$

We have completed the analysis of member CFG. It is possible to continue the application of the equilibrium relations to the member ABCDE to obtain results for $R_B$ and $R_C$ that will enable you to check the results from the analysis of member CFG.

## The Effect of Rigid Joints in Frames

Let's explore the effect of converting the pinned joints to rigid ones on stability of a structure. A rectangular frame with pinned joints is illustrated in Fig. 8.1a. Clearly, the application of a horizontal force at the top left corner of the frame causes it to rotate about its pinned joints. The slightest horizontal force component will cause the rectangular frame to collapse. However, if the pins at the upper corners are replaced with gusset plates, that are sufficiently rigid to constrain rotation, as shown in Fig 8.1b, the rectangular frame is stable.

Fig. 8.1 Stability of a rectangular frame is achieved by making the joints rigid.

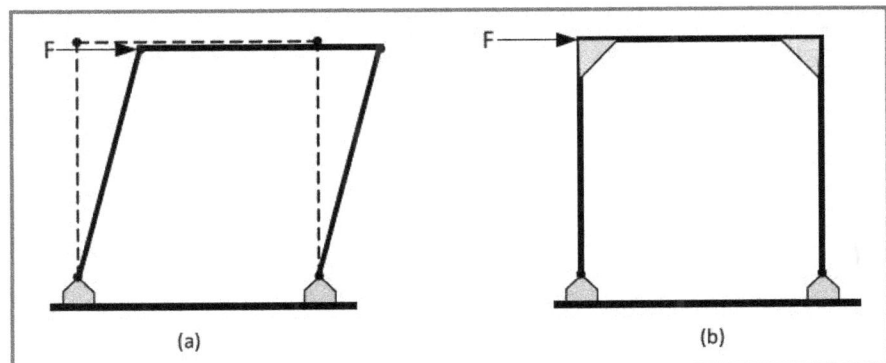

(a)

(b)

The forces applied to a frame need not be restricted to the location of the pins, and can be applied at any point along the length of any member. Of course the application of an additional force along its length produces a multiforce member as described in the previous section. A rectangular frame with a uniformly distributed force system is presented in Fig. 8.2.

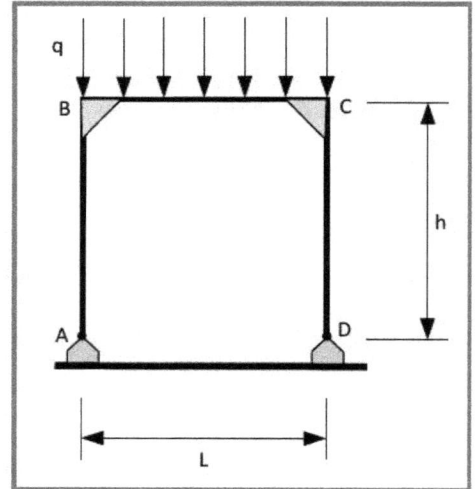

Fig. 8.2 Rectangular frame supporting a uniformly distributed force q.

Let's analyze this rectangular frame to determine the internal forces acting within its members.

## EXAMPLE 8.2

Determine the internal forces and moments for the three members in the rectangular frame presented in Fig. 8.2.

**Solution:**

**Step 1:** The procedure for determining the internal forces and moments is identical to that previously established when analyzing trusses. However, it is critical to recognize that some of the members in the frame will be multiforce members. We begin with a FBD of the entire frame, presented in Fig. E8.2, and then write the appropriate equations of equilibrium.

**Step 2:** The force system in Fig. E8.2 is coplanar and non-concurrent. Accordingly, we write:

$$\Sigma F_x = R_{Ax} - R_{Dx} = 0$$

$$\Sigma F_y = R_{Ay} + R_{Dy} - qL = 0 \qquad\qquad (a)$$

$$\Sigma M_A = R_{Dy} L - qL(L/2) = 0$$

Solving Eqs. (a) yields:

$$R_{Ay} = R_{Dy} = qL/2 \qquad \Rightarrow \qquad R_{Ax} = R_{Dx} \qquad\qquad (b)$$

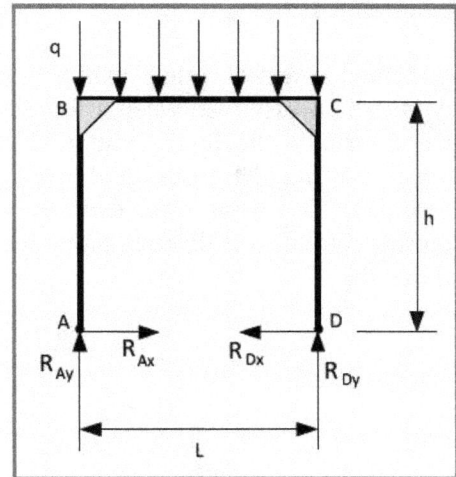

Fig. E8.2

**Step 3:** Let's draw FBDs for each of the members in the frame, as shown in Fig. E8.2a. In preparing each of these FBDs, we make use of the law of action and reaction. Examination of the FBDs again indicates that the force system is coplanar and non-concurrent for each of its three members.

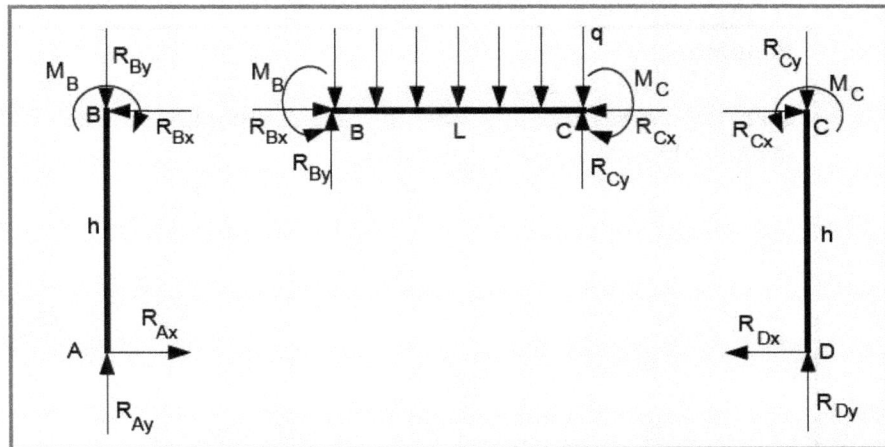

Fig. E8.2a

**Step 4:** Writing the equations of equilibrium and recognizing the symmetry of both the loading and the geometry yields:

$$M_B = M_C \qquad \Rightarrow R_{Bx} = R_{Cx} = R_{Ax} = R_{Dx} \qquad \Rightarrow R_{By} = R_{Cy} = R_{Ay} = R_{Dy} = qL/2 \qquad (c)$$

Consider the vertical member and write moments about point B to give $R_{Ax}$ as:

$$R_{Ax} = M_B/h \qquad (d)$$

We have employed all of the equations of equilibrium for each of the three members; however, we have not determined the magnitude of the reactions in the x direction or the magnitude of the moment produced by the rigid constraint at points B and C. We have established the relation given in Eq. (d), but not the magnitude. The problem is statically indeterminate. To resolve the indeterminacy, it is necessary to consider the deformed shape of the frame, which is beyond the scope of this textbook.

## EXAMPLE 8.3

A manual hydraulic pump with a 2.00 in. diameter piston is illustrated in Fig. E8.3. The fluid in the cylinder is pressurized when a force F is applied to the lever arm ADB. If the cylinder pressure is 500 psi, determine the force F applied to the lever arm and the force in the link CD. Note that CD is a two-force member, so the internal force $F_{CD}$ is directed along the axis of the link.

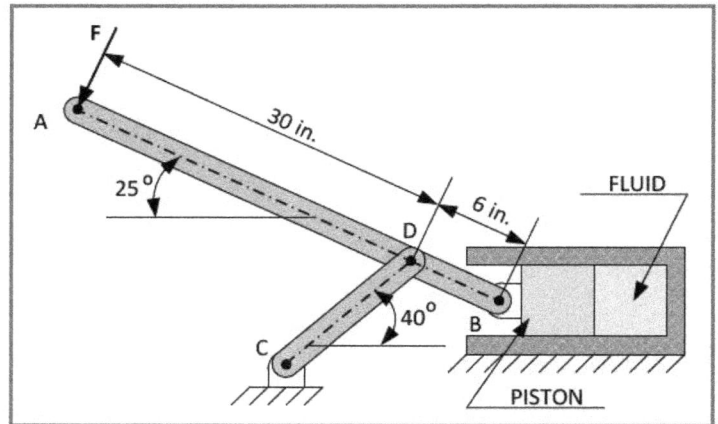

Fig. E8.3

**Solution:**

**Step 1:** We begin with a FBD of the entire pump, presented in Fig. E8.3a, and then write the appropriate equations of equilibrium.

Fig. E8.3a

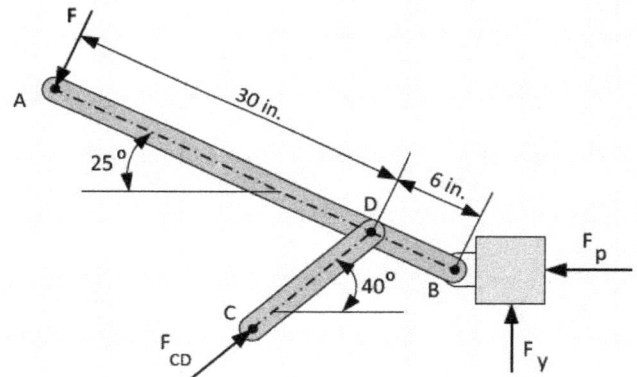

**Step 2:** The force $F_p$ acting on the piston is given by:

$$F_p = pA = 500\ (\pi d^2/4) = 500\ [\pi\ (2)^2/4] = 550\pi = 1571\ \text{lb} \qquad \text{(a)}$$

**Step 3:** The force system in Fig. E8.3a is coplanar and non-concurrent. Accordingly, we write:

$$\Sigma\ F_x = F_{CD} \cos (40°) - F \sin (25°) - F_p = 0 \qquad \text{(b)}$$

$$\Sigma\ F_y = F_y + F_{CD} \sin (40°) - F \cos (25°) = 0 \qquad \text{(c)}$$

We cannot write a meaningful moment equation for the FBD in Fig. 8.3a because the location of the force $F_y$ along the piston is not known.

Substituting Eq. (a) into Eq. (b) yields:

$$(0.7660) F_{CD} - (0.4226) F - 1571 = 0 \qquad\qquad (d)$$

**Step 4:** Let's draw FBDs for the two members in the pump's lever system, as shown in Fig. E8.3b. In preparing each of these FBDs, the law of action and reaction is used. Examination of the FBD for the lever arm indicates that the force system is coplanar and non-concurrent.

Fig. 8.3b

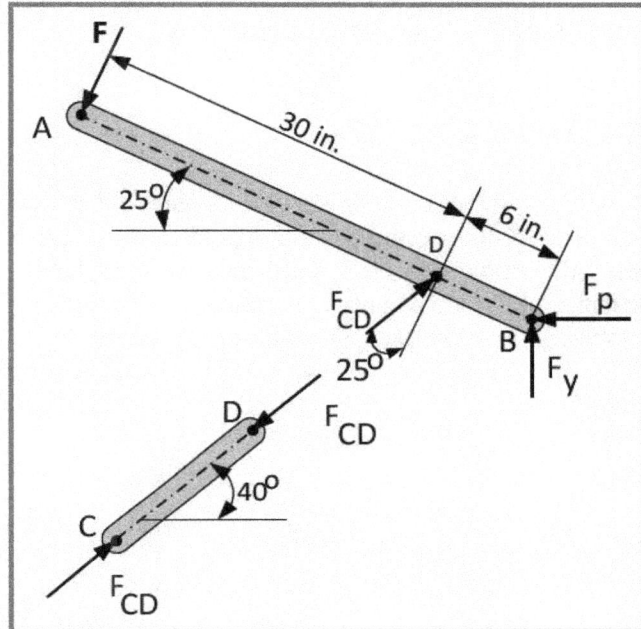

**Step 5:** Writing the moment equation of equilibrium for the lever arm about point B yields:

$$\Sigma M_B = (36.00) F - (6.00)[F_{CD} \cos (25°)] = 0$$

$$F = (1/6) F_{CD} \cos (25°) = 0.1511 F_{CD} \qquad\qquad (e)$$

Combining Eq. (d) and Eq. (e) gives:

$$F_{CD} = \frac{1571}{[0.7660 - (0.4226)(0.1511)]} = 2237 \text{ lb} \qquad\qquad (f)$$

$$F = (0.1511)(2237) = 338.0 \text{ lb}$$

The force required on the lever to develop the pressure of 500 psi is very large for a manual pump. Clearly, most people would not be capable of generating a pump pressure 500 psi. It would be advisable to suggest a redesign of the lever to decrease the force required.

## 8.3 MACHINES—EXAMPLES AND SOLUTIONS

Machines are similar to frames in that they usually contain one or two multiforce members. The difference is that one or more of the members comprising the machine moves. The movement is necessary to amplify or attenuate a force or to change its direction. The equations employed in the analysis of machines are identical to those used in solving problems related to trusses and frames—the equilibrium relations. The motion that occurs in these machines is considered to be without acceleration; hence, the static equations of equilibrium apply. As with frames, the analysis usually requires a number of carefully prepared and accurate FBDs.

### EXAMPLE 8.4

A pair of pliers clamps a small diameter rubber cylinder as shown in Fig. E8.4. If opposing forces F, equal to 30 lb, are applied to the handles of the pliers determine the force acting on the rubber cylinder. Also determine the work performed if each handle moves though a distance of 0.3 in. Finally, determine the diametrical compression of the rubber cylinder by the application of the forces.

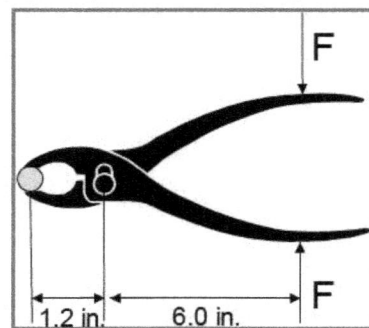

Fig. E8.4

**Solution:** Let's consider each member of the pliers individually and prepare a FBD for each as illustrated in Fig. E8.4a. The force acting on the rubber cylinder is R and the force acting at the pivot pin on the pliers is P.

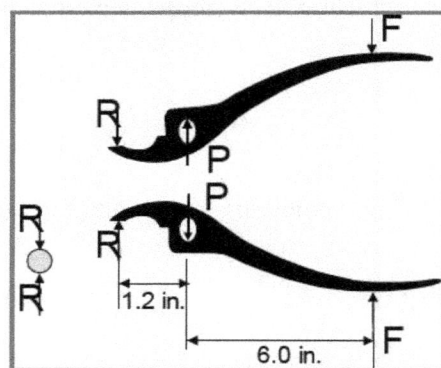

Fig. E8.4a

Three forces are applied to each handle of the pliers; hence the handles are considered as multiforce members. Writing the moment equation about point P, we obtain:

$$\Sigma M_P = (1.2)\, R - (6.0)\, F = 0 \qquad \Rightarrow \qquad R = (6.0/1.2)\, F = (5)(30) = 150\ \text{lb} \qquad (a)$$

The pliers exhibit a mechanical advantage of 5 since the output force R is five times the input force F. The mechanical advantage **MA** is given by:

$$MA = R/F \tag{8.1}$$

The work **W** performed is given by:

$$\mathbf{W} = F\,d \tag{8.2}$$

where d is the distance that the force F is moved along its line of action by a machine.

In this case, each of the two forces is moved a distance 0.3 in. along the line of action of F; hence:

$$\mathbf{W} = 2Fd = 2\,(30)(0.3) = 18 \text{ in.-lb} \tag{b}$$

Each handle of the pliers is a lever (curved to better fit the hand but a lever nevertheless). The pin holding the two handles together is the fulcrum. The diametrical compression of the rubber cylinder $\Delta d$ is given by:

$$\Delta d = (1/MA)s_h \tag{8.3}$$

$$\Delta d = (1/5)(2)(0.3) = 0.12 \text{ in.} \tag{c}$$

where $s_h$ is the distance through which each handle moves.

## EXAMPLE 8.5

Determine the reaction force at point G and the mechanical advantage of the toggle mechanism illustrated in Fig. E8.5. The dimensions are a = 310 mm, b = 220 mm, c = 160 mm, d = 100 mm, e = 140 mm and f = 120 mm. The input force F = 200 N. The contact at point G is made with a roller and pins are inserted at joints B, C, D and E.

Fig. E8.5

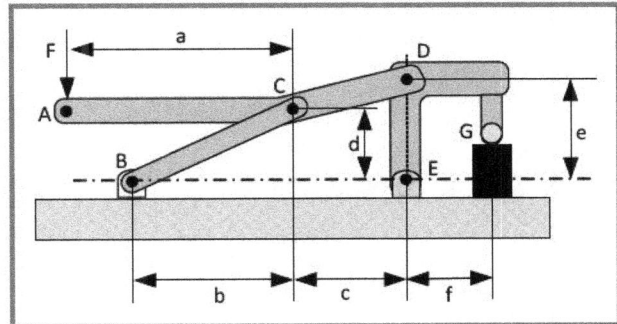

**Solution:**

**Step 1:** Identify the link BC as a two-force member and ACD and EDG as multiforce members.

**Step 2:** Prepare FBDs of the three members as presented in Fig. E8.5a. Note that we have used the fact that the link BC is a two-force member. The force $R_B$ acts along its axis. Also the reaction at pin C is equal and opposite to the reaction at pin B. We have used the laws of action and reaction in assigning forces and directions at all of the pins. Select member ACD for analysis because only three unknown forces are acting on it.

**Step 3:** Determine the angle $\theta$ and then apply the equations of equilibrium to member ACD.

$$\theta = \tan^{-1}(d/b) = \tan^{-1}(100/220) = 24.44° \tag{a}$$

$$\Sigma F_x = R_B \cos\theta - R_{Dx} = 0 \qquad \Rightarrow \qquad R_{Dx} = 0.9104\, R_B \tag{b}$$

$$\Sigma F_y = R_B \sin\theta - R_{Dy} - F = 0 \qquad \Rightarrow \qquad R_{Dy} = (0.4138)\, R_B - 200 \tag{c}$$

$$\Sigma M_C = F\, a - R_{Dy}\, c + R_{Dx}\, (e - d) = 0 \tag{d}$$

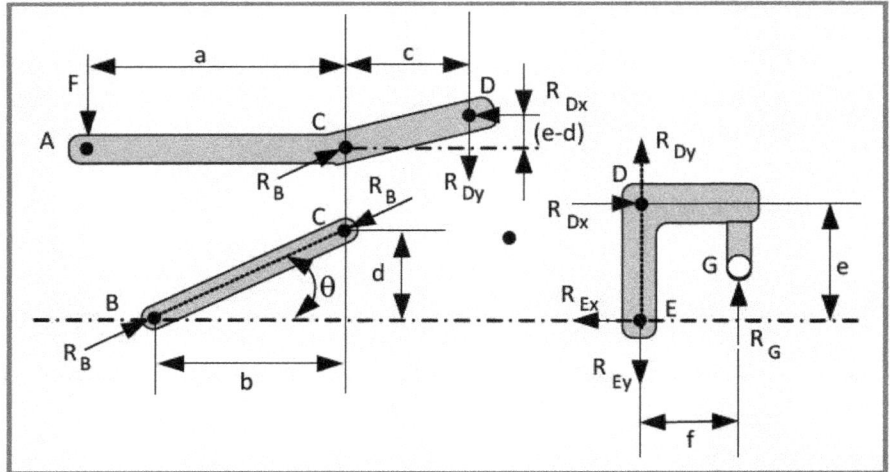

Substituting the results from Eqs. (b) and (c) into Eq. (d) yields:

$$\Sigma M_C = (200)(310) - [(-200) + (0.4138)\, R_B](160) + (0.9104) R_B\, (40) = 0 \tag{e}$$

Solving Eq. (e) for $R_B$ and substituting the result into Eqs. (b) and (c) yields:

$$R_B = 3,155\text{ N} \qquad \Rightarrow \qquad R_{Dx} = 2,873\text{ N} \quad \Rightarrow \qquad R_{Dy} = 1,106\text{ N} \tag{f}$$

**Step 4:** Apply one of the equations of equilibrium to member EDG.

$$\Sigma M_E = -R_{Dx}\, e + R_G\, f = 0$$

$$R_G = (e/f)\, R_{Dx} = (140/120)(2873) = 3,352\text{ N} \tag{g}$$

**Step 5:** Determine the mechanical advantage from Eq. (8.1).

$$\mathbf{MA} = R/F = 3352/200 = 16.76 \tag{h}$$

The toggle mechanism has amplified the input force of 200 N to an output force of 3.352 kN. The mechanical advantage of a toggle mechanism is significant. For this reason toggles are often used when input forces are limited and high output forces are necessary.

## 8.4 CONSTRUCTION EQUIPMENT

Construction equipment provides several real examples of machines that are used to modify forces. The backhoe, shown in Fig. 8.3, involves two large arms that support a bucket. Two hydraulic cylinders are employed to rotate the arms and another hydraulic cylinder is used to rotate the bucket. Both ends of each hydraulic cylinder are pinned, and the end of each of the large arms arm is pinned. The bucket rotation involves a linkage arrangement.

Fig. 8.3 Hydraulic cylinders actuate the arms and bucket on a backhoe.

## EXAMPLE 8.6

A backhoe in action is illustrated in Fig. E8.6. Prepare a drawing showing the bucket, its linkage, the large arm supporting the bucket, and the hydraulic cylinder, which actuates the bucket. Draw FBDs of this machine showing the forces at the pins. If the bucket contains 800N of sand, determine the force exerted by the hydraulic cylinder. Dimensions of the components are presented in the FBDs given in the solution.

Fig. E8.6 A backhoe bucket containing 800 N of sand.

**Solution:**

**Step 1**: Prepare a drawing showing the arm and linkages supporting the bucket as presented in Fig. E8.6a.

Fig. E8.6a Note the bucket arm and the bucket are welded together.

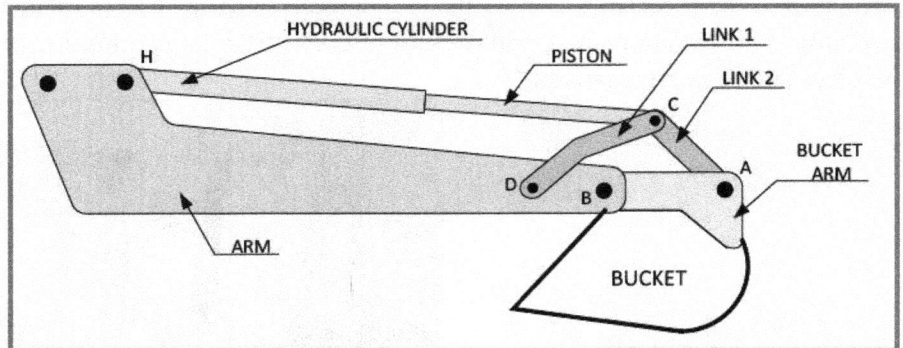

An examination of this drawing shows that Link 1 and 2 are both two force members and that the hydraulic actuator is also a two-force member. This fact will be instrumental in solving for the force generated by the hydraulic cylinder in supporting the loaded bucket.

**Step 2**: Prepare a FBD of the backhoe bucket as shown in Fig. E8.6b.

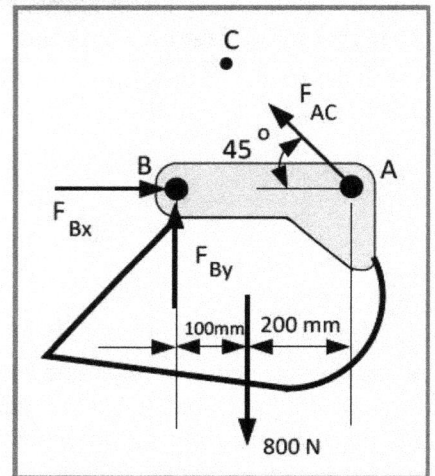

**Step 3**: Apply the equilibrium relations and solve for the three unknown forces.

$$\Sigma M_B = (300) \, F_{AC} \sin (45°) - (100)(800) = 0 \qquad \text{(a)}$$

$$\Sigma F_y = F_{By} + F_{AC} \sin (45°) - 800 \qquad \text{(b)}$$

$$\Sigma F_x = F_{Bx} - F_{AC} \cos (45°) = 0 \qquad \text{(c)}$$

Solving Eqs. (a), (b) and (c) for the three unknown forces yields:

$$F_{AC} = 377.1 \text{ N} \qquad F_{By} = 533.4 \text{ N} \qquad F_{Bx} = 266.7 \text{ N} \qquad \text{(d)}$$

**Step 4**: Prepare a FBD for the pin at point C where links 1 and 2 join the piston end of the hydraulic actuator as indicated in Fig. E8.6c.

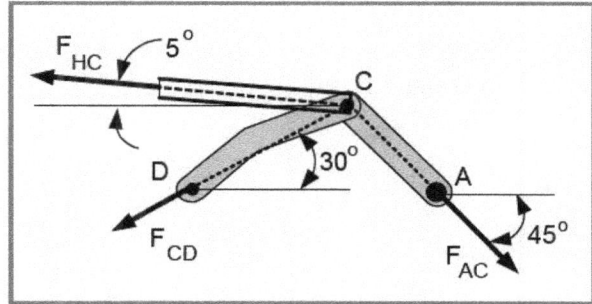

Fig. E8.6c

**Step 5**: Apply the equilibrium relations and solve for the two unknown forces.

$$\Sigma F_y = F_{HC} \sin (5°) - F_{AC} \sin (45°) - F_{CD} \sin (30°) = 0 \qquad (e)$$

$$\Sigma F_x = - F_{HC} \cos (5°) + F_{AC} \cos (45°) - F_{CD} \cos (30°) = 0 \qquad (f)$$

Combining the results in Eq. (d) with Eqs. (e) and (f) gives:

$$F_{HC} (0.08716) - F_{CD} (0.5000) = (377.1)(0.7071) = 266.65 \qquad (g)$$

$$F_{HC} (0.9962) + F_{CD} (0.8660) = (377.1)(0.7071) = 266.65 \qquad (h)$$

Solving these two equations for the two unknowns yields:

$$F_{CD} = - 422.5 \text{ N} \qquad\qquad F_{HC} = 635.0 \text{ N}$$

Of course, the negative sign for $F_{CD}$ indicates the force is compressive.

## EXAMPLE 8.7

The Loadall is a versatile piece of construction equipment that can be used for several different purposes. The unit shown in Fig. E8.7 is fitted with a pair of forks for lifting pallets. These forks are fixed to the end of a telescoping boom, and are positioned over a dumpster in the photograph of Fig E8.7. If a pallet weighing 1200 lb were lifted from a truck bed with these forks, determine the actuator force necessary to rotate the boom and the forces at the bracket hinge pin.

Fig. E8.7 A Loadall unit equipped with a pair of forks for lifting pallets.

**Solution:**

**Step 1**: Prepare a drawing showing the telescoping arm and system supporting the forks as presented in Fig. E8.7a.

Fig. E8.7a

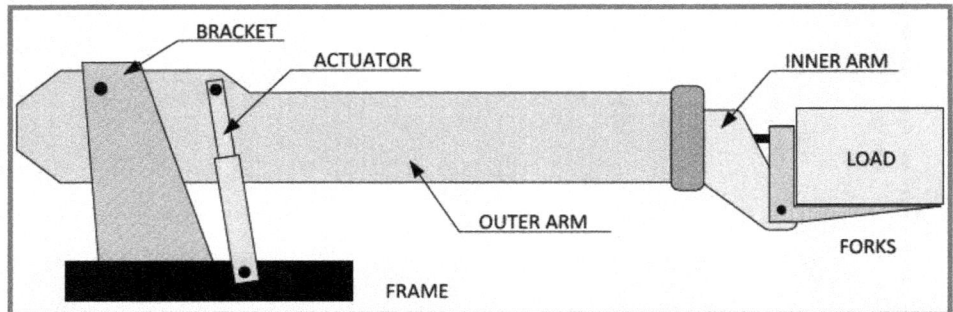

**Step 2**: Prepare a FBD of the Loadall mechanism as shown in Fig. E8.7b. Dimensions of the points where forces are applied are shown in the illustration. The weight of the arms acts through the center of gravity as shown in Fig. E8.7b.

Fig. E8.7b

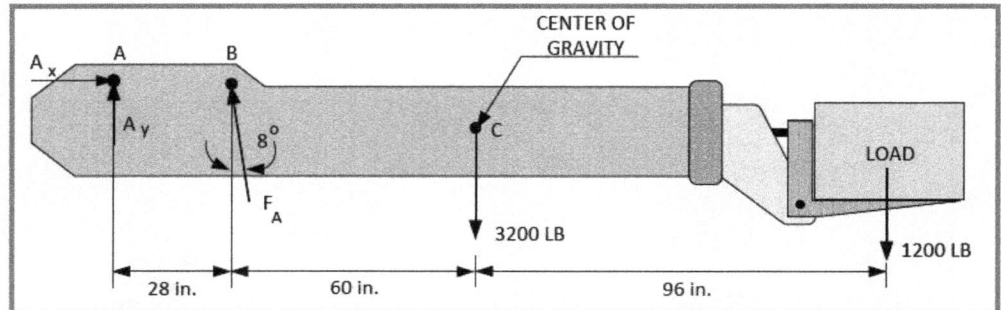

**Step 3**: Apply the equilibrium relations and solve for the three unknown forces—$A_x$, $A_y$ and $F_A$.

$$\Sigma M_A = [F_A \cos (8°)](28) - (3{,}200)(28 + 60) - (1{,}200)(28 + 60 + 96) = 0$$

$$F_A = 18{,}120 \text{ lb} \qquad \text{(a)}$$

$$\Sigma F_x = A_x - F_A \sin (8°) = 0 \qquad A_x = (18{,}120)(0.1392) = 2{,}522 \text{ lb} \qquad \text{(b)}$$

$$\Sigma F_y = A_y + F_A \cos (8°) - 3{,}200 - 1{,}200 = A_y + (18{,}120)(0.9903) - 4{,}400 = 0$$

$$A_y = -13{,}540 \text{ lb} \qquad \text{(c)}$$

While the Loadall appears to be a complex machine capable of lifting and moving loads, it is relatively easy to analyze. The arm is a multiforce member, but if the FBD is prepared correctly, the solution for the actuator force $F_A$ and the forces $A_x$ and $A_y$ at the hinge pin is straightforward.

## 8.5 SUMMARY

The method of analysis of frames and machines was introduced. Frames and machines contain one or more members that have three or more externally applied forces. For these multiforce members, the internal force does not coincide with the axis of the member. This fact is extremely important since it implies that two internal forces (P and V) and an internal moment M must be applied when making a section cut through a multiforce member. Because we introduce so many unknowns, section cuts through a multiforce member are usually avoided when analyzing frames and machines. The approach is to construct FBDs of the individual members of the structure and to apply the law of action and reaction as well as the equations of equilibrium to each member. Examples for both frames and machines were presented to demonstrate the method of analysis. Examples of construction equipment were described to show that complex machines could be analyzed using these relatively straightforward techniques.

## PROBLEMS

8.1  For the hoist shown in the figure to the right, determine all of the external forces on member CFG as point D is moved along member AE so that the angle θ varies from 0 to 45°. We suggest that you use a spreadsheet in preparing this solution. Note the angle β = 60° and F = 580 N. The dimensions are: CF = 200 mm, FG = 320 mm and the pulley radius r = 10 mm.

8.2  For the hoist shown in the figure to the right, determine all of the external forces on member CFG as point B is moved along member AE so that the angle β varies from 35 to 75°. We suggest that you use a spreadsheet in preparing this solution. Note the angle θ = 30° and F = 240 N. The dimensions CF = 200 mm, FG = 320 mm and the pulley radius r = 10 mm.

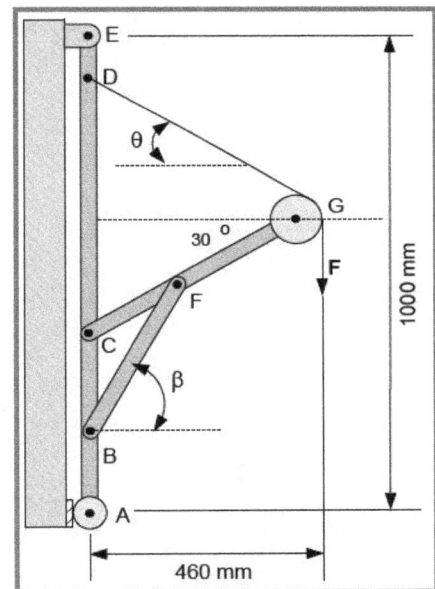

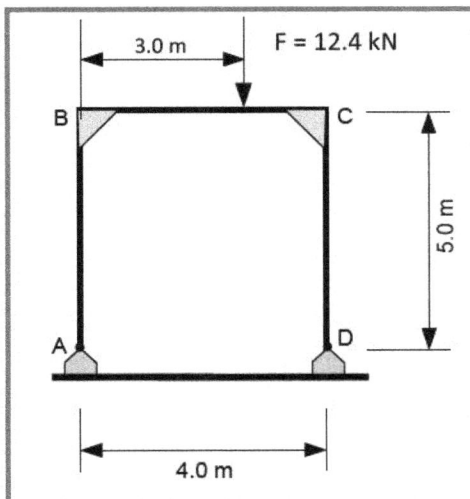

8.3  Prepare a FBD for the entire rectangular frame, shown in the figure to the left, if it is subjected to a concentrated force of 12.4 kN located a distance of 3.0 m from the left end of the horizontal member.

8.4    Determine the force F required to develop a pressure p = 2.1 MPa in the cylinder of the pump shown in the figure to the right. The piston area for the pump is 300 mm$^2$.

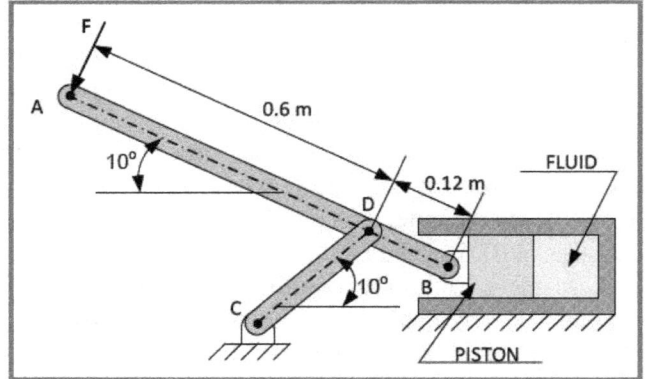

8.5    Determine the reaction force at point G and the mechanical advantage of the toggle mechanism illustrated in the figure to the right. A force F of 300 N is applied to the lever at point A. The contact at point G is made with a roller, and the links are connected with pins inserted at points B, C, D and E.

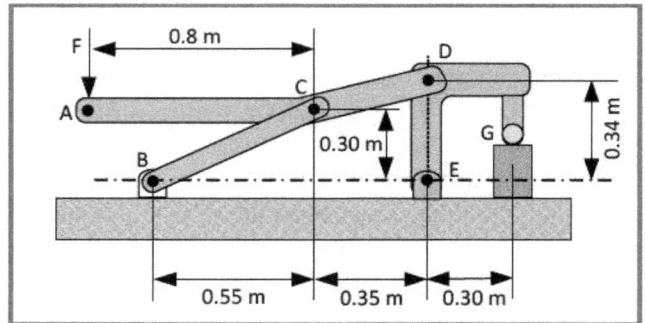

8.6    Determine the reaction force at point G and the mechanical advantage of the toggle mechanism illustrated in the figure to the right. The input force F is now 500 N. The contact at point G is made with a roller and the links are connected with pins inserted at points B, C, D and E.

8.7    Your manager believes that the dimension (e − d) for the toggle mechanism shown in the figure to the left is a critical design parameter. She asks you to determine the mechanical advantage of this mechanism if the dimension d is fixed and the dimension e is modified so that (e − d) varies from 5 mm to 50 mm. You may consider using a spreadsheet for this analysis.

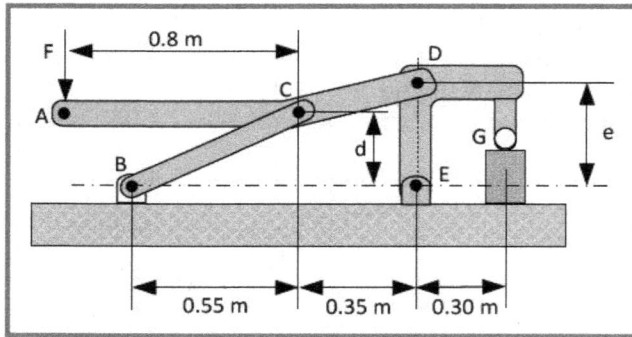

8.8    If the spring constant of the block under point G is 2.0 kN/mm, determine the vertical displacement of point G in Problem 8.5.

8.9    If the spring constant of the block under point G is 2.0 kN/mm, determine the work output of the toggle mechanism for the conditions described in Problem 8.5.

8.10   If the spring constant of the block under point G is 3.8 kN/mm, determine the vertical displacement of point G in Problem 8.6.

8.11 If the spring constant of the block under point G is 3.8 kN/mm, determine the work output of the toggle mechanism for the conditions described in Problem 8.6.

8.12 For the compaction press, shown in the figure to the right, determine the compressive force developed by the sliding platen if a force of 5 ton is applied to the toggle mechanism at point B. Also determine the mechanical advantage.

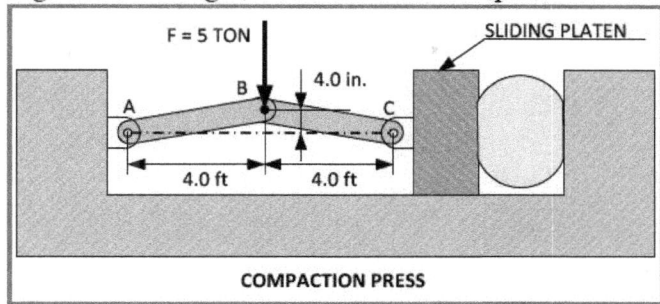

8.13 A pair of pliers clamps a small diameter rubber cylinder as shown in the figure to the right. If opposing forces of 30 lb are applied to the handles of the pliers, determine the reaction forces acting on the cylinder. Also determine the work performed if each handle moves though a distance d = 0.14 in. Finally, determine the amount that the cylinder is squeezed by the application of the forces. The dimensions of the pliers are a = 1.0 in. and b = 5.5 in.

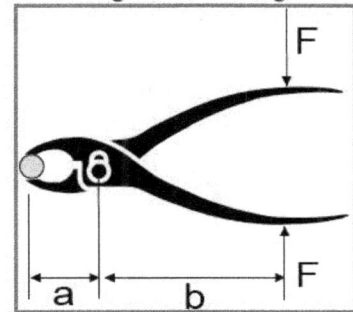

8.14 For the frame, shown in the figure to the left, determine the forces and moment at the fixed support at A, the forces at pin C and the force in link BD.

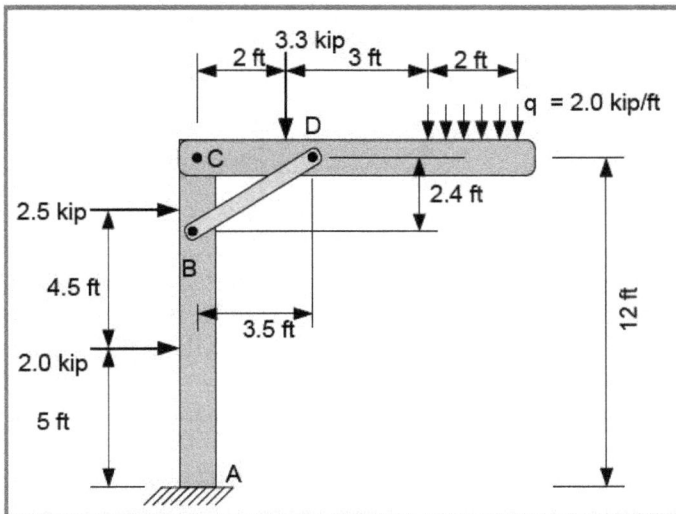

For the frame, illustrated in the figure to the right, determine all the forces acting on member ABCDE. Also determine the internal force in member AF. The attached weight W is 4.8 kN and the pulley radius is 150 mm. The dimensions in the figure are given in mm.

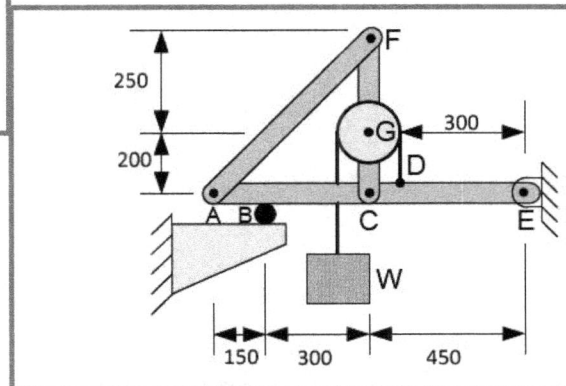

8.15 For the frame illustrated in the figure to the right, determine all the forces acting on member ABCDE. Also determine the internal force in member AF. The attached weight W is 4.8 kN and the pulley radius is 100 mm. The dimensions in the figure are given in mm.

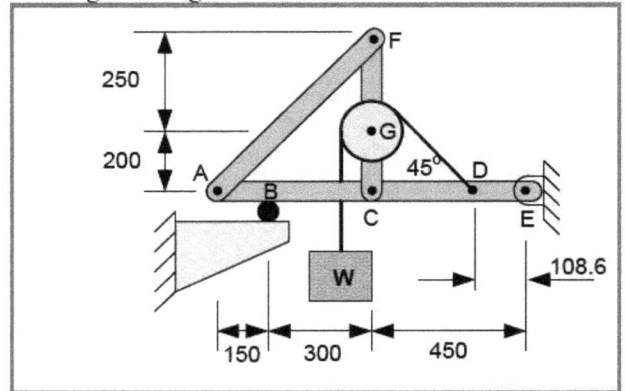

8.16 For the arm of the backhoe, shown in the figure below, determine the force that must be exerted by the hydraulic actuator AC to maintain the bucket loaded with a weight of 1.0 kN in equilibrium. Also find the forces in links BC and CE and the force acting on pin D. The bucket and the crosshatched appendage are welded together.

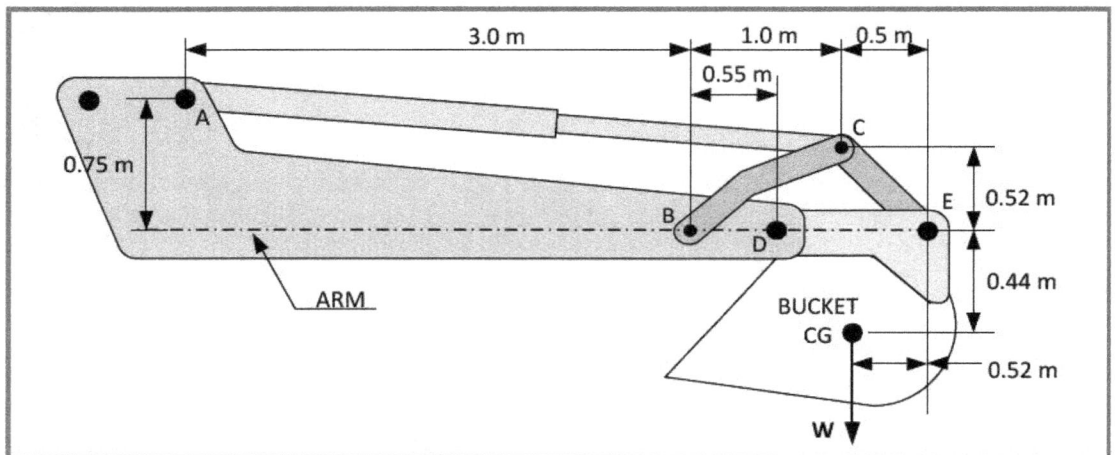

8.17 For the "Loadall" arm, presented in the figure below, determine the forces exerted by the hidden actuator and the visible actuator BC to maintain equilibrium. Also determine the forces acting at pins A and D. The arm weighs 1.5 kN and its center of gravity is located 1.8 m to the right of point A. All of the dimensions are given in meters.

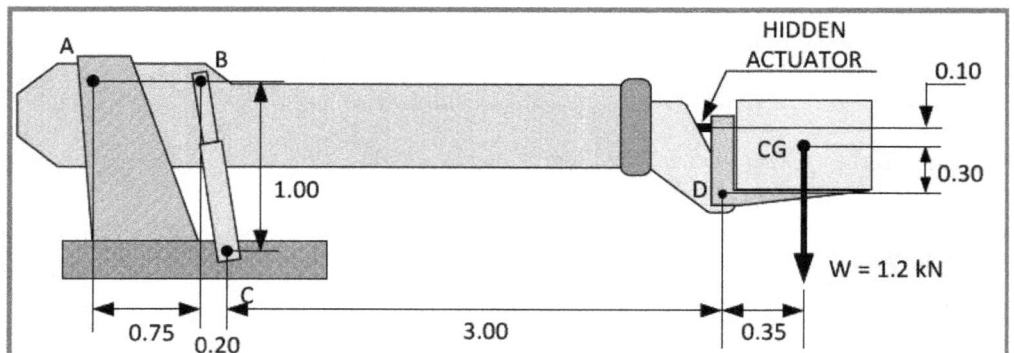

8.18 Levers are well known machines used for either amplifying or attenuating forces. When used in scales to weigh heavy objects, the levers are often arranged to compound the attenuation. Such an arrangement is illustrated in the figure below. If the pins are frictionless, show that the relation between the known scale weight w and the unknown scale weight W is:

$$W = w \left[ \frac{(a+b+c+d-x)(a+b+c)b}{aec} \right]$$

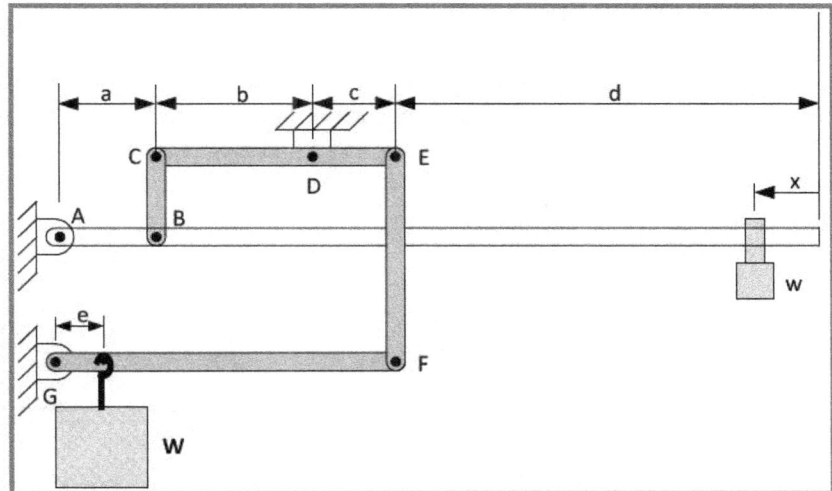

8.19 Select dimensions a through e if the scale, illustrated in the figure above, is to measure a weight W = 5.0 kN with a small sliding weight w = 10 N. The value of x = 200 mm is fixed in this design analysis.

8.21 The horizontal boom of a construction crane is counter balanced with a 15 kip weight that is centered at point B. A cable BFE anchored at points B and E supports the horizontal boom. The cable is maintained at a constant tension over its length with a small pulley located at point F. The horizontal boom is attached to the tower with a pin located at point C. The crane is lifting a load W = 10 kip from a hoist located at point D. Determine all of the forces acting on the horizontal boom, the cable tension, all of the reactions at point A and the force acting on the pin at point F. The weight of the boom is 5.0 kip and the weight of the tower member is 4.0 kip.

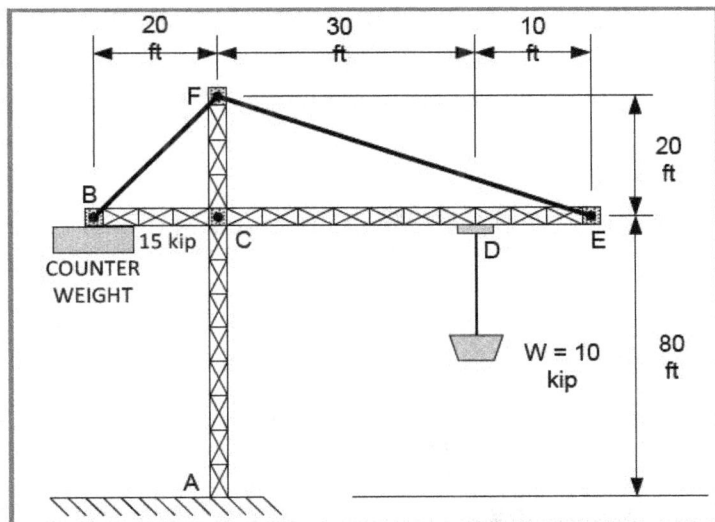

# CHAPTER 9

# FRICTION

## 9.1 INTRODUCTION

**Friction** occurs when two bodies are in contact and forces are applied to one or both bodies. Frictional forces develop at the contacting surfaces that tend to inhibit the sliding of one body relative to the other. For example, it is friction between the tires and road that keeps a car from sliding off when it rounds a turn. This ability of a tire to maintain contact with the road is often referred to as **traction**. We depict this situation with the illustration given in Fig. 9.1.

Fig. 9.1 Two bodies in contact with a force applied to the upper block.

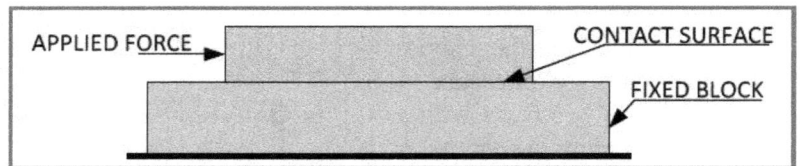

If we slowly increase the lateral force applied to the upper block, we observe that it remains stationary. The block remains in equilibrium even with an applied force of significant magnitude. Is this behavior consistent with Newton's second law, which states $\Sigma \mathbf{F} = \mathbf{ma}$? Let's look more closely at the upper block in Fig. 9.1 by preparing a FBD showing **all** the forces that act on the body. The FBD for both the upper and lower blocks is shown in Fig. 9.2. Since the upper block is in equilibrium, it is clear that the frictional force must act in the opposite direction to the applied force. The frictional force on the fixed (lower) block is the same magnitude, but it is in the opposite direction because of the law of action and reaction. Normal forces also develop between the two contacting surfaces. The normal forces are perpendicular to the contacting surfaces, equal in magnitude and opposite in direction.

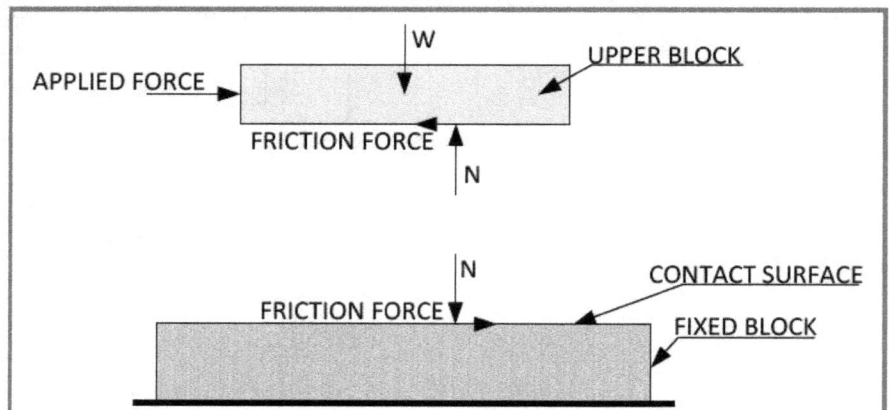

Fig. 9.2 Free body diagram showing the friction forces developed on both bodies at the contact surface.

If we continue to increase the applied force, the upper block will eventually begin to slide over the surface of the lower block, indicating some upper limit on the frictional forces. Before we explore this upper limit, let's consider several factors that influence friction. We know that friction only occurs with two bodies in contact so it is reasonable to examine in detail the two contact surfaces. Consider a magnified view of the surfaces presented in Fig. 9.3.

We note in Fig. 9.3 that smooth surfaces are not really smooth when magnified sufficiently. Instead, the surfaces are made up of valleys and peaks, which we call **asperities**. Contact occurs only when two opposing asperities are aligned. Since the alignment of two opposing peaks is rare, the actual area of contact $A_c$ is small compared to the area A of the entire surface. For reasonably smooth surfaces, the ratio of $A_c/A$ is:

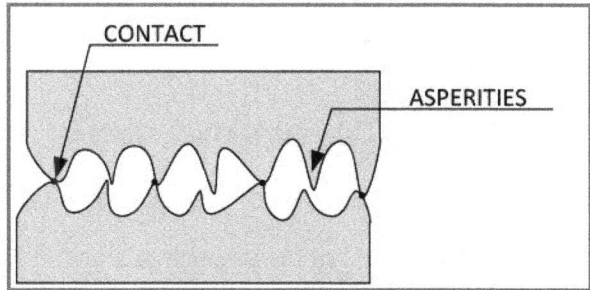

Fig. 9.3 Schematic illustration of contacting surfaces.

$$A_c/A \leq 0.001 \qquad\qquad (9.1)$$

When a lateral force is applied to one of the bodies a resistive force (friction) develops because some of the asperities are coupled together. In some cases, these asperities mechanically interfere with each other, and must be sheared off before motion between the surfaces may occur. In other cases, the contacting asperities bond together, and this bond must be fractured before sliding may occur. The shearing of asperities and the fracture of the contact bonds produce wear of sliding surfaces.

In the study of mechanics, we ignore the microscopic characteristics of the surfaces and treat friction from a more pragmatic view point. We simply distinguish between dry and fluid friction. In **dry** or **Coulomb**[1] **friction** the asperities are in contact, and appreciable friction forces develop to resist sliding motion between the contacting surfaces. In **fluid friction**, we introduce a lubricant that separates the two surfaces as illustrated in Fig. 9.4.

When a lubricant is present the frictional forces are reduced significantly. This is the reason for using oil to lubricate the engines on our automobiles. The resistance is no longer due to breaking asperities free to permit sliding. Instead, much smaller forces are required to shear the lubricant. In this chapter we will be concerned with only dry or Coulomb friction. Lubrication theory is a topic introduced later in the Mechanical Engineering curriculum.

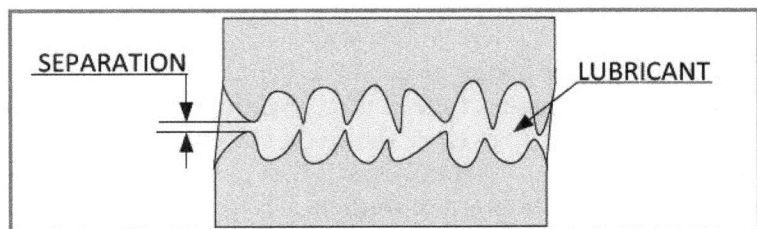

Fig. 9.4 Lubricant separates the surfaces preventing contact of the asperities.

## 9.2 STATIC AND DYNAMIC FRICTION

Let's return to the model depicted in Fig. 9.1 with two blocks in contact, and examine the conditions required to maintain the upper block in equilibrium. The body is two-dimensional and three equations of equilibrium must be satisfied, namely $\Sigma F_x = 0$, $\Sigma F_y = 0$ and $\Sigma M_O = 0$. The FBD of the upper block presented in Fig. 9.5, shows the applied lateral force F, the weight of the block W, the vertical reaction of the supporting block N, and the friction force $F_f$. The equilibrium relations yield:

---

[1] Charles Augustine Coulomb, a military engineer, developed the general theory of friction and published his work in a lengthy memoir. Théorie des Machines Simple, En ayant égard au frottement de leurs parties, et a la roideur des Cordages, Mémoires de Mathématique et de Physique, vol. X. 1785.

$$\Sigma F_y = 0 \Rightarrow N = W \qquad\qquad\qquad (a)$$

$$\Sigma F_x = 0 \Rightarrow F = F_f \qquad\qquad\qquad (b)$$

The normal force N is located along the contact surface at position s, to satisfy the moment equation of equilibrium.

From the relation $\Sigma M_O = 0$, we may determine the location s for the normal force N as:

$$s = (Ww - Fh)/2W \qquad\qquad\qquad (9.2)$$

We note that s = w/2 in the absence of a lateral force F. When F increases the value of s decreases as the normal force shifts to the right. When $F = (wW)/h$, s = 0, and the block is on the verge of tipping. We will consider whether the block slides or tips later in this chapter.

Fig. 9.5 FBD of a block prior to the initiation of sliding.

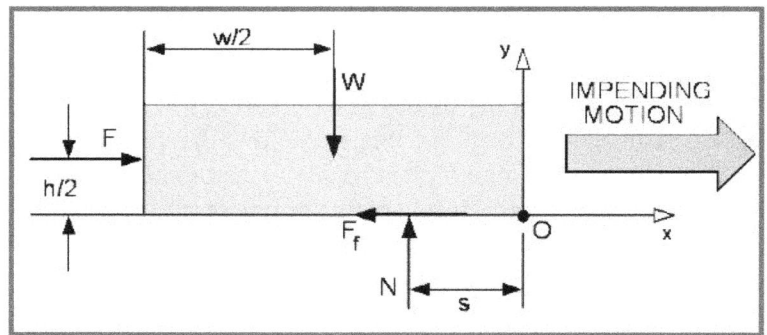

The magnitude of the frictional force is variable. The frictional force becomes as large as necessary to prevent sliding up to some maximum force when the asperities break free. The maximum frictional force is approximated by:

$$(F_f)_{max} = \mu N \qquad\qquad\qquad (9.3)$$

where $\mu$ is the coefficient of friction.

The coefficient of friction depends on the materials in contact and to a significant degree on the surface finish and cleanliness of the contact surfaces. A listing of the coefficient of friction for a number of different materials is given in Table 9.1; however, it is important to note the very wide variations in $\mu$ for all of the material combinations. The variation is so large that it is advisable to conduct simple experiments to measure the coefficient of friction of the materials involved in a design of a machine component.

The maximum frictional force increases linearly with the normal force N and is considered to be independent of the area of the surfaces in contact providing that $A_c/A$ remains very small, as indicated in Eq. (9.1). When the normal force becomes extremely large the asperities undergo plastic deformation and the ratio $A_c/A$ increases markedly. Another situation leading to significant increases in $A_c/A$ is when one of the two bodies is fabricated from soft materials such as rubber or plastics. In these two cases, the area of the surfaces is important, and Eq. (9.3) is no longer valid.

**Table 9.1**
**Approximate coefficient of friction for contact between different materials.**

| Surface Materials | Coefficient of Friction (static) |
|---|---|
| Metal on metal | 0.15 – 0.50 |
| Metal on stone or concrete | 0.30 – 0.70 |
| Metal on leather | 0.30 – 0.60 |
| Metal on wood | 0.20 – 0.60 |
| Wood on wood | 0.30 – 0.70 |
| Wood on stone or concrete | 0.30 – 0.70 |
| Rubber on concrete | 0.60 – 0.90 |
| Metal on brake pad | 0.20 – 0.40 |
| Metal on Ice | 0.03 – 0.05 |

For normal conditions Eq. (9.3) is valid, and the friction force increases linearly with the applied lateral force F as shown in Fig. 9.6 until motion (sliding or tipping) occurs.

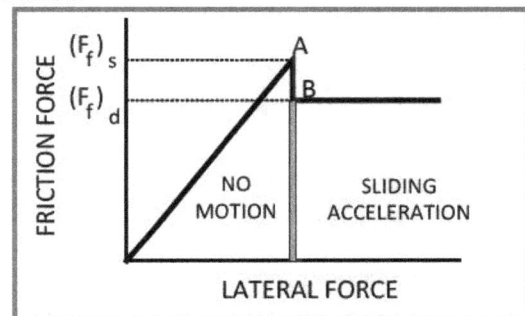

Fig. 9.6 Graph of frictional force with the applied lateral force.

The frictional force is exactly equal to the applied lateral force until the sliding initiates due to the failure of asperities. At this instant, the frictional force decreases abruptly (from points A to B in Fig. 9.6). After the onset of sliding, the frictional force remains constant at a value of $(F_f)_d$ with further increases in the applied lateral force. The friction force that develops under sliding conditions is smaller than the friction under static conditions. The mechanical interference of the asperities still occurs, but the time required for adhesion (bonding) of the asperities is not available when one surface is sliding relative to the other. Under sliding conditions, the kinetic friction force, $(F_f)_d$ is approximated by:

$$(F_f)_d = \mu_d\, N \qquad\qquad (9.4)$$

where $\mu_d$ is the dynamic or kinetic coefficient of friction, which is usually approximated by multiplying the static value by 0.75 to 0.80.

## 9.3 MEASURING THE COEFFICIENT OF FRICTION

The coefficient of friction $\mu$ exhibits significant variation due to affinity of different surfaces, surface roughness, and the cleanliness of the two mating surfaces. For these reasons it is advisable to measure the coefficient of friction of the materials to be employed in the design of machine components and structures. The measurement of $\mu$ is relatively easy to accomplish using the simple experimental arrangement shown in Fig. 9.7.

A block of one material is placed on an inclined plane with a surface fabricated with the other material. The angle of inclination of the plane $\theta$ is slowly increased until the block begins to slide. The angle required to initiate movement (sliding) is known as the angle of impending motion $\theta_s$. It is measured with a protractor. Let's perform an equilibrium analysis to show the relation between $\mu$ and $\theta_s$.

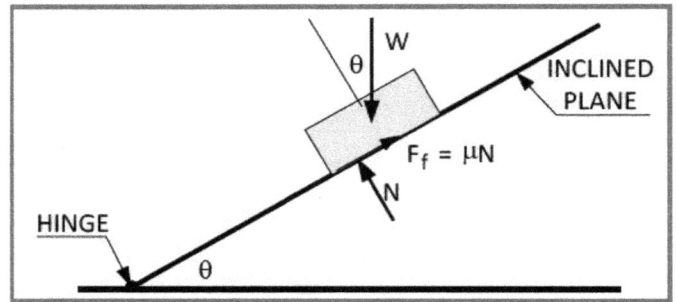

Fig. 9.7 Illustration of a method for measuring the static coefficient of friction.

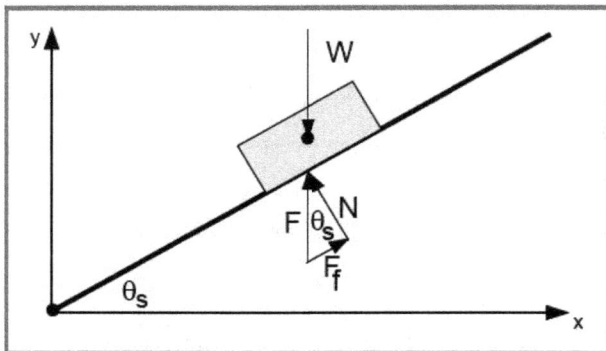

Fig. 9.8 FBD of the block on the incline plane at the position of impending motion.

We begin the equilibrium analysis by drawing a FBD of the block where the inclined plane is at the position of impending motion as shown in Fig. 9.8. We have rearranged the forces acting on the block to show the resultant force **F**, which is given by:

$$\mathbf{F} = \mathbf{N} + \mathbf{F_f} \tag{a}$$

Next, consider the equilibrium relation $\Sigma F_y = 0$, and write:

$$\Sigma F_y = F - W = 0 \quad \Rightarrow \quad F = W \tag{b}$$

We also note that:

$$N = F \cos \theta_s = W \cos \theta_s \tag{c}$$

$$F_f = F \sin \theta_s = W \sin \theta_s \tag{d}$$

Recall Eq. (9.3) and use Eqs. (c) and (d) to write:

$$\mu = F_f / N = \sin \theta_s / \cos \theta_s = \tan \theta_s \tag{9.5}$$

Clearly, we can measure the coefficient of friction quickly and efficiently with this simple experiment. The angle $\theta_s$ is related to the coefficient of friction through the tangent function. Also, the angle $\theta_s$ is sometimes referred to as the angle of repose, which we will show is equal to the angle of friction $\phi$.

Let's consider a block resting on the horizontal surface. Apply a lateral force F to the block until is just ready to slide (**impending motion**). We illustrate the FBD for the block under these loading conditions in Fig. 9.9.

To analyze this FBD note that the angle of friction is defined as:

$$\tan \phi = F_f / N \tag{a}$$

But with the block subject to **impending motion**, we may write $F_f = \mu\, N$. Substituting this relation into Eq. (a) leads to:

$$\tan \phi = \mu \qquad\qquad (9.6)$$

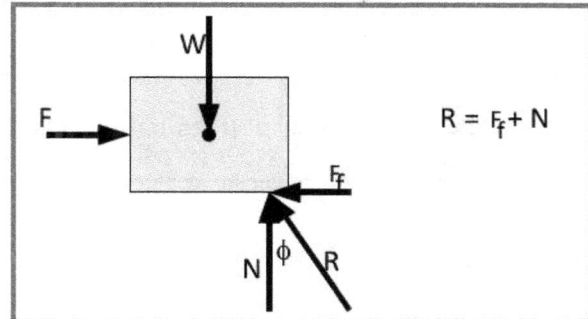

Fig. 9.9 FBD of a block subjected to applied forces associated with impending motion.

Combine Eqs. (9.5) and (9.6) to yield:

$$\theta_s = \phi \qquad\qquad (9.7)$$

From these derivations, we have proven that the angle of repose $\theta_s$ is equal to the angle of friction $\phi$.

## EXAMPLE 9.1

A sales representative states that the coefficient of friction $\mu$ for a flat neoprene-fabric belt is 0.6 when the belt is operated against a smooth metal pulley. You obtain a sample of the belt, and plan to conduct an inclined plane experiment to measure $\mu$. As a part of your plan determine the expected angle of repose.

**Solution:**

Recall Eq. (9.5) and write:

$$\theta_s = \tan^{-1} \mu = \tan^{-1} (0.6)$$

$$\theta_s = 31.0°$$

Your plan for the friction experiment should permit the inclined plane to tilt to at least 40°, and the protractor for measuring the angle of repose should be accurate to 1/2°.

## EXAMPLE 9.2

In conducting the experiments with the sample of flat neoprene/fabric belt, you record five different values of the angle $\theta_s$ including 29.5, 31.5, 31.0, 32.5, and 28.0. Determine the average value of $\mu$ and its range.

**Solution:**

Recall Eq. (9.5) and write:

$$\mu = \tan \theta_s \qquad\qquad (a)$$

The results from Eq. (a) are given in the table below:

| $\theta_s$ | $\mu$ |
|------|-------|
| 29.5 | 0.566 |
| 31.5 | 0.613 |
| 31.0 | 0.601 |
| 32.5 | 0.637 |
| 28.0 | 0.532 |

The average value of $\mu$ is $\Sigma\mu/n = 2.949/5 = 0.590$; and the range in the measurements of $\mu$ is $\mu_{max} - \mu_{min} = 0.637 - 0.532 = 0.105$

## EXAMPLE 9.3

You are attending a county fair and decide to watch an exhibition involving a harnessed team of mules that are pulling a sled with a load of weights. The load is pulled only short distances (about two meters). You are amazed at the size of the cast iron blocks that are placed on the sled, and decide to conduct an analysis to determine the force F with which the team pulls. Draw the FBD necessary to conduct the analysis, and list the information necessary to determine the force F.

**Solution:**

Let's represent the sled, its weight, and the weight of the cast iron blocks with a single rectangular block as shown in Fig. E9.3.

Fig. E9.3

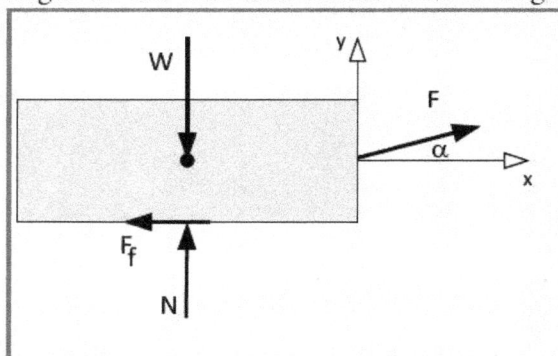

This FBD represents a model of the mule team, sled and the load of weights. The mule team pulls with a force F that is oriented at an angle $\alpha$ relative to the ground. The friction force $F_f$ provides the resistance in the x direction.

To determine the force F, we must gather information defining the angle $\alpha$, the friction coefficient, and the weight W. Suppose the sled together with the added weights total 5200 lb. The harness on the mules is shortened as much as possible to maximize the angle $\alpha$ at a value of 15°. The steel runners on the sled interact with a flat, bare, dry, clay field. If we assume that the coefficient of friction of metal (steel) on dry clay is 0.45, we have collected all of the information needed to determine F. Let's solve for the pull F provided by the team of mules.

Begin by writing the equilibrium relations $\Sigma F_x = 0$ and $\Sigma F_y = 0$, which yield:

$$\Sigma F_x = F \cos \alpha - F_f = 0 \qquad (a)$$

$$\Sigma F_y = F \sin \alpha + N - W = 0 \qquad (b)$$

Next recall Eq. (9.3) and note the maximum load pulled by the mule team occurs when motion of the sled is impending. In this instance Eq. (9.3) is valid and we may write:

$$(F_f)_{max} = \mu N$$

Substituting Eq. (9.3) into Eq. (a) yields:

$$F = \mu N/\cos \alpha \qquad\qquad\qquad (c)$$

Substituting Eq. (c) into Eq. (b), and solving for the unknown force N leads to:

$$\mu N \tan \alpha + N - W = 0$$

$$N = W/(1 + \mu \tan \alpha) \qquad\qquad\qquad (9.8)$$

Finally, substitute Eq. (9.8) into Eq. (c) to obtain:

$$F = \mu W/[\cos \alpha(1 + \mu \tan \alpha)] \qquad\qquad\qquad (9.9)$$

We have numbered the relations for the two unknown forces N and F in the sled problem as Eqs. (9.8) and (9.9). These relations may be used in the solution of N and F for any block that is moved by sliding providing $\mu$, W, and $\alpha$ are known. Let's complete the solution and determine the capability of our mule team. Substituting into Eq. (9.9) gives:

$$F = [(0.45)(5200 \text{ lb})]/[\cos 15°(1 + (0.45)\tan 15°)] = 2162 \text{ lb}$$

## EXAMPLE 9.4

Consider a wooden crate that is resting on a level concrete floor of a warehouse. A worker attempts to slide the crate a short distance as shown in Fig. E9.4. If the crate weighs 400 N and the coefficient of friction between the wooden crate and the concrete floor is $\mu = 0.5$, determine the force that must be applied to slide the crate. Note that the crate is short and the worker must lean over to apply a force at an angle of 30° relative to the floor.

Fig. E9.4

**Solution:**

Begin the solution by constructing a FBD to model the physical situation. We recognize this problem is identical to the problem in Example 9.3, except the worker is pushing with a force that has components in the negative x and y directions. We account for this fact by deriving Eqs. (9.8) and (9.9) again. Following the same procedure, we obtain:

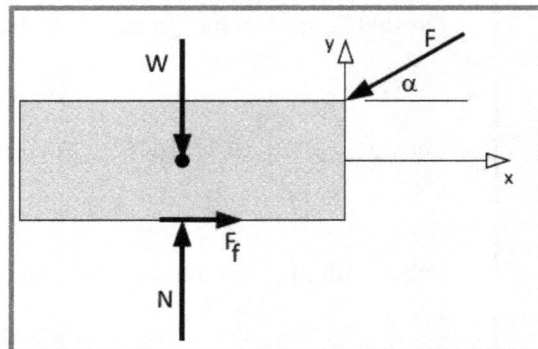

Fig. E9.4a

$$N = W/(1 - \mu \tan \alpha) \qquad\qquad (9.8a)$$

$$N = 400/[1 - (0.5) \tan (30°)] = 562.3 \text{ N}$$

$$F = \mu W/[\cos \alpha(1 - \mu \tan \alpha)] \qquad\qquad (9.9a)$$

$$F = (0.5)(400)/ \{\cos (30°)[1 - (0.5) \tan(30°)]\} = 325 \text{ N}$$

The result of 325 N is much greater than a force of 200 N, which would have been required if the worker had applied the force F in the direction of motion. There are two reasons for the significant increase in the required force. First, the worker is pushing with a component of force downward on the crate, which increases the normal and friction forces. Second, only the horizontal component of the applied force is useful in sliding the crate.

## EXAMPLE 9.5

Suppose you are driving in a storm that is depositing a mixture of snow and ice on the highways. As a prudent driver, you reduce your speed significantly, and are making slow but steady progress toward your destination. However, you encounter a long hill with a grade of 4% as shown in Fig. E9.5. Unfortunately part way up the hill your front wheel drive car loses traction and stops. Your attempts to move the car forward fail because the tires spin as you depress the accelerator to increase the power applied to the wheels. In fact, your efforts cause the car to slip a short distance down the hill.

Determine the coefficient of friction that exists between the surface of the tire and the roadbed. Also explain why your attempts to move the car forward (up the hill) are futile.

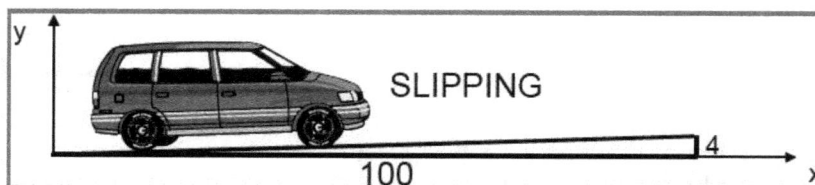

**Solution:**

Let's begin our analysis by drawing a FBD of one of the front wheels as shown in Fig. E9.5a.

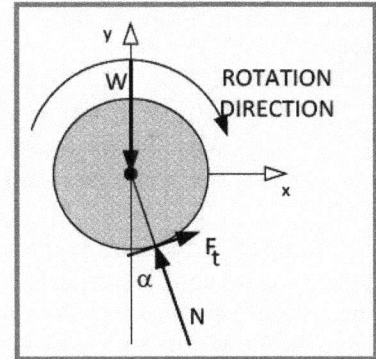

Fig. E9.5a

Clearly, we select a front wheel for the FBD because the car applies the traction force $F_t$ to the pavement through the front wheels. Begin the analysis by writing the equilibrium relations.

$$\Sigma F_x = F_t \cos\alpha - N \sin \alpha = 0 \qquad\qquad (a)$$

$$\Sigma F_y = F_t \sin \alpha + N \cos \alpha - W = 0 \qquad\qquad (b)$$

Since the car slid down the hill, a condition of impending motion existed prior to the slide. Hence, we may write:

$$F_t = F_f = \mu N \qquad\qquad (c)$$

We have equated the traction force $F_t$ with the friction force $F_f$ in this equation.

Next, let's substitute Eq. (c) into Eq. (a) to give:

$$\mu N \cos \alpha - N \sin \alpha = 0 \qquad\qquad (d)$$

$$\mu = \tan \alpha = 0.04 \qquad\qquad (e)$$

The result in Eq. (e) shows that the friction coefficient is independent of the normal force N, and is equal to the grade of the hill. This fact is clear because N cancels out of Eq. (d). We will return to this point in the discussion following this example.

To continue the analysis, substitute Eq. (9.3) into Eq. (b) and write:

$$\mu N \sin \alpha + N \cos \alpha - W = 0 \qquad\qquad (f)$$

$$N = W/ (\mu \sin \alpha + \cos \alpha) \qquad\qquad (g)$$

We have determined the relation for N to present a complete solution. The coefficient of friction is independent of both N and the weight W acting on the wheel.

Attempts to get the car moving are almost always unsuccessful because the wheels begin to spin as power is applied to the drive train. As the wheels spin, heat is generated melting the snow and ice mixture. Water is formed that better lubricates the surface of the ice and snow coating the highway, which reduces still further the coefficient of friction.

If you have ever driven much in snow and ice and have mastered the art of successfully negotiating slippery hills when all about you get stuck, the solution in Example 9.5 should certainly raise two very important questions.

1. Why do front wheel drive cars perform better in the snow and ice than rear drive cars?
2. Why do we spend significant amounts of money and put up with excessive road noise to equip our cars with snow tires?

Front wheel drive cars perform better in the snow than rear wheel drive cars, because more weight is placed over the drive wheels. The analysis showed that the friction coefficient was independent of the weight on the wheel only because we did not consider the possibility of the tire tread being forced into the snow and ice mixture covering the pavement. Higher wheel forces may in many cases produce enough penetration of the tread into the coating of snow to give some degree of interlocking at the tire-snow interface. The analysis in Example 9.5 considered the surfaces of the wheel and the pavement to be perfectly rigid, which is not true.

Snow tires are also beneficial because the tread is designed to penetrate snow and gain traction by interlocking with a coating of even relatively hard packed snow. However, the benefit gained by a tread that penetrates a coating of snow largely vanishes if the pavement is coated with a rigid layer of ice.

## 9.4 FRICTION AND STABILITY

In many engineering applications friction is a culprit. It results in a loss of power, a loss of efficiency, and the generation of undesirable heat. However, in other applications friction is a benefit that enables us to move about with control and to maintain stability. The next series of examples demonstrate the role of friction in stability, in preventing slipping, in inducing tipping and in promoting rolling.

### 9.4.1 Stability

When you climb a ladder, you want it to remain stable or otherwise you will fall. The stability is maintained because friction keeps the ladder from slipping and sliding as you climb it. Often ladders are designed with rotating footpads with a rubber surface that improves frictional contact, increases the coefficient of friction, and enables these ladders to remain stable at smaller angles of inclination.

### EXAMPLE 9.6

Consider the ladder that rests on a level concrete surface and leans against a vinyl-clad wall as shown in Fig. 9.10. A man with a mass of 81.5 kg climbs ¾ of the way up the ladder. We question the stability of the ladder. Is it stable or will it slip and fall?

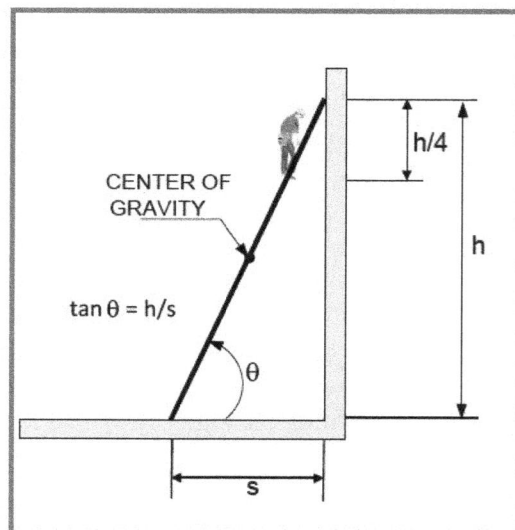

Fig. 9.10 Illustration of climbing a ladder.

**Solution:**

Before considering the stability of the ladder, we need to gather much more information about the ladder, its interactions with the floor and the wall, and its weight. A FBD of the ladder shown in Fig. E9.6 provides guidance before we seek the additional information necessary to determine if the ladder is stable.

Fig. E9.6a

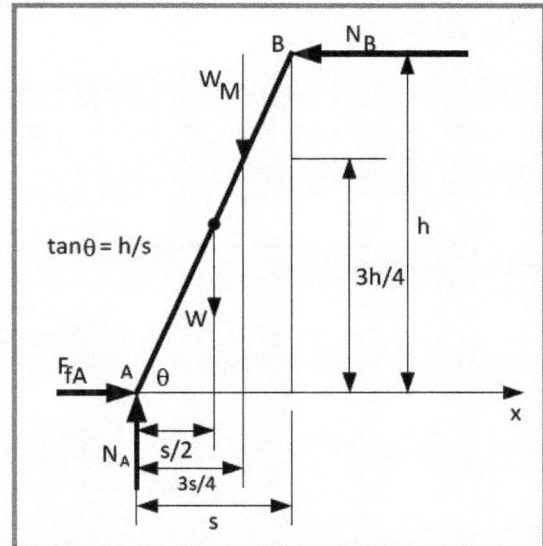

Let's introduce the parameters that control the stability of the ladder.

$$s = 3\text{m}; \ h = 7 \text{ m}; \ \theta = \tan^{-1} (h/s) = \tan^{-1} (7/3) = 66.80°; \ W = 250 \text{ N};$$

$$W_M = (81.5)(9.81) = 800 \text{ N}; \ \mu_A = 0.6 \text{ and } \mu_B = 0.$$

Note that we have assumed that $\mu_B = 0$ because a wet vinyl clad surface exhibits very little frictional restraint.

In performing a stability analysis, we assume that the ladder is about to slip (implying impending motion). This assumption is important because it permits us to determine the maximum friction force that can be developed to resist motion (instability).

Let's apply the three equilibrium relations that must be satisfied if the ladder is stable.

$$\Sigma F_y = N_A - W - W_M = 0$$

$$N_A = W + W_M = 250 + 800 = 1050 \text{ N} \tag{a}$$

$$\Sigma M_A = N_B \, h - W s/2 - 3 W_M \, s/4 \tag{b}$$

$$N_B = (s)(2W + 3W_M)/4h = (3)(500 + 2400) /(4)(7) = 310.7 \text{ N} \tag{c}$$

$$\Sigma F_x = F_{fA} - N_B = 0 \tag{d}$$

$$F_{fA} = N_B = 310.7 \text{ N} \tag{e}$$

This result indicates that the friction force at the base of the ladder $F_{fA}$ must be 310.7 N to maintain the ladder in a state of stable equilibrium. The maximum friction force that can develop at Point A is given by Eq. (9.3) as:

$$(F_{fA})_{max} = \mu_A \, N_A = (0.6)(1050) = 630 \text{ N} \qquad (f)$$

Since $(F_{fA})_{max} = 630 \text{ N} \geq F_{fA} = 310.7 \text{ N}$, the ladder will be stable under the conditions described above. The magnitude of the friction force that may develop at the base of the ladder is larger than that necessary to maintain equilibrium.

## EXAMPLE 9.7

Let's consider the ladder shown in Fig. 9.10, and determine its stability if all the conditions are the same as described in Example 9.6 except for the following parameters:

$$h = 6 \text{ m}; \; s = 5 \text{ m}; \; \mu_A = 0.4, \text{ and } \mu_B = 0$$

**Solution:**

Utilizing the FBD presented in Example 9.6, we explore the stability question by writing the equilibrium relations:

$$\Sigma F_y = N_A - W - W_M = 0 \qquad (a)$$

$$N_A = W + W_M = 250 + 800 = 1050 \text{ N}$$

$$\Sigma M_A = N_B \, h - Ws/2 - 3W_M \, s/4$$

$$N = (s)(2W + 3W_M)/(4h) = (5)(500 + 2400)/(4)(6) = 604.2 \text{ N} \qquad (b)$$

$$\Sigma F_x = F_{fA} - N_B = 0 \qquad (c)$$

Substituting the results from Eq. (b) into Eq. (c) gives:

$$F_{fA} = N_B = 604.2 \text{ N} \qquad (d)$$

Next, we must determine if the friction force $F_{fA} = 604.2$ N at the base of the ladder can be developed. From Eq. (9.3), $(F_{fA})_{max}$ is given by:

$$(F_{fA})_{max} = \mu_A \, N_A = (0.4)(1050) = 420 \text{ N} \qquad (e)$$

Clearly, the ladder is unstable and the base will slide away from the wall since:

$$F_{fA} = 604.2 \text{ N} > (F_{fA})_{max} = 420 \text{ N} \qquad (f)$$

## EXAMPLE 9.8

Let's consider a slightly more complex question pertaining to the stability of the ladder shown in Fig. 9.10. The new parameters describing the condition of the ladder are:

$$h = 7\text{m}; \; s = 4\text{m}; \; \mu_A = 0.4; \text{ and } \mu_B = 0.3$$

All of the other parameters are the same as those described in Example 9.6.

**Solution:**

Since the coefficient of friction at the point of contact of the ladder with the wall is not zero, we must modify the FBD by adding a friction force $F_{fB}$ as shown in Fig. E9.8. Examination of the FBD indicates that we encounter four unknown forces, namely $N_A$, $N_B$, $F_{fA}$, and $F_{fB}$. However, we only have three equations of equilibrium to apply in our stability analysis. To resolve this difficulty, we assume that the ladder is in a condition of impending motion at the wall (point B), which enables us to use Eq. (9.3) and express $F_{fB}$ in terms of $N_B$. This assumption reduces the number of unknown forces to three, which is equal to the number of relevant equilibrium relations.

Fig. E9.8

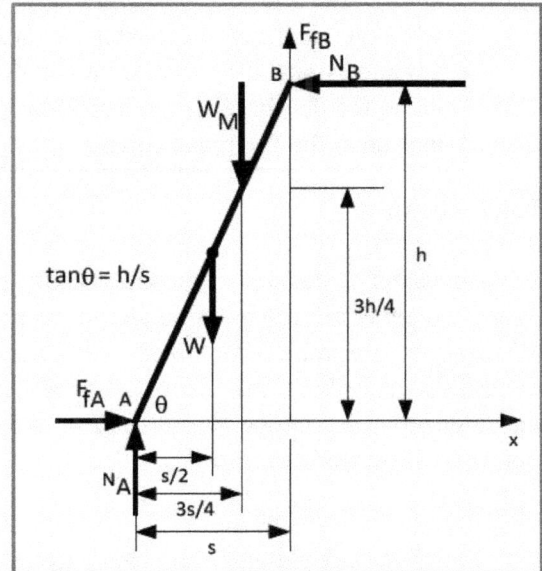

Next we write the equation for the friction force at point B as:

$$F_{fB} = (F_{fB})_{max} = \mu_B N_B \qquad (a)$$

Applying the equilibrium relations yields:

$$\Sigma M_A = -Ws/2 - 3W_M s/4 + F_{fB} s + N_B h = 0 \qquad (b)$$

Substituting Eq. (a) into Eq. (b), and solving for $N_B$ gives:

$$N_B = (s/4)(2W + 3W_M)/(\mu_B s + h) \qquad (c)$$

$$N_B = (4/4)(500 + 2400)/[(0.3)(4) + 7] = 354 \text{ N}$$

$$\Sigma F_y = N_A - W - W_M + \mu_B N_B = 0 \qquad (d)$$

Solving Eq. (d) for $N_A$ yields:

$$N_A = W + W_M - \mu_B N_B = 250 + 800 - (0.3)(354) = 943.8 \text{ N} \qquad (e)$$

Next, we employ the final equilibrium relation to determine the friction force at the base of the ladder required to insure stability.

$$\Sigma F_x = F_{fA} - N_B = 0 \qquad (f)$$

$$F_{fA} = N_B = 354 \text{ N} \qquad (g)$$

From Eq. (9.3) and the condition of impending motion at point A, we write:

$$(F_{fA})_{max} = \mu_A \, N_A = (0.4)(943.8) = 377.5 \text{ N} \tag{h}$$

Then we compare the maximum friction force possible with the required friction force as indicated below:

$$(F_{fA})_{max} = 377.5 \text{ N} > F_{fA} = 354 \text{ N} \tag{i}$$

We conclude that the ladder is stable, but with only a modest margin of safety. If we had assumed impending motion at the floor rather than the wall, the solution would give a negative value for the force $F_{fB}$. Clearly the wall cannot pull on the ladder the assumption of impending motion at the floor was correct.

## 9.4.2 Tipping

Let's consider a different type of stability problem that involves sliding objects along some surface. When we apply a force to the object to overcome the frictional forces, the object may slide or it may tip. We seek to determine if the object will slide or tip for a given set of conditions describing the object. We will introduce this topic by considering a rectangular box resting on a horizontal surface as the object as illustrated in Fig. 9.11.

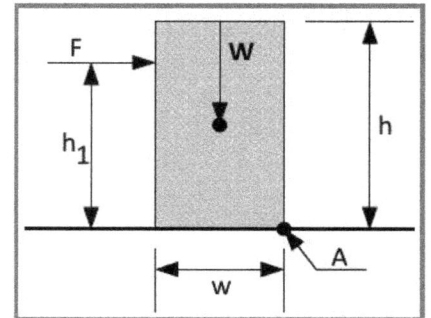

Fig. 9.11 A rectangular box being moved by sliding along a horizontal surface.

## EXAMPLE 9.9

Consider a crate that weighs 400 lb resting on a level concrete floor. The coefficient of friction $\mu = 0.4$ between the floor and the box. The dimensions of the quantities defined in Fig. 9.11 are h = 4 ft, $h_1$ = 3 ft, and w = 1 ft. Determine the force required moving the box, and establish if it will slide or tip when the force F is applied.

**Solution:**

Begin with a FBD of the crate as shown in Fig. E9.9. Next, apply the equations of equilibrium.

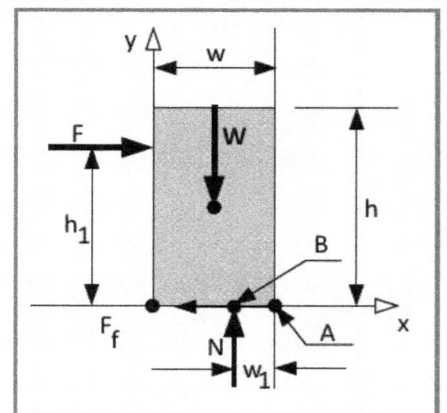

Fig. E9.9

$$\Sigma F_y = N - W = 0 \qquad \text{(a)}$$

$$N = W = 400 \text{ lb}$$

$$\Sigma F_x = F - F_f = 0 \qquad \text{(b)}$$

$$\Sigma M_B = W[(w/2) - w_1] - Fh_1 = 0 \qquad \text{(c)}$$

At this stage, we have solved for one unknown (N); however, three unknowns remain (F, F and w) with only two applicable equations remaining. We must consider each case of impending motion separately to determine if the crate will tip or slide.

First, assume that the crate slides with $F_{Slide} = F_f$. Then the following relation holds:

$$F_{Slide} = F_f = \mu N = (0.4)(400) = 160 \text{ lb} \qquad \text{(d)}$$

Now, let's consider the argument for tipping. When the crate begins to tip, the point of application of N (point B) shifts to the right hand edge of the crate (point A). For this situation, $w_1 = 0$ and Eq. (c) becomes:

$$F_{Tip} = (Ww)/(2h) = [(400)(1)]/[(2)(3)] = 66.67 \text{ lb} \qquad \text{(e)}$$

Clearly, the crate will tip before it slides because $F_{Tip} < F_{Slide}$.

## EXAMPLE 9.10

The crate in Example 9.9 tipped because we applied the force F near the top of the crate. For the same conditions, determine the largest value of $h_1$ for applying the force F, which will result in the crate sliding without tipping.

**Solution:**

To solve this problem, recognize that tipping becomes possible when the point of application of the normal force N shifts to point A in Fig. 9.11, or when $w_1 = 0$. Recall Eq. (c) from Example 9.9, let $w_1 = 0$, and then solve for $h_1$.

$$h_1 = (W/F)[(w/2 - w_1) \qquad \text{(a)}$$

$$h_1 = (400/160)[1/2 - 0] = 1.25 \text{ ft} \qquad \text{(b)}$$

For h < 1.25 ft, the crate will slide.

### 9.4.3 Rolling

Let's next consider the role of friction in determining if a cylindrical object rolls or slides. To illustrate this situation, refer to the roller positioned adjacent to a step as shown in Fig. 9.12.

As we apply a force F to the cable wound about the smaller cylinder, the larger roller contacts the step. If the friction force developed at point A is sufficiently large, the cylinders will roll up the vertical face, and the assembly will negotiate the step. On the other hand, if the friction force is insufficient, the cylinders will slip and remain at the bottom of the step.

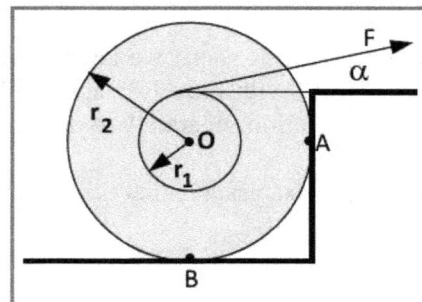

Fig. 9.12 Rolling a cylinder over a step.

Rolling friction can be both beneficial and detrimental in the same application. For example, when designing wheels for a vehicle it is necessary to maintain enough rolling friction to provide the thrust forces that are generated during acceleration of the vehicle. At the same time, the rolling friction also reduces the fuel efficiency of the car by resisting the forward motion of the vehicle when it travels at a constant velocity.

### EXAMPLE 9.11

Consider a cylindrical arrangement similar to that shown in Fig. 9.12 with a force F applied at an angle $\alpha = 30°$. If the coefficient of friction at the contact points A and B is $\mu_A = \mu_B = 0.4$, determine if the cylinder will roll up and over the step or slip and remain in the corner as the force is increased. Also determine the force required to roll a cylindrical arrangement that weighs 600 N up and over the step. Note that $r_1 = 100$ mm, and $r_2 = 300$ mm.

**Solution:** In constructing the FBD shown in Fig. E9.11, we have assumed that the cylindrical assembly has been lifted from the horizontal surface ($N_B = 0$) and is beginning to roll up the step. To resolve the roll or slip question, write the equilibrium relations to obtain:

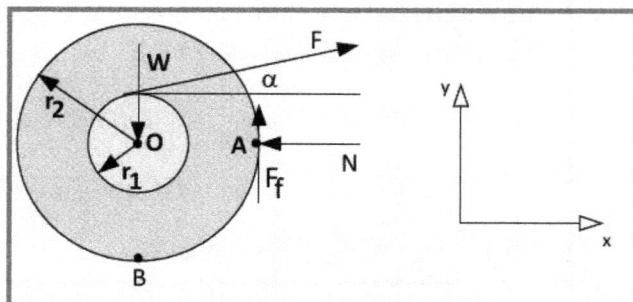

Fig. E9.11

$$\Sigma M_O = F_f r_2 - F r_1 \cos \alpha = 0 \qquad \text{(a)}$$

$$F_f = (r_1/r_2)F \cos \alpha = (1/3)F(0.8660) = 0.2886\ F \qquad \text{(b)}$$

$$\Sigma F_x = F \cos \alpha - N = 0; \quad \Rightarrow \quad N = 0.866 \, F \tag{c}$$

The maximum possible value of the friction force at point A is given by Eq. (9.3) as:

$$(F_f)_{max} = \mu N = (0.4)(0.8660)F = 0.346 \, F \tag{d}$$

A comparison of the results of Eq. (b) and Eq. (d) shows that:

$$(F_f)_{max} > F_f \tag{e}$$

Clearly, the cylinder will roll up and over the step.

Finally, let's determine the force F required.

$$\Sigma F_y = F \sin \alpha - W + F_f = 0 \tag{f}$$

Substituting the results of Eq. (b) into Eq. (f) yields:

$$(0.5)F + 0.2886 \, F = 600 \quad \Rightarrow \quad F = 760.8 \, N$$

Because $(F_f)_{max} > (F_f)$, the cylindrical assembly will roll up the step without slipping.

## 9.5 FRICTION EFFECTS ON WEDGES

A wedge is classified as a simple machine because it can be used to magnify a force if the effects of friction are not too large. The lifting of a block with a weight W by a tapered wedge is illustrated in Fig. 9.13. For the wedge to act as a machine, the force F used to drive the wedge under the block must be less that the weight W.

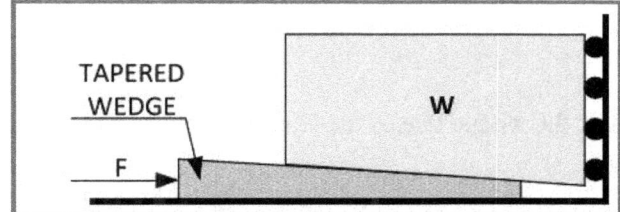

Fig. 9.13 A tapered wedge acts as a machine to lift a weight.

The use of a wedge as a simple machine has three modes of operation:

- Lifting
- Holding
- Lowering

Friction plays an important role in all three of these operations.

### 9.5.1 Lifting with a Wedge

Let's first consider the lifting operation and explore the role of friction. As usual, we begin with the FBD diagram of the block and the wedge shown in Fig. 9.14. Note that the force F, in Fig. 9.14, acts to the right in lifting the block. Also observe that we have assumed that the rollers eliminated the friction force $F_{f3}$ on the vertical surface of the block.

Let's apply the equations of equilibrium to the block and then to the wedge to determine the relation between the applied force F and the weight W.

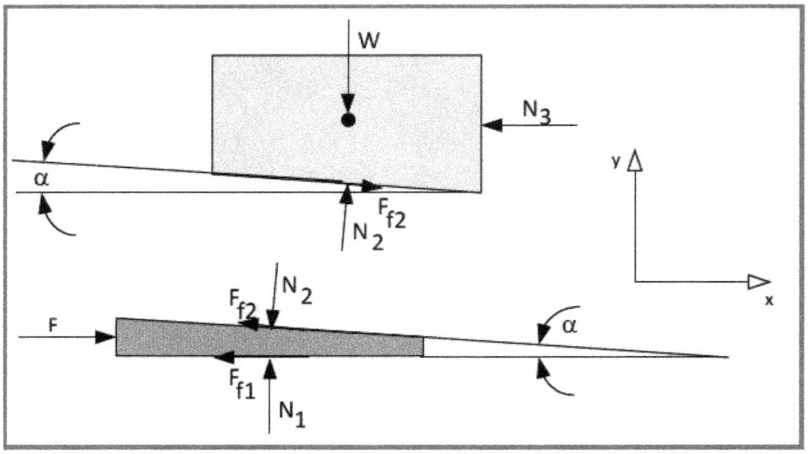

Fig. 9.14 FBD of a block and wedge in a lifting operation.

For the block, we write:

$$\Sigma F_y = N_2 \cos \alpha - F_{f2} \sin \alpha - W = 0 \tag{a}$$

Since motion occurs to move the wedge, we may write:

$$F_{f1} = \mu N_1 \quad \text{and} \quad F_{f2} = \mu N_2 \tag{b}$$

Combining Eq. (a) and Eq. (b) and solving for $N_2$ yields:

$$N_2 = W/(\cos \alpha - \mu \sin \alpha) \tag{c}$$

For the wedge one of the equilibrium equations gives:

$$\Sigma F_y = N_1 + F_{f2} \sin \alpha - N_2 \cos \alpha = 0 \tag{d}$$

Combining Eqs. (b), (c) and (d) yields:

$$N_1 = W \tag{e}$$

Applying the final relevant equilibrium relation for the wedge gives:

$$\Sigma F_x = F - F_{f1} - F_{f2} \cos \alpha - N_2 \sin \alpha = 0$$

$$F = \mu W + \mu N_2 \cos \alpha + N_2 \sin \alpha \tag{f}$$

$$F = W\left[\frac{2\mu + (1 - \mu^2) \tan \alpha}{1 - \mu \tan \alpha}\right] = CW \tag{9.10}$$

where C is the coefficient for the wedge acting as a simple machine. It is given by:

$$C = \frac{2\mu + (1 - \mu^2)\tan\alpha}{1 - \mu\tan\alpha} \qquad (9.11)$$

For the wedge to be effective as a machine and lift a heavy weight W with a small force F, we seek to minimize the coefficient C. This minimization is accomplished by reducing the friction coefficient and by using small wedge angles. A graph showing the machine coefficient C as a function of the wedge angle for several different values of the friction coefficient is presented in Fig. 9.15.

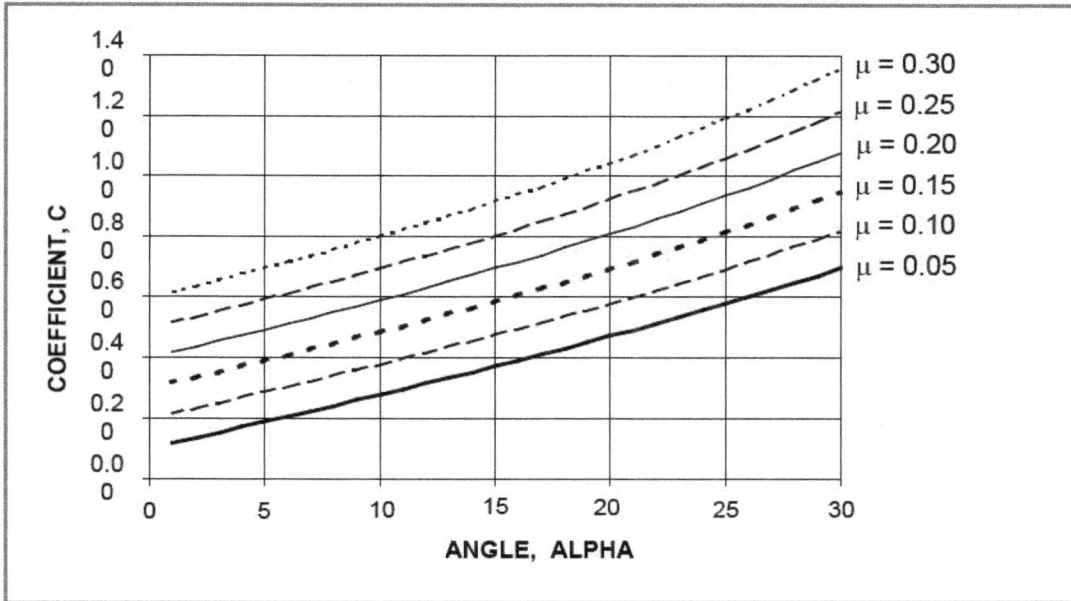

Fig. 9.15 Small forces lift heavy weights with wedges when $\alpha$ is small and $\mu$ is low.

## EXAMPLE 9.12

You are attempting to level a heavy machine bed with a mass of 20,000 kg. To lift one end of the bed, you insert a wedge with an angle $\alpha = 10°$. If the coefficient of friction between the wedge and the machine bed is $\mu = 0.2$, determine the force necessary to lift the machine bed by 120 mm.

**Solution:**

The weight of the machine bed is given by:

W = (9.81)(20,000) = 196.2 kN

Recall Eq. (9.10)

F = C W = (0.590)(196.2) = 115.8 kN

where C = 0.590 is determined from the graph in Fig. 9.15 or by evaluating Eq. (9.11).

Suppose that the hydraulic jack that is available only has a capacity of 10 tons (89.0 kN). Clearly this jack cannot drive the wedge with the conditions listed above. What can you do to resolve this situation?

There are three possible solutions.

- Purchase a larger capacity jack. This option is expensive and will delay the project.
- Find a wedge with a smaller angle. This is a valid solution if a smaller angle wedge exists. If not ordering a new wedge will cost money and delay the project.
- Lubricate the wedge with a heavy grease which will reduce the friction coefficient to $\mu = 0.11$. This is the best solution because grease is readily available at a low cost.

If we apply a coating of grease to both sides of the wedge, the friction coefficient is reduced to $\mu = 0.11$. Then for a wedge angle of $\alpha = 10°$, the machine coefficient is C = 0.401. Then:

$$F = C\,W = (0.401)(196.2\ kN) = 78.68\ kN$$

This result for F is equivalent to 8.84 tons so the 10-ton capacity jack, which is available, provides sufficient force to drive the wedge.

## 9.5.2 Holding with a Wedge

Suppose the wedge is driven to raise the weight as illustrated in Fig. 9.13. The question is then whether or not the wedge will stay in place after the force F is removed. The weight acting downward is resisted by a normal force that has a component tending to push the wedge outward. Frictional forces must be sufficiently large to prevent the wedge from being forced out. If the friction forces are insufficient, the weight will fall, and the situation could be dangerous.

Let's explore the conditions under which the wedge will hold the load in the absence of the driving force F. As usual we begin with a free body diagram that is shown in Fig. 9.16.

Fig. 9.16 FBD under holding conditions with no force F applied to the wedge.

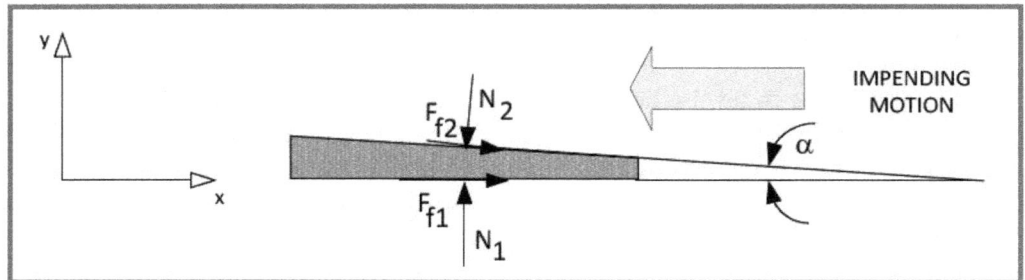

If we consider the wedge and the weight coupled together as a single body, the equilibrium relation $\Sigma F_y = 0$ yields:

$$\Sigma F_y = 0 \quad \Rightarrow \quad W = N_1 \tag{a}$$

For the wedge, we may write:

$$\Sigma F_y = W - N_2 \cos \alpha - F_{f2} \sin \alpha = 0 \tag{b}$$

Since the wedge slipping implies impending motion, we may use Eq. (9.3) and write:

$$F_{f1} = \mu\,N_1 \quad \Rightarrow \quad F_{f2} = \mu\,N_2 \tag{c}$$

$$\Sigma F_x = N_2 \sin \alpha - F_{f2} \cos \alpha - F_{f1} = 0 \tag{d}$$

Combining Eqs. (a) through (d) leads to:

$$\tan \alpha = 2\mu/(1 - \mu^2) \tag{9.12}$$

This relation obviously needs some interpretation because the quantities $\alpha$ and $\mu$ are usually specified conditions and are not physically related. We have developed a relation in Eq. (9.12) that describes the condition for the wedge to begin to slip from under the weight. For stability or instability of the weight we explore the inequalities on either side of Eq. (9.12), and write:

$$\tan \alpha < 2\mu/(1 - \mu^2) \tag{9.13}$$

This relation defines the locking angle for a wedge; the wedge remains fixed after it is driven if this equation is satisfied.
On the other hand the wedge will slip, and the weight will fall (unstable) if:

$$\tan \alpha > 2\mu/(1 - \mu^2) \tag{9.14}$$

## EXAMPLE 9.13

A wedge with an angle $\alpha = 20°$ is used to lift a 1200 lb weight. The wedge is lubricated to facilitate the lifting operation, and its coefficient of friction with the mating surfaces is only 0.12. Determine if the wedge will slip when the driving force is removed.

---

**Solution:** Check for stability by employing Eq. (9.13).

$$\tan \alpha < 2\mu/(1 - \mu^2) \quad \Rightarrow \quad \text{stability condition}$$

$$\tan 20° = 0.3640 \quad \text{and} \quad 2\mu/(1 - \mu^2) = (2)(0.12)/[1 - (0.12)^2] = 0.2435$$

Since $\tan 20° = 0.3640 > [2\mu/(1 - \mu^2)] = 0.2436$, Eq. (9.14) is satisfied and the instability condition exists. The wedge will slip, and the weight will fall.

---

## 9.5.3 Removing the Wedge

Suppose that the wedge is a tapered key that is used to fix a gear to a shaft. In some instances, the gear must be removed from the shaft to perform maintenance on the gear, shaft, or bearings. In these cases, we must apply a force to pull the wedge from the keyway. Let's determine the force $F_R$ required to pull the key. As usual, we begin with the free body diagram presented in Fig. 9.17 to model the physical situation and to guide us in writing the appropriate relations.

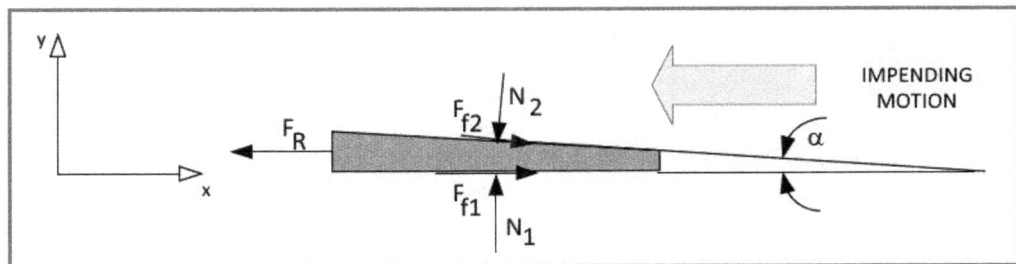

Fig. 9.17 FBD of a wedge removed from a keyway.

Let's assume the pressure of the gear acting on the wedge-like key produces a force W that is equal to force $N_1$ as defined in Fig. 9.17.

$$W = N_1 \qquad\qquad (a)$$

For the wedge, we may write:

$$\Sigma F_y = W - N_2 \cos \alpha - F_{f2} \sin \alpha \qquad\qquad (b)$$

Since pulling the wedge implies impending motion, we may use Eq. (9.3) and write:

$$F_{f1} = \mu\, N_1 \quad \Rightarrow \quad F_{f2} = \mu\, N_2 \qquad\qquad (c)$$

$$\Sigma F_x = N_2 \sin \alpha - F_{f2} \cos \alpha - F_{f1} + F_R = 0 \qquad\qquad (d)$$

Combining Eqs. (a) through (d) leads to:

$$F_R = W\,[2\mu - (1 - \mu^2)\tan\alpha]\,/\,[1 + \mu \tan\alpha] \qquad\qquad (9.15)$$

## EXAMPLE 9.14

Determine the force required to pull a key if the coefficient of friction $\mu = 0.8$ and the angle $\alpha$ of the key is 6°. The pressure between the gear and the key produces a force W of 2500 lbs.

Solution:

We employ Eq. (9.15) and write:

$$F_R = W\,[2\mu - (1 - \mu^2)\tan\alpha]\,/\,[1 + \mu \tan\alpha]$$

$$F_R = 2500\,[2(0.8) - (1 - 0.64)\tan(6°)]\,/\,[1 + (0.8)\tan(6°)]$$

$$F_R = 3602\ \text{lb}$$

The result may seem high, but the large coefficient of friction may imply that the key has corroded in the keyway.

## 9.6 FRICTION EFFECTS ON SCREWS

Screws are used as fasteners to hold two or more components together in an assembly. They are also used in jacks to lift bodies. In all probability, you have a screw jack in your automobile that is used in lifting one corner of a car in changing a flat tire. Screws are also used in positioning and measuring devices. For example, a screw is used to move the carriage that supports the cutting tool on a lathe, and a screw is used to control the position of the head on a micrometer.

A screw is very much like a wedge or an inclined plane that is wrapped around a cylinder as shown in Fig. 9.18. The inclined plane is wrapped around the cylinder in the from of a helix with an angle $\alpha$ to form the thread. A nut, not shown in Fig. 9.18, is fitted over the thread. As the thread is rotated, the nut is driven in one direction or the other depending on the direction of rotation. In one rotation, the nut is driven a distance of p, which is the pitch of the thread. In driving the carriage of a lathe, the thread rotates in a fixed position and the nut drives the carriage along the length of the lathe bed. In a typical jack, the nut is fixed to the support structure for the jack and the end (or both ends) of the screw moves to lift an auto.

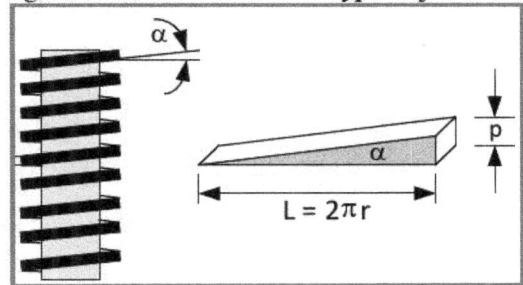

Fig. 9.18 Representing a screw thread with an inclined plane.

We apply a moment to turn the screw in either lifting or lowering a load. The frictional forces and the magnitude of the load determine the magnitude of the moment applied to the screw. Let's determine the magnitude of the moment required to lift a load.

### 9.6.1 Lifting a Load with a Screw

Suppose we are lifting a load of W with a screw jack. In designing the jack, we must determine the moment required to lift the load. Let's begin this analysis by modeling the thread on the cylinder with an inclined plane and the nut with a block as shown in the FBD of Fig. 9.19.

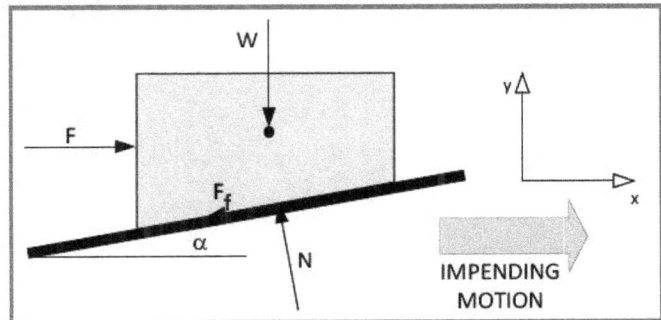

Fig. 9.19 FBD of a model of a screw and nut for lifting a load W.

We apply the equilibrium relations to determine the moment required to lift the load.

$$\Sigma F_y = N \cos \alpha - F_f \sin \alpha - W = 0 \qquad \text{(a)}$$

$$N = W/(\cos \alpha - \mu \sin \alpha)$$

$$\Sigma F_x = F - N \sin \alpha - F_f \cos \alpha = 0 \qquad \text{(b)}$$

$$F = N(\sin \alpha + \mu \cos \alpha)$$

$$F = W(\tan \alpha + \mu)/(1 - \mu \tan \alpha) \qquad \text{(c)}$$

However, the relation of the moment M to the force F is:

$$M = F r \qquad (9.16)$$

where r is the radius of the screw.

Then from Eqs (c) and (9.16), we write:

$$M = W \, r(\tan \alpha + \mu)/(1 - \mu \tan \alpha) \tag{9.17}$$

## 9.6.2 Lowering a Load with a Screw

The analysis of the screw employed to lower a load is almost identical to that shown above. The only differences are the direction of the frictional force $F_f$ and the applied force F as shown in Fig. 9.20.

We apply the equilibrium relations to determine the moment required to lower the load.

$$\Sigma F_y = N \cos \alpha + F_f \sin \alpha - W = 0 \tag{a}$$

$$N = W/(\cos \alpha + \mu \sin \alpha)$$

$$\Sigma F_x = - F - N \sin \alpha + F_f \cos \alpha = 0 \tag{b}$$

$$F = N(- \sin \alpha + \mu \cos \alpha)$$

$$F = W \, (\mu - \tan \alpha)/(1 + \mu \tan \alpha) \tag{c}$$

Fig. 9.20 The FBD of a screw and nut used in lowering the load W.

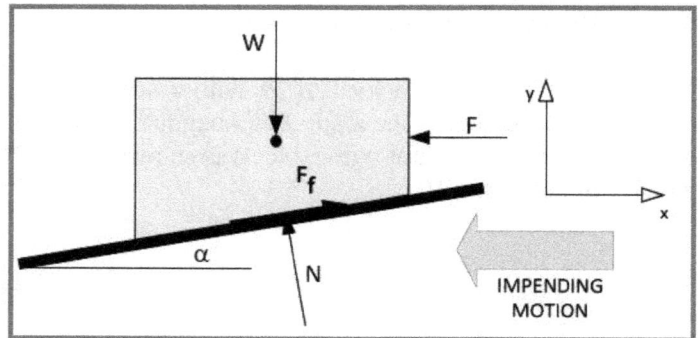

Then by substituting Eq. (c) into (9.16), we obtain:

$$M = W \, r \, (\mu - \tan \alpha)/(1 + \mu \tan \alpha) \tag{9.18}$$

## 9.6.3 Holding a Load with a Screw

The analysis of the screw employed to hold the load in a fixed position is also very similar to the two previous studies. The friction force $F_f$ opposes the impending motion and the applied force F is absent. We show the FBD for the new model, which represents lowering the load in Fig. 9.21.

Fig. 9.21 The FBD for a screw holding a load without an applied moment (torque).

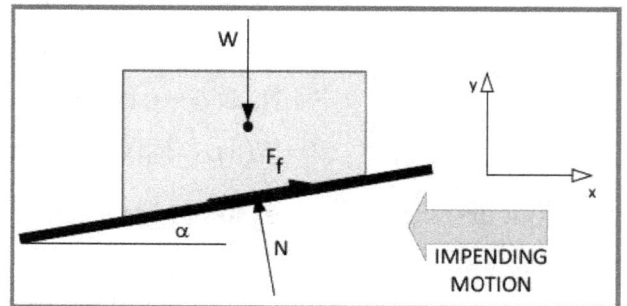

Apply the equilibrium relations to determine if the jack will be stable without a moment applied to the screw. This is an important consideration as self locking screws are essential for the safe operation of a lifting device. No one wants to have the load crashing down as soon as the moment is released from the screw.

Let's begin by writing the equations of equilibrium for the block shown in Fig. 9.21.

$$\Sigma F_x = - N \sin \alpha + F_f \cos \alpha = 0 \qquad \text{(a)}$$

We again assume that motion is impeding and substitute Eq. (9.3) into Eq. (a) to obtain:

$$\mu = \sin \alpha / \cos \alpha = \tan \alpha \qquad \text{(b)}$$

Equation (b) describes the equilibrium condition for the screw with the moment M = 0. However, to insure stability, we write the inequality condition as:

$$\tan \alpha < \mu \qquad \text{(9.19)}$$

Substituting Eq. (9.6) into Eq. (9.19) gives:

$$\tan \alpha < \tan \phi \qquad \text{(9.20)}$$

Clearly, Eq. (9.20) shows that the helix angle $\alpha$ of the screw must be less than the friction angle $\phi$ for the screw to be self-locking.

Another method for showing the self-locking condition for a screw is to set M = 0 in Eq. (9.18). Then it is clear that $(\mu - \tan \alpha) = 0$ with M = 0 and the conditions for stability given in Eq. (9.19) are obtained.

## 9.7 SUMMARY

Friction occurs when we have two bodies in contact and attempt to move one relative to another. Although motion may not occur, friction forces develop that must be considered in any equilibrium analysis of the individual bodies. Friction effects may be beneficial when friction forces are essential in maintaining stability and in providing traction. However, in many instances friction forces are detrimental because they result in losses of power, energy, and efficiency, and in the generation of heat.

Friction is due to the interactions of the asperities on both contacting surfaces. With dry friction, the frictional forces increase until they become sufficiently large to break the bonds between adjoining asperities and/or shear the interlocking asperities. Lubricating the surfaces markedly reduces the frictional forces because the lubricants physically separate the asperities. The small friction forces developed under fully lubricated conditions are due to shearing of the fluid lubricant.

The magnitude of the friction force is variable. It increases up to some maximum value to prevent relative motion between the two contacting surfaces. The maximum friction force is related to the shearing force necessary to destroy the bonds between the contacting asperities. It is approximated by:

$$\boxed{(F_f)_{max} = \mu N \qquad \text{(9.3)}}$$

A table providing approximate values for the coefficients of friction, $\mu$ for different material pairs is presented; however, a typical coefficient exhibits a large variation; therefore it is advisable to measure $\mu$ with a simple experiment involving an inclined plane as illustrated in Fig. 9.7.

There is a difference between the static and dynamic coefficients of friction. The static coefficient of friction describes the maximum friction force that may be developed before motion occurs.

The dynamic coefficient describes a constant value of the friction force after motion begins. The dynamic coefficient is usually 20 to 25% lower than the static coefficient.

We often describe the coefficient of friction with either the friction angle or the angle of repose. These angles are equal. The angle of repose $\theta_s$ is measured with an inclined plane, and the friction angle $\phi$ is the angle between the resultant force and the friction force as shown in Fig. 9.9. The tangent of both of these angles is equal to the coefficient of friction.

$$\mu = \tan \theta_s = \tan \phi \qquad\qquad (9.5 - 9.7)$$

Many examples are presented to illustrate procedures useful in solving "friction" problems including:

- Construction of a FBD to model the physical situation and to show all of the forces acting on the body and their dimensions.
- Application of the appropriate equations of equilibrium.
- Execution of the mathematics required for the solution of these equations.

Friction dependent stability was demonstrated by considering a number of different examples involving the stability of a ladder. We assume a condition of impending motion and derive an inequality relation that indicates stability or the lack thereof:

$$F_f \le (F_f)_{max} \qquad \Rightarrow \quad \text{for stability.}$$

$$F_f \ge (F_f)_{max} \qquad \Rightarrow \quad \text{for instability.}$$

Examples showing the role of friction in tipping versus sliding were considered. The determination involves locating the point of application of the normal reactive force. If this normal force is positioned to the right of point A in Fig. 9.11, the crate tips rather than slides.

One example was introduced to show the conditions for rolling or slipping. The determination of whether a cylinder rolls or slips is identical to the stability condition for the ladder.

$$F_f \le (F_f)_{max} \qquad \Rightarrow \quad \text{the rolling condition.}$$

$$F_f \ge (F_f)_{max} \qquad \Rightarrow \quad \text{the slipping condition.}$$

We examined the role of friction relative to both the wedge and the screw, which are simple machines. In both components, three situations must be analyzed—lifting, holding and lowering. All of the equations necessary to determine the driving force and or moment (torque) required to perform these three functions for both the wedge and the screw are summarized below:

The force required to lift a weight W with a wedge is:

$$F = W\left[\frac{2\mu + (1 - \mu^2)\tan \alpha}{1 - \mu \tan \alpha}\right] = CW \qquad (9.10)$$

$$C = \frac{2\mu + (1 - \mu^2)\tan \alpha}{1 - \mu \tan \alpha} \qquad (9.11)$$

With no force applied a wedge will hold if the inequality below is satisfied:

$$\tan \alpha < 2\mu/(1 - \mu^2) \qquad (9.13)$$

On the other hand the wedge will slip, and the weight will fall (unstable) if:

$$\tan \alpha > 2\mu/(1 - \mu^2) \qquad (9.14)$$

The force $F_R$ required to remove a wedge is given by:

$$F_R = W \, [2\mu - (1 - \mu^2) \tan \alpha] \, / \, [1 + \mu \tan \alpha] \qquad (9.15)$$

The moment that must be applied to a screw to lift a load W is given by:

$$M = W \, r \, (\tan \alpha + \mu)/(1 - \mu \tan \alpha) \qquad (9.17)$$

The moment that must be applied to a screw to lower a load W is given by:

$$M = W \, r \, (\mu - \tan \alpha)/(1 + \mu \tan \alpha) \qquad (9.18)$$

Finally the condition for stability of a screw in holding a load is given by:

$$\tan \alpha < \tan \phi = \mu \qquad (9.20)$$

The conditions for holding a wedge or screw in place when the load or moment is removed are particularly important, because if a screw or wedge does not hold when the load or moment is removed, the machine is unstable. The load will fall and injury to persons or property may result.

## REFERENCES

1. Ling, F. F. and C. H. T. Pan, *Approaches to Modeling Friction and Wear*, Springer-Verlag, New York, 1986.

## PROBLEMS

9.1 Write an engineering brief describing the factors affecting friction forces acting between two bodies.

9.2 Write an engineering brief explaining the reasons for the difference between static and dynamic friction coefficients.

9.3 Derive Eq. (9.2).

9.4 Explain the conditions that must exist before you apply the equation $(F_f)_{max} = \mu N$.

9.5 Suppose you develop a fixture consisting of an inclined plane and a block of some solid material to measure the material's coefficient of friction. You measure and record the angle of the inclined plane at the instant motion is initiated. Determine the average value of the friction coefficient from the five separate tests and its range using the data given in the table below.

| Problem No. | Angle | Angle | Angle | Angle | Angle |
|-------------|-------|-------|-------|-------|-------|
| 9.5a | 22.0° | 22.2° | 24.3° | 21.8° | 22.5° |
| 9.5b | 24.0° | 23.2° | 26.3° | 24.8° | 23.5° |
| 9.5c | 32.0° | 31.2° | 33.3° | 32.8° | 32.7° |
| 9.5d | 37.0° | 38.2° | 37.9° | 39.0° | 39.2° |

9.6 Derive Eq. (9.7).

9.7 Determine the force F required to move the sled, illustrated in the figure to the right, if the coefficient of friction between the sled and the surface is $\mu = 0.28$.

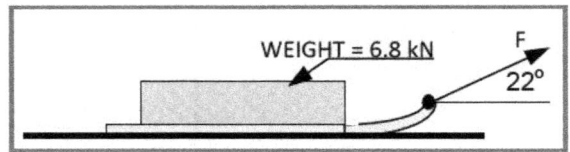
WEIGHT = 6.8 kN    F    22°

9.8 Determine the force required to slide a box, illustrated in the figure to the left, across a level floor if the force is applied at an angle $\alpha = 10°$. The coefficient of friction between the box and the floor is $\mu = 0.25$.

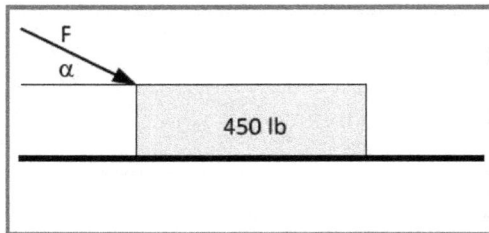
F
α
450 lb

9.9 A worker attempts to slide a wooden crate along a level concrete floor in a warehouse as shown in the figure to the right. The weight of the crate is 80 lb and the worker weighs 120 lb. The coefficient of friction between the crate and the floor $\mu_c = 0.55$ and between the worker and the floor $\mu_w = 0.70$. Note that the crate is short and the worker must lean over to apply a force at an angle of 30° relative to the floor. Is it possible for the worker to move the crate? Justify your answer by showing all the relevant calculations.

9.10 It snowed on the day of your final examination for this course, and you find the roads covered with a coating of ice and snow. In your attempt to drive to the University, you cannot negotiate a hill with a grade of 4°. Estimate the coefficient of friction between the road and the tire for your instructor, as you explain the reason for missing the final examination.

9.11 Two blocks rest on a surface as illustrated in the figure below. If $F_2 = 400$ lb, determine the maximum force $F_1$ that can be applied before one of the blocks moves. The coefficient of friction between the two blocks is 0.28 and between block B and the surface it is 0.22. Block A weighs 1,200 lb and Block B weighs 2,000 lb.

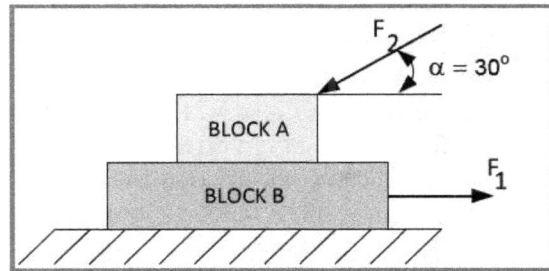

9.12   Determine if the ladder, shown in the figure to the left, is stable or not.  The man on the ladder weighs 145 lb and the ladder weighs 52 lb.  The coefficient of friction at the base of the ladder is 0.30 and the coefficient of friction between the ladder and the wall is 0.15.

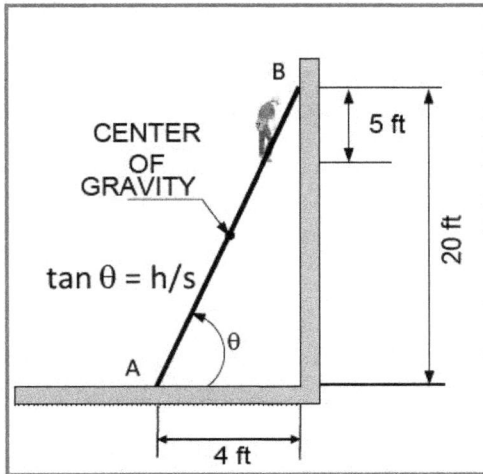

9.13   Determine the force required to move the crate, shown in the figure to the right, and indicate whether it will slide or tip.  The crate weighs 720N and the coefficient of friction between it and the floor is $\mu = 0.42$.

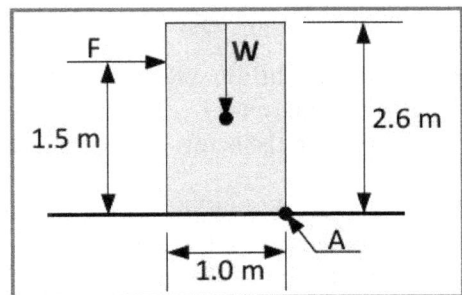

9.14   A cylindrical assembly is pulled by a force F to roll up and over a step as shown in the figure to the left.  The cylinder weighs 420 N, and the coefficient of friction between the cylinder and the surfaces at points A and B is 0.43. The radii $r_1 = 0.5$ m and $r_2 = 1.0$ m. Determine if the cylinder will roll up and over the step or slip and remain in the corner.  Show all of your work and justify your answer.

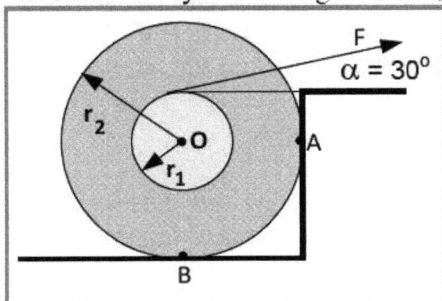

9.15 A wedge inserted at point A is used to lift one end of the machine bed shown in the figure to the right. If the wedge has an included angle $\alpha = 12°$, determine the force necessary to drive the wedge if the coefficient of friction with both surfaces $\mu = 0.35$, and the weight per unit area of the machine bed is 4 kN/m$^2$.

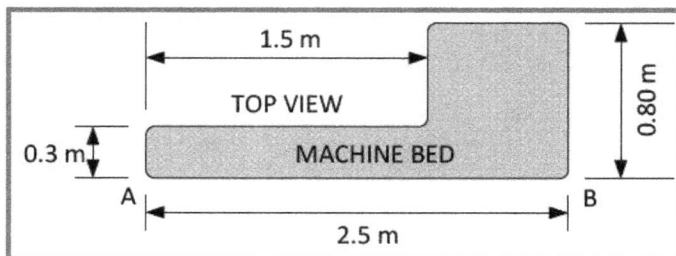

9.16 Reconsider Problem 9.15; however, the wedge is to lift the machine bed from point B and the angle of the wedge is reduced from 12° to 6°.

9.17 Determine the machine coefficient for the wedge, which measures its effectiveness as a simple lifting machine for the results obtained in Problem 9.15. Discuss the parameters leading to the best machine coefficient. Recall, work is defined as **F • d**.

9.18 If the machine bed in Problem 9.15 is lifted through a distance of 60 mm, determine the work expended in driving the wedge. Also determine the work accomplished in lifting the weight of the machine bed. Recall that work is defined as **F • d**.

9.19 If the machine bed in Problem 9.16 is lifted through a distance of 30 mm, determine the work expended in driving the wedge. Also determine the work accomplished in lifting the weight of the machine bed. Recall that work is defined as **F • d**.

9.20 For the conditions described in Problem 9.15, determine if the wedge will hold when the driving force is removed. Also determine the force $F_R$ required to remove the wedge.

9.21 For the conditions described in Problem 9.16, determine if the wedge will hold when the driving force is removed. Also determine the force $F_R$ required to remove the wedge.

9.22 You are lifting one corner of a car with a screw jack. The car weighs 3800lb, and the screw has a helix angle of 4°. The screw is lubricated with a thick, heavy grease and the coefficient of friction between the screw and the nut is $\mu = 0.11$. Determine the torque necessary to turn the screw.

9.23 You are lifting one corner of a heavy plate of steel with a screw jack. The plate weighs 10,000 N, and the screw has a helix angle of 3°. The screw is lubricated with a thick, heavy grease and the coefficient of friction between the screw and the nut is $\mu = 0.11$. Determine the torque necessary to turn the screw.

9.24 You are lifting one corner of a truck with a screw jack. The truck weighs 25 kN, and the screw has a helix angle of 4.5°. The screw is lubricated with a heavy grease and the coefficient of friction between the screw and the nut is $\mu = 0.11$. Determine the torque necessary to turn the screw.

9.25 Determine the locking angle $\alpha$ for helical threads if the coefficient of friction is 0.11.

9.26 Write an engineering brief describing the dangers associated with using a screw with a helical angle of more than 6° in the design of a jack to be used in lifting heavy weights. The screw is made from steel and the nut is made from brass.

9.27 A barrel with a diameter of 0.7 m and weighing 500 N is being rolled up an inclined ramp as shown in the figure to the right. The ramp weighs 250 N. If the coefficient of friction is 0.24 between all contacting surfaces, determine the distance x through which the barrel can be rolled before slippage occurs.

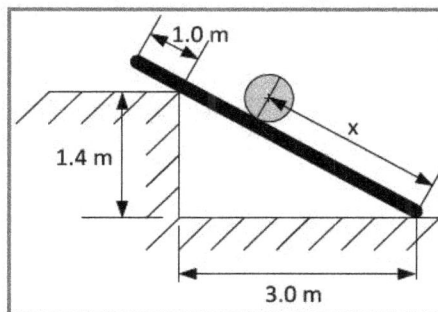

9.28 Block B with a mass $m_B$ = 19 kg sits on top of a larger block A with a mass $m_A$ = 32 kg. A horizontal force F is applied to block B as shown in the figure to the left. If the coefficient of friction between all of the surfaces is 0.28, determine the force F required to initiate motion. Describe the motion that occurs.

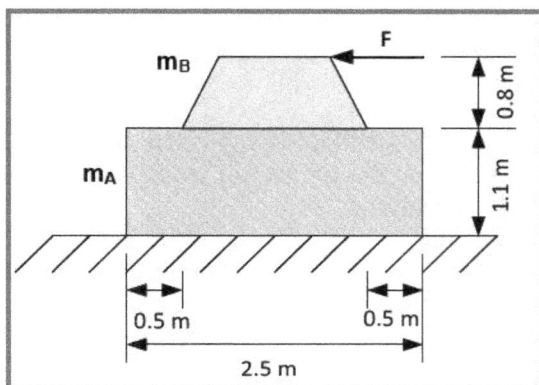

9.29 The links in the toggle mechanism, depicted in the figure to the right, are sufficiently light in weight to be neglected. The pins are frictionless. The coefficient of friction between the block and the surface is 0.34, and the block weighs 3.8 kN. Determine the angle $\theta$ that permits the application of the largest force F for which motion of the block does not occur.

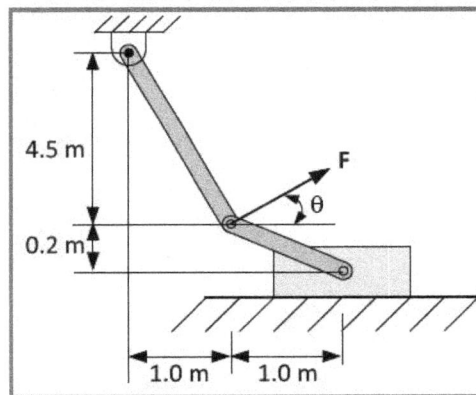

9.30 A carton weighing $W_A$ = 85 lb has tipped and is resting against another carton weighing $W_B$ = 125 as shown in the figure to the left. The coefficient of friction between the floor and the first (tipped) carton is $\mu_A$ = 0.45 and the coefficient of friction between the floor and the second (upright) carton is $\mu_B$ = 0.40. Determine if the boxes are in equilibrium in the position shown. Neglect the effect of friction between the cartons at the contact point.

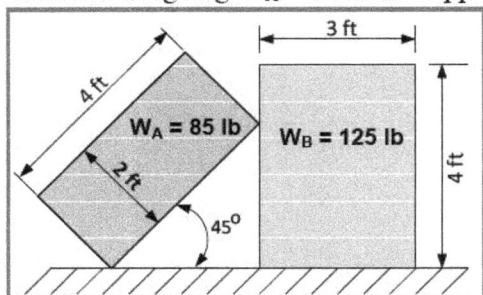

9.31    A crate with a weight W = 2.5 kN is connected to two opposing weights $W_1$ and $W_2$ by thin cables that pass over frictionless pulleys as shown in the figure to the right.    Determine the minimum and maximum values for the weight $W_1$ so that motion does not occur. The coefficient of friction between the crate and the floor is 0.28.

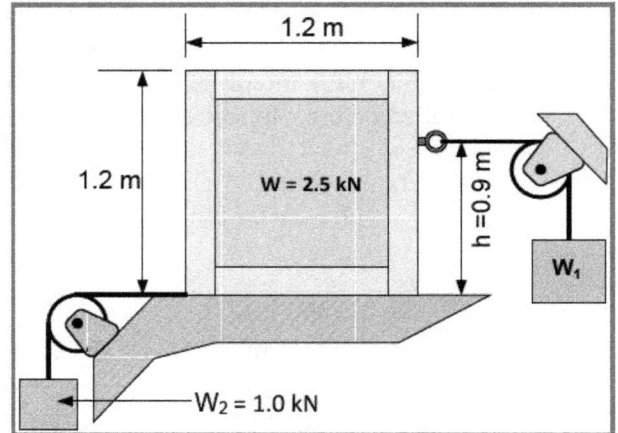

9.32    Consider the crate, shown in the figure to the right and above that is subjected to the conditions given in Problem 9.31 except for the height h.  Determine the product of h times the weight $W_1$ that produces motion by slipping and tipping simultaneously.

# APPENDIX A

## WIRE AND SHEET METAL GAGES
## DIMENSIONS ARE SHOWN IN INCHES

| Gage No. of Wire | B & S[1] Non Ferrous Metals | American S. & W Steel Wire | American S. & W Music Wire | Steel Manufactures' Sheet | Gage No. of Wire |
|---|---|---|---|---|---|
| 7-0s | 0.651354 | 0.4900 | ...... | ...... | 7-0s |
| 6-0s | 0.580049 | 0.4615 | 0.004 | ...... | 6-0s |
| 5-0s | 0.516549 | 0.4305 | 0.005 | ...... | 5-0s |
| 4-0s | 0.460 | 0.3938 | 0.006 | ...... | 4-0s |
| 000s | 0.40964 | 0.3625 | 0.007 | ...... | 000s |
| 00 | 0.3648 | 0.3310 | 0.008 | ...... | 00 |
| 0 | 0.32486 | 0.3065 | 0.009 | ...... | 0 |
| 1 | 0.2893 | 0.2830 | 0.010 | ...... | 1 |
| 2 | 0.25763 | 0.2625 | 0.011 | ...... | 2 |
| 3 | 0.22942 | 0.2437 | 0.012 | 0.2391 | 3 |
| 4 | 0.20431 | 0.2253 | 0.013 | 0.2242 | 4 |
| 5 | 0.18194 | 0.2070 | 0.014 | 0.2092 | 5 |
| 6 | 0.16202 | 0.1920 | 0.016 | 0.1943 | 6 |
| 7 | 0.14428 | 0.1770 | 0.018 | 0.1793 | 7 |
| 8 | 0.12849 | 0.1620 | 0.020 | 0.1644 | 8 |
| 9 | 0.11443 | 0.1483 | 0.022 | 0.1495 | 9 |
| 10 | 0.10189 | 0.1350 | 0.024 | 0.1345 | 10 |
| 11 | 0.090742 | 0.1205 | 0.026 | 0.1196 | 11 |
| 12 | 0.080808 | 0.1055 | 0.029 | 0.1046 | 12 |
| 13 | 0.071961 | 0.0915 | 0.031 | 0.0897 | 13 |
| 14 | 0.064084 | 0.0800 | 0.033 | 0.0747 | 14 |
| 15 | 0.057068 | 0.0720 | 0.035 | 0.0763 | 15 |
| 16 | 0.05082 | 0.0625 | 0.037 | 0.0598 | 16 |
| 17 | 0.045257 | 0.0540 | 0.039 | 0.0538 | 17 |
| 18 | 0.040303 | 0.0475 | 0.041 | 0.0478 | 18 |
| 19 | 0.03589 | 0.0410 | 0.043 | 0.0418 | 19 |
| 20 | 0.031961 | 0.0348 | 0.045 | 0.0359 | 20 |
| 21 | 0.028462 | 0.0317 | 0.047 | 0.0329 | 21 |
| 22 | 0.025347 | 0.0286 | 0.049 | 0.0299 | 22 |
| 23 | 0.022571 | 0.0258 | 0.051 | 0.0269 | 23 |
| 24 | 0.0201 | 0.0230 | 0.055 | 0.0239 | 24 |
| 25 | 0.0179 | 0.0204 | 0.059 | 0.0209 | 25 |
| 26 | 0.01594 | 0.0181 | 0.063 | 0.0179 | 26 |
| 27 | 0.014195 | 0.0173 | 0.067 | 0.0164 | 27 |
| 28 | 0.012641 | 0.0162 | 0.071 | 0.0149 | 28 |
| 29 | 0.011257 | 0.0150 | 0.075 | 0.0135 | 29 |
| 30 | 0.010025 | 0.0140 | 0.080 | 0.0120 | 30 |
| 31 | 0.008928 | 0.0132 | 0.085 | 0.0105 | 31 |
| 32 | 0.00795 | 0.0128 | 0.090 | 0.0097 | 32 |
| 33 | 0.00708 | 0.0118 | 0.095 | 0.0090 | 33 |
| 34 | 0.006304 | 0.0104 | ...... | 0.0082 | 34 |
| 35 | 0.005614 | 0.0095 | ...... | 0.0075 | 35 |
| 36 | 0.005 | 0.0090 | ...... | 0.0067 | 36 |
| 37 | 0.004453 | 0.0085 | ...... | 0.0064 | 37 |
| 38 | 0.003965 | 0.0080 | ...... | 0.0060 | 38 |
| 39 | 0.003531 | 0.0075 | ...... | ...... | 39 |
| 40 | 0.003144 | 0.0070 | ...... | ...... | 40 |

[1] Courtesy of Brown and Sharpe Manufacturing Company.

# APPENDIX B-1

## PHYSICAL PROPERTIES OF COMMON STRUCTURAL MATERIALS

| MATERIAL | ELASTIC MODULUS, E | | SHEAR MODULUS, G | | POISSON'S RATIO, ν | THERMAL EXPANSION COEFFICIENT, α | |
|---|---|---|---|---|---|---|---|
| | Mpsi | GPa | Mpsi | GPa | — | $\times 10^{-6}/°F$ | $\times 10^{-6}/°C$ |
| METAL | | | | | | | |
| Aluminum Alloy | 10.4 | 72 | 3.9 | 27 | 0.32 | 12.9 | 23.2 |
| Brass, Bronze | 16 | 110 | 6.0 | 41 | 0.33 | 11.1 | 20.0 |
| Copper | 17.5 | 121 | 6.6 | 46 | 0.33 | 9.4 | 16.9 |
| Cast Iron - Gray | 15 | 103 | 6.0 | 41 | 0.26 | 6.7 | 12.1 |
| Cast Iron - Malleable | 25 | 170 | 9.9 | 68 | 0.26 | 6.7 | 12.1 |
| Magnesium Alloy | 6.5 | 45 | 2.4 | 17 | 0.35 | 14.4 | 25.9 |
| Nickel Alloy | 30 | 207 | 11.5 | 79 | 0.30 | 7.8 | 14.0 |
| Steel | 30 | 207 | 11.5 | 79 | 0.30 | 6.3 | 11.3 |
| Stainless Steel | 27.5 | 190 | 10.6 | 73 | 0.30 | 9.6 | 17.3 |
| Titanium Alloy | 16.5 | 114 | 6.2 | 43 | 0.33 | 4.9 | 8.8 |
| WOOD[1] | | | | | | | |
| Douglas Fir | 1.9 | 13 | 0.1 | 0.7 | — | 2.1 | 3.8 |
| Sitka Spruce | 1.5 | 10 | 0.07 | 0.5 | — | 2.1 | 3.8 |
| Western White Pine | 1.5 | 10 | — | — | — | 2.1 | 3.8 |
| White Oak | 1.8 | 12 | — | — | — | 2.1 | 3.8 |
| Red Oak | 1.8 | 12 | — | — | — | 2.1 | 3.8 |
| Redwood | 1.3 | 9 | — | — | — | 2.1 | 3.8 |
| CONCRETE | | | | | | | |
| Medium Strength | 3.6 | 25 | — | — | — | 5.5 | 9.9 |
| High Strength | 4.5 | 31 | — | — | — | 5.5 | 9.9 |
| PLASTIC | | | | | | | |
| Nylon Type 6/6 | 0.4 | 2.8 | — | — | — | 80 | 144 |
| Polycarbonate | 0.35 | 2.4 | — | — | — | 68 | 122 |
| Polyester, PBT | 0.35 | 2.4 | — | — | — | 75 | 135 |
| Polystyrene | 0.45 | 3.1 | — | — | — | 70 | 125 |
| Vinyl, Rigid PVC | 0.45 | 3.1 | — | — | — | 75 | 135 |
| STONE | | | | | | | |
| Granite | 10 | 70 | 4 | 28 | 0.25 | 4 | 7.2 |
| Marble | 8 | 55 | 3 | 21 | 0.33 | 6 | 10.8 |
| Sandstone | 6 | 40 | 2 | 14 | 0.50 | 5 | 9.0 |
| GLASS | 9.6 | 65 | 4.1 | 28 | 0.17 | 44 | 80 |
| RUBBER | $0.22^2$ | 0.0015 | $0.073^2$ | 0.0005 | 0.50 | 125 | 225 |

The values for the properties given above are representative. Because processing methods and exact composition of the material influence the properties to some degree, the exact values may differ from those presented here.

1. Wood is an orthotropic material with different properties in different directions. The values given here are parallel to the grain.

2. The modulus for rubber is given in ksi.

# APPENDIX B-2

## TENSILE PROPERTIES OF COMMON STRUCTURAL MATERIALS

| MATERIAL | ULTIMATE TENSILE STRENGTH, $S_u$ | | YIELD STRENGTH, $S_y$ | | DENSITY, $\rho$ | |
|---|---|---|---|---|---|---|
| | ksi | MPa | ksi | MPa | lb/in.$^3$ | Mg/m$^3$ |
| CARBON & ALLOY STEELS | | | | | | |
| 1010 A | 44 | 303 | 29 | 200 | 0.284 | 7.87 |
| 1018 A | 49.5 | 341 | 32 | 221 | 0.284 | 7.87 |
| 1020 HR | 66 | 455 | 42 | 290 | 0.284 | 7.87 |
| 1045 HR | 92.5 | 638 | 60 | 414 | 0.284 | 7.87 |
| 1212 HR | 61.5 | 424 | 28 | 193 | 0.284 | 7.87 |
| 4340 HR | 151 | 1041 | 132 | 910 | 0.283 | 7.84 |
| 52100 A | 167 | 1151 | 131 | 903 | 0.284 | 7.87 |
| STAINLESS STEELS | | | | | | |
| 302 A | 92 | 634 | 34 | 234 | 0.286 | 7.92 |
| 303 A | 87 | 600 | 35 | 241 | 0.286 | 7.92 |
| 304 A | 83 | 572 | 40 | 276 | 0.286 | 7.92 |
| 440C A | 117 | 807 | 67 | 462 | 0.286 | 7.92 |
| CAST IRON | | | | | | |
| Gray | 25 | 170 | — | — | 0.260 | 7.20 |
| Malleable | 50 | 340 | 32 | 220 | 0.266 | 7.37 |
| ALUMINUM ALLOYS | | | | | | |
| 1100-0 | 12 | 83 | 4.5 | 31 | 0.098 | 2.71 |
| 2024-T4 | 65 | 448 | 43 | 296 | 0.100 | 2.77 |
| 6061-T6 | 38 | 260 | 35 | 240 | 0.098 | 2.71 |
| 7075-0 | 34 | 234 | 14.3 | 99 | 0.100 | 2.77 |
| 7075 T6 | 86 | 593 | 78 | 538 | 0.100 | 2.77 |
| MAGNESIUM ALLOYS | | | | | | |
| HK31XA-0 | 25.5 | 176 | 19 | 131 | 0.066 | 1.83 |
| HK31XA-H24 | 36.2 | 250 | 31 | 214 | 0.066 | 1.83 |
| NICKEL ALLOYS | | | | | | |
| Monel 400 A | 80 | 550 | 32 | 220 | 0.319 | 8.83 |
| Cupronickel A | 53 | 365 | 16 | 110 | 0.323 | 8.94 |
| COPPER ALLOYS | | | | | | |
| Oxygen-free (99.9%) A | 32 | 220 | 10 | 70 | 0.322 | 8.91 |
| 90-10 Brass A | 36.4 | 251 | 8.4 | 58 | 0.316 | 8.75 |
| 80-20 Brass A | 35.8 | 247 | 7.2 | 50 | 0.316 | 8.75 |
| 70-30 Brass A | 44.0 | 303 | 10.5 | 72 | 0.316 | 8.75 |
| Naval Brass | 54.5 | 376 | 17 | 117 | 0.316 | 8.75 |
| Tin Bronze | 45 | 310 | 21 | 145 | 0.318 | 8.80 |
| Aluminum Bronze | 90 | 620 | 40 | 275 | 0.301 | 8.33 |
| TITANIUM ALLOY | | | | | | |
| Annealed | 155 | 1070 | 135 | 930 | 0.167 | 4.63 |

A = Annealed and HR = Hot Rolled

# APPENDIX B-3

## TENSILE PROPERTIES OF NON METALLIC MATERIALS

| MATERIAL | ULTIMATE TENSILE STRENGTH, $S_u$ | | YIELD STRENGTH, $S_y$ | | DENSITY, $\rho$ | |
|---|---|---|---|---|---|---|
| | ksi | MPa | ksi | MPa | lb/in.$^3$ | Mg/m$^3$ |
| WOOD | | | | | | |
| Douglas Fir | 15 | 100 | — | — | 0.017 | 0.470 |
| Sitka Spruce | 8.6 | 60 | — | — | 0.015 | 0.415 |
| Western White Pine | 5.0 | 34 | — | — | 0.014 | 0.390 |
| White Oak | 7.4 | 51 | — | — | 0.025 | 0.690 |
| Red Oak | 6.8 | 47 | — | — | 0.024 | 0.660 |
| Redwood | 9.4 | 65 | — | — | 0.015 | 0.415 |
| CONCRETE[1] | | | | | | |
| Medium Strength | 4.0 | 28 | — | — | 0.084 | 2.32 |
| High Strength | 6.0 | 40 | — | — | 0.084 | 2.32 |
| PLASTIC | | | | | | |
| Nylon Type 6/6 | 11 | 75 | 6.5 | 45 | 0.0412 | 1.14 |
| Polycarbonate | 9.5 | 65 | 9 | 62 | 0.0433 | 1.20 |
| Polyester, PBT | 8 | 55 | 8 | 55 | 0.0484 | 1.34 |
| Polystyrene | 8 | 55 | 8 | 55 | 0.0374 | 1.03 |
| Vinyl, Rigid PVC | 6.5 | 45 | 6 | 40 | 0.0520 | 1.44 |
| STONE[1] | | | | | | |
| Granite | 35 | 240 | — | — | 0.100 | 2.77 |
| Marble | 18 | 125 | — | — | 0.100 | 2.77 |
| Sandstone | 12 | 85 | — | — | 0.083 | 2.30 |
| GLASS[1] | | | | | | |
| 98% Silica | 7 | 50 | — | — | 0.079 | 2.19 |
| RUBBER | | | | | | |
| Natural, Vulcanized | 4 | 28 | — | — | 0.034 | 0.95 |

1. The tensile strength of concrete, stone and bulk glass is negligible. The compressive strength for these materials is reported in this table.

# APPENDIX C

## PROPERTIES OF AREAS

### C.1 AREA

The cross sectional area, A of structural members plays an extremely important role in the efficiency and the adequacy of any member in the design of the structure. For example, the stress in a uniaxial tension or compression member is given by:

$$\sigma = P/A$$

The force P is in the numerator and the area A of the cross section is in the denominator of the relation for the stresses. To lower the stresses we have only two options—decrease the force P or increase the area A. The area A of the cross section depends on its shape as indicated in Fig. C.1:

Fig. C.1 Dimensions and areas for common cross-sections.

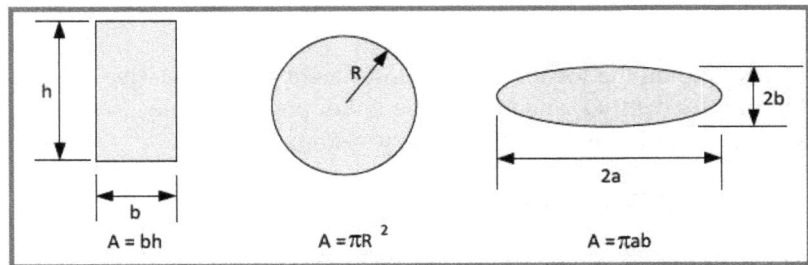

For areas of arbitrary shape, we determine their area by integration as indicated in Fig. C.2:

Fig. C.2 An arbitrary area A.

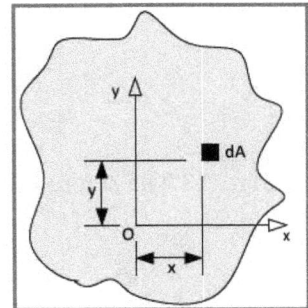

The area, A is determined by summing the incremental area dA in an integration process:

$$A = \int_A dA \qquad\qquad (C.1)$$

If the boundaries of the area are not known in terms of well-defined mathematical expressions, it is not possible to integrate to determine the area A. However, it is always possible to divide the area into many small squares or rectangles each with an area $\Delta A$. These squares or rectangles follow the boundary of the shape in question and completely fill the interior region. The area is then given by:

$$A = \sum_{i=1}^{N} \Delta A_i = N\Delta A \qquad (C.2)$$

where N is the number of small squares or rectangles.

## C.2 FIRST MOMENT OF AN AREA

The first moment of an area is important because it is useful in locating the position of the centroid of a given cross sectional area. Let's consider an arbitrary area with the coordinate system Oxy as shown in Fig. C.2. The first moment of the area A about the x-axis is defined as:

$$Q_x = \int_A y\,dA \qquad (C.3)$$

Also, the first moment of the area A with respect to the y-axis is defined as:

$$Q_y = \int_A x\,dA \qquad (C.4)$$

Depending on the location of the coordinate system relative to the area, the numerical values obtained for the first moment $Q_x$ and $Q_y$ may be either positive or negative. The units for $Q_x$ and $Q_y$ are $mm^3$ in the SI system or $in^3$ in the U. S. Customary system.

## C.3 CENTROID OF THE AREA A

The centroid of an area A is defined by point C located relative to an arbitrary coordinate system Oxy in Fig. C.3. Indeed, the definition of a centroid is the point, which is considered the center of gravity of a line, an area or a volume.

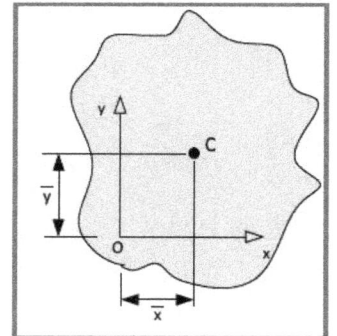

Fig. C.3 The centroid C of an area A is located with coordinates x and y.

The coordinates $\bar{x}$ and $\bar{y}$ of a centroid of an area are determined from the first moments as:

$$Q_y = \int_A x\,dA = A\bar{x}$$

$$Q_x = \int_A y\,dA = A\bar{y} \qquad (C.5)$$

where $\bar{x}$ and $\bar{y}$ are dimensions locating the centroid as shown in Fig. C.3.

Let's illustrate the method for determining the first moment of the area and the location of a centroid by considering a few simple shapes the examples presented below.

## EXAMPLE C.1

Consider the rectangular area illustrated in Fig C.4, with the origin of the Oxy coordinate system positioned at its corner. Determine the first moments of the rectangular area and the location of its centroid relative to this coordinate system.

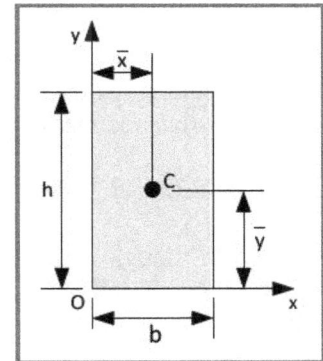

Fig C.4 A rectangle with the coordinate system located along its edges.

**Solution:**

For the rectangle area presented in Fig. C.4, the first moments of the area about are given by Eq. (C.5) as:

$$Q_x = A\bar{y} = (bh)\frac{h}{2} = \frac{bh^2}{2}$$

$$Q_y = A\bar{x} = (bh)\frac{b}{2} = \frac{b^2h}{2}$$

t was possible to quickly solve for the first moments, $Q_x$ and $Q_y$, because we recognized the location of the centroid for the rectangular area. When an axis of symmetry exists for a given area, the centroid is located somewhere on this axis of symmetry. With the rectangle, two axes of symmetry exist, and the location of the centroid is obviously at the intersection of the axes.

For the circular cross section shown in Fig. C.5, the center of the circle clearly locates the centroid. The center also serves as the origin for the centroidal axes $x_c$ and $y_c$.

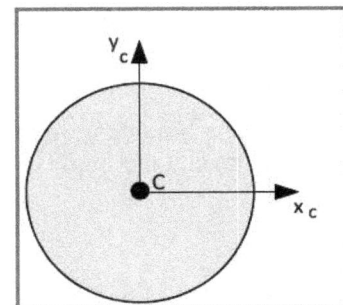

Fig. C.5 The centroid serves as the origin for the centroidal axes $Cx_cy_c$.

For cross sectional shapes such as ellipses, circles, squares and rectangles, the center may be located by inspection because these geometries have two axes of symmetry. However, for non-symmetric figures, such as triangles, portions of circles, parabolic areas, etc., locating the center of the area is not obvious. We will show methods for determining the location of the centroid of areas that do not exhibit two axes of symmetry.

With respect to a centroidal coordinate system, the first moment of the area must vanish for both axes. Therefore:

$$Q_{\bar{x}} = \int_A y\,dA = 0 \qquad\qquad Q_{\bar{y}} = \int_A x\,dA = 0 \qquad\qquad (C.6)$$

These relations are employed to locate the centroid of an area of any shape providing its boundary is defined with some mathematical function.

## EXAMPLE C.2

For a right triangle, determine the first moment of the area about its base and vertical side and the position of its centroid relative to its two sides. The right triangle with a base b and a height h is illustrated in Fig. C.6:

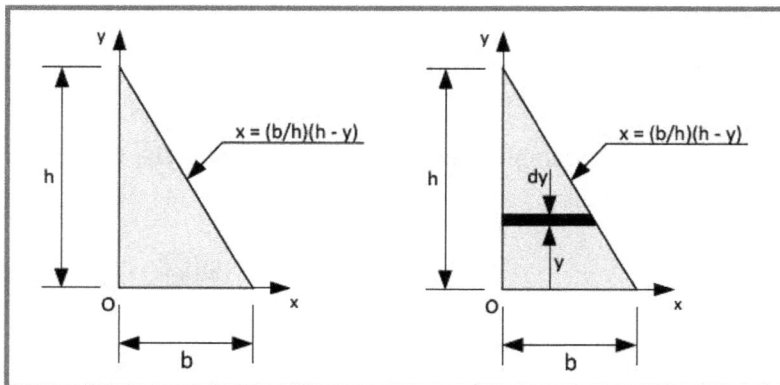

Fig C.6 A right triangle with a coordinate system coincident with its base and vertical side.

**Solution:**

To begin, let's determine the first moment of the area, A of the right triangle relative to the x-axis (its base side). Writing Eq. C.3 gives:

$$Q_x = \int_A y\,dA = \int_y xy\,dy \qquad\qquad (a)$$

Note, the equation for the diagonal boundary of the triangle is given by:

$$x = (b/h)(h - y) \qquad\qquad (b)$$

The limits on the integral go from 0 to h to encompass the area of the triangle. We substitute Eq. (b) and the limits on the integral into Eq. (a), and write:

$$Q_x = \frac{b}{h}\int_0^h (h - y)y\,dy \qquad\qquad (c)$$

Integrating Eq. (c) gives:

$$Q_x = \frac{b}{h}\left[\frac{hy^2}{2} - \frac{y^3}{3}\right]_0^h \qquad\qquad (d)$$

Completing the integration gives:

$$Q_x = bh^2/6 \qquad\qquad (C.7)$$

By using Eq. C.4 and following the same procedure, we find the first moment about the side of the triangle is given by:

$$Q_y = b^2 h/6 \qquad\qquad (C.8)$$

Equation C.7 gives the first moment of the area of a right triangle about its base. This is an interesting exercise in calculus, but what does it have to do with determining the location of the centroid of the right triangle? The result presented in Eq. (C.7) is an intermediate step. We continue the solution by combining the results of Eqs. (C.7) and (C.8) with Eqs. (C.5) to obtain:

$$Q_x = bh^2/6 = A\,\bar{y} = (bh/2)\,\bar{y} \qquad\qquad (e)$$

$$Q_y = b^2h/6 = A\,\bar{x} = (bh/2)\,\bar{x} \qquad\qquad (f)$$

where $\bar{x}$ and $\bar{y}$ locate the $Cx_c y_c$ coordinates relative to the Oxy coordinates (see Fig. C.7).

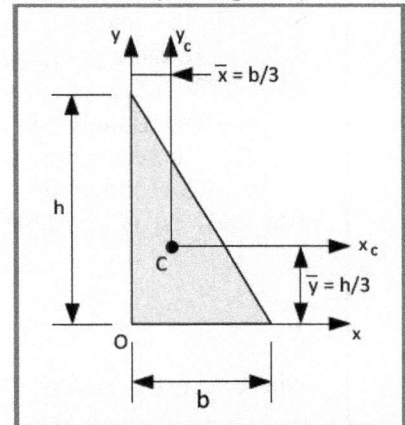

Fig. C.7 A right triangular area with a base axes Oxy and centroidal axes $Cx_c y_c$.

To determine the position of the centroid, let's solve Eqs. (e) and (f) for x and y to obtain:

$$\bar{y} = h/3 \qquad\qquad \bar{x} = b/3 \qquad\qquad (C.9)$$

We employ Eqs. (C.3) and (C.4) to determine the first moment of the area Q relative to either the x or y axes. The location of the centroid is then established from Eq. (C.5). The location of the centroid for common shapes has been determined, and the results are presented together with the definition of the shape of the area in Fig. C.8.

## C.4 LOCATING THE CENTROID OF A COMPOSITE AREA

In many cases, the shape of a cross section is unusual and differs from the common geometries described in Fig. C.8. To analytically locate the centroid of areas with uncommon shapes, we divide its area into several different common shapes for which we have solutions for the location of the centroid. Then we combine the product of these individual areas and their centroid locations to give the location of the centroid of the composite. Let's consider the uncommon shape defined in Fig. C.9, and locate the position of its centroid by employing the composite area technique.

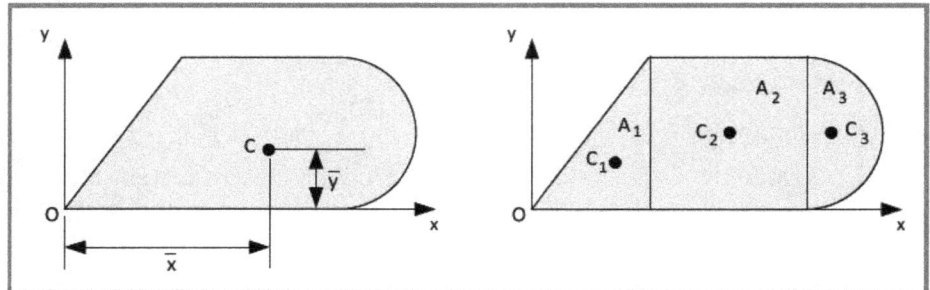

Fig. C.9 The unusual shaped area on the left is divided into common shaped areas on the right.

### EXAMPLE C.3

Determine the location of the centroid of the odd shape defined in Fig. C.9.

**Solution:**

To begin the solution, we divide the area into three sub-regions—$A_1$, $A_2$ and $A_3$ as shown in Fig. C.9. Note the sub-regions are a right triangle, rectangle and semicircle. The dimensions of the three different shapes are:
- The triangle—base b = 4 units and height h = 6 units.
- The rectangle—width w = 8 units and height h = 6 units.
- The semicircle—radius R = 3 units.

We apply Eq. (C.5) to the composite area, and write:

$$\Sigma Q_x \quad \Rightarrow \quad \overline{Y} A_t = \Sigma \overline{y}_n A_n \qquad (C.10)$$

$$\Sigma Q_y \quad \Rightarrow \quad \overline{X} A_t = \Sigma \overline{x}_n A_n \qquad (C.11)$$

where $A_t$ is the total area of the composite shape.

Let's first determine $\overline{Y}$ from Eq. (C.10):

$$\overline{Y} (A_1 + A_2 + A_3) = \overline{y}_1 A_1 + \overline{y}_2 A_2 + \overline{y}_3 A_3 \qquad (a)$$

Solving for $\overline{Y}$ yields:

$$\overline{Y} = (\overline{y}_1 A_1 + \overline{y}_2 A_2 + \overline{y}_3 A_3)/(A_1 + A_2 + A_3) \qquad (b)$$

Substituting results from Fig. C.8 into Eq. (b) gives:

$$\overline{Y} = \frac{\left( \dfrac{h}{3} \dfrac{bh}{2} + \dfrac{h}{2} wh + R \dfrac{\pi R^2}{2} \right)}{\left( \dfrac{bh}{2} + wh + \dfrac{\pi R^2}{2} \right)} \qquad (c)$$

Substituting b = 4, h = 6, w = 8 and R = 3 units into Eq. (c) yields:

$$\overline{Y} = 2.838 \text{ units}$$

This result is slightly less than h/2 as we anticipated. The presence of the triangle shifts the location of the centroid downward from the centerline of the rectangle and semi-circle.

Next, let's determine the position of the centroid in the direction of the x-axis. We begin by using Eq. (C.11), and write:

$$\overline{X}(A_1 + A_2 + A_3) = \overline{x}_1 A_1 + \overline{x}_2 A_2 + \overline{x}_3 A_3 \qquad (d)$$

Solving for $\overline{X}$ yields:

$$\overline{X} = (\overline{x}_1 A_1 + \overline{x}_2 A_2 + \overline{x}_3 A_3)/(A_1 + A_2 + A_3) \qquad (e)$$

From the information listed in Fig. C.8, we determine the centroid location for each of the shapes in the composite area as indicated below:

$$\overline{X} = \frac{\dfrac{2b}{3}\dfrac{bh}{2} + \left(b + \dfrac{w}{2}\right)wh + \left(b + w + \dfrac{4R}{3\pi}\right)\dfrac{\pi R^2}{2}}{\dfrac{bh}{2} + wh + \dfrac{\pi R^2}{2}} \qquad (f)$$

Substituting b = 4, h = 6, w = 8 and R = 3 units into Eq. (f) yields:

$$\overline{X} = 8.142 \text{ units}$$

We note that the location of the centroid of the unusual area is slightly to the right of the center of the rectangular area. This position is to be expected because the orientation of the right triangle with its area concentrated toward the right side tends to shift the centroid to the right.

## TRIANGULAR AREA

$\bar{x} = b/3$

$\text{AREA} = bh/2$

$I_{xc} = \dfrac{bh^3}{36}$

$\bar{y} = h/3$

$I_x = \dfrac{bh^3}{12}$

## SEMICIRCULAR AREA

$I_{xc} = \dfrac{\pi r^4}{8} - \dfrac{8r^4}{9\pi}$

$\text{AREA} = \pi r^2/2$

$\bar{y} = \dfrac{4r}{3\pi}$

$I_x = \dfrac{\pi r^4}{8}$

## PARABOLIC AREA

$\text{AREA} = 4ah/3$

$\bar{y} = 3h/5$

## SEMIELLIPTICAL AREA

$\text{AREA} = \pi ab/2$

$\bar{y} = \dfrac{4b}{3\pi}$

## CIRCULAR SECTOR

$\text{AREA} = \alpha r^2$

$\bar{x} = 2r\sin\alpha/(3\alpha)$

$I_x = \dfrac{r^4}{4}\left(\alpha - \dfrac{1}{2}\sin 2\alpha\right)$

$I_y = \dfrac{r^4}{4}\left(\alpha + \dfrac{1}{2}\sin 2\alpha\right)$

## GENERAL SPANDREL

$\text{AREA} = (ah)/(n+1)$

$y = Kx^n$

$\bar{x} = (n+1)a/(n+2)$

$\bar{y} = (n+1)h/(4n+2)$

$I_x = K\, a^{3n+1}\, \dfrac{3n-2}{3(3n+1)}$

## RECTANGULAR AREA

$\text{AREA} = bh$

$\bar{y} = h/2$

$I_x = \dfrac{bh^3}{3}$

$I_{xc} = \dfrac{bh^3}{12}$

## CIRCULAR AREA

$\text{AREA} = \pi r^2$

$\bar{y} = r$

$I_x = \dfrac{5\pi r^4}{4}$

$I_{xc} = \dfrac{\pi r^4}{4}$

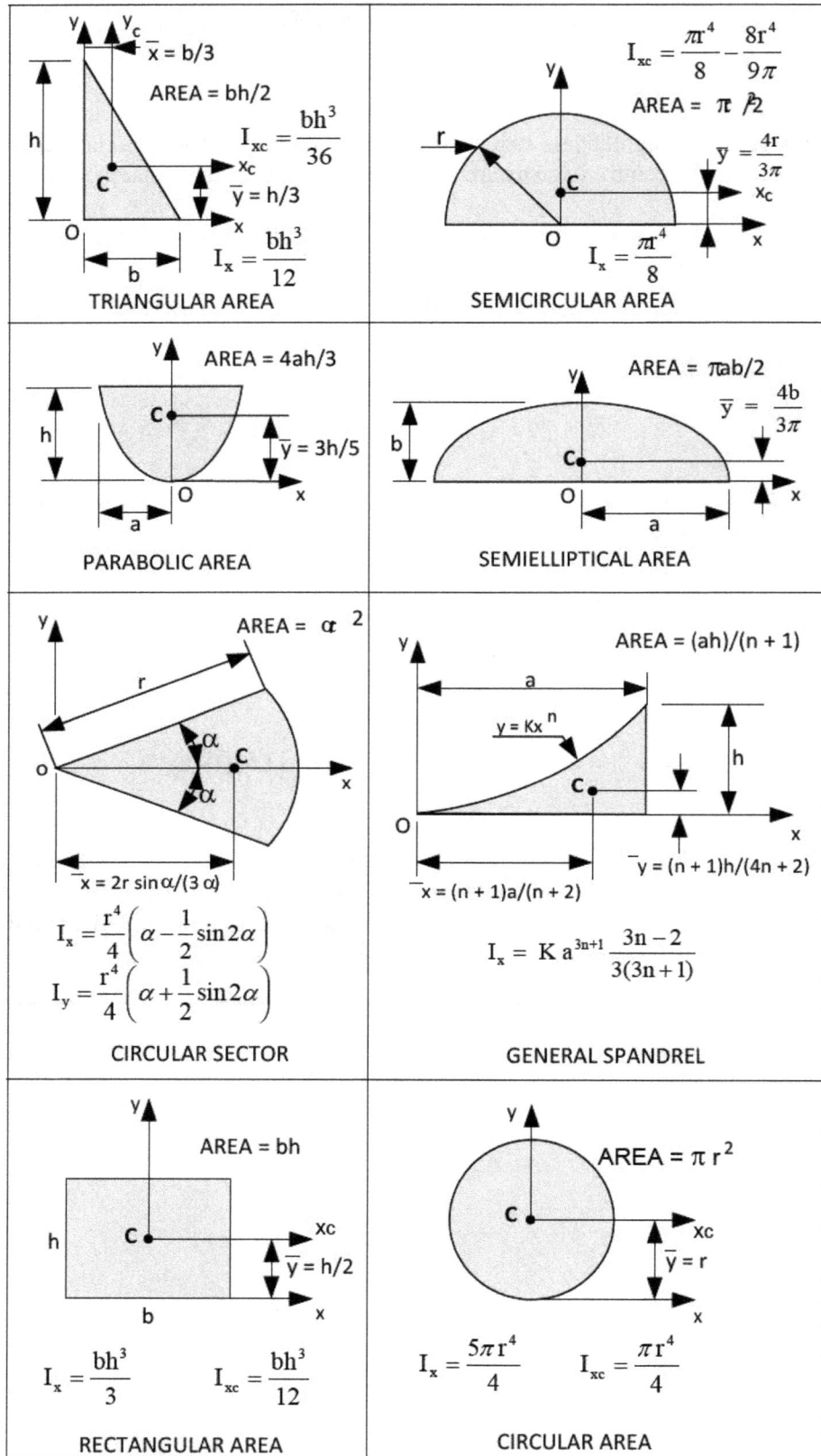

Fig. C.8 Area properties of some common shapes.

# C.5 SECOND MOMENT OF THE AREA

The second moment of the area is also know as the moment of inertia. In a discussion of beam theory in Mechanics of materials, we indicated the importance of the moment of inertia $I_z$ in determining the stresses produced by internal moments. Several different equations for the second moments of an area A, illustrated in Fig. C.10, are defined in this Section.

The second moment of the area, often called the moment of inertia, is referenced to one or both of the coordinate axes. The moment of inertia relative to the y and z axes is defined by:

$$I_z = \int_A y^2 dA \qquad\qquad I_y = \int_A z^2 dA \qquad\qquad (C.12)$$

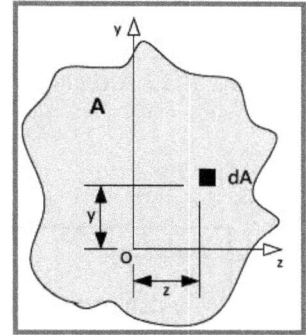

Fig. C.10 An elemental area dA is used when integrating to determine the second moment of the area A.

We also define a polar moment of inertia of the area A relative to the origin O of the coordinate system as indicated in Fig. C.11 as:

$$J_O = \int_A r^2 dA = \int_A (z^2 + y^2) dA = \int_A z^2 dA + \int_A y^2 dA$$

$$(C.13)$$

$$J_O = I_y + I_z$$

The units for the second moments of the area are in.$^4$ in the U. S. Customary system and m$^4$ or mm$^4$ in the SI system.

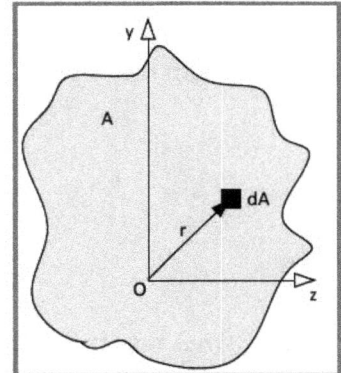

Fig. C.11 Coordinate system for the polar moment of inertia $J_O$.

Finally, the radius of gyration of an area A with respect to its axes is defined by:

$$I_z = r_z^2 A \qquad\qquad I_y = r_y^2 A \qquad\qquad J_O = r_O^2 A \qquad\qquad (C.14)$$

The symbols $r_z$ and $r_y$ reference the radii of gyration relative to the z and y-axes, respectively. The radius of gyration for the polar moment of inertia is $r_O$.

Let's consider a few examples to demonstrate the integration of Eq. (C.12) to determine the second moment of the area.

## EXAMPLE C.4

For the rectangular cross sectional area and coordinate system as defined in Fig C.12, determine the equations for $I_z$ and $I_y$.

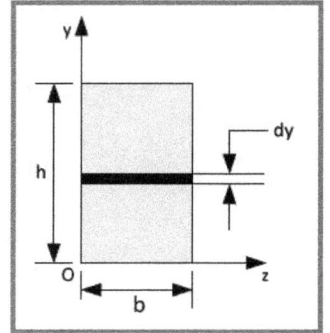

Fig. C.12 A rectangular area with a coordinate system along its edges.

**Solution:**

To determine $I_z$, we write Eq. (C.12) and observe that $dA = b\,dy$. Then:

$$I_z = \int_A y^2 dA = b\int_0^h y^2 dy = b\left[\frac{y^3}{3}\right]_0^h = \frac{bh^3}{3} \tag{C.15}$$

Similarly for $I_y$, we write Eq. C.12 and observe that $dA = h\,dz$. Then:

$$I_y = \int_A z^2 dA = h\int_0^b z^2 dz = h\left[\frac{z^3}{3}\right]_0^b = \frac{hb^3}{3} \tag{C.16}$$

To show the importance of the location of the coordinate system in determining the second moment of the area, let's shift the origin of the coordinate system to the center of the rectangle as shown in Fig C.13.

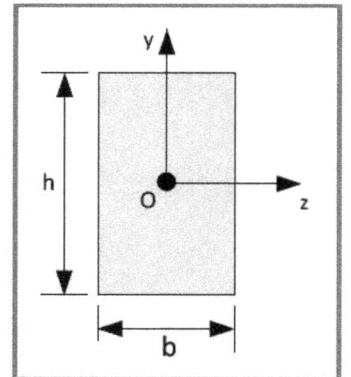

Fig. C.13 A rectangular area with a centrally located coordinate system.

## EXAMPLE C.5

Determine the moment of inertia of a rectangular area relative to its centroidal axes.

**Solution:**

To determine the moment of inertia $I_z$, we write Eq. (C.12) and note that $dA = b\,dy$.

$$I_z = \int_A y^2 dA = b\int_{-h/2}^{h/2} y^2 dy = b\left[\frac{y^3}{3}\right]_{-h/2}^{h/2} = \frac{bh^3}{12} \qquad (C.17)$$

To determine the moment of inertia $I_y$, we write Eq. (C.12) and note that $dA = h\,dz$.

$$I_y = \int_A z^2 dA = h\int_{-b/2}^{b/2} z^2 dz = h\left[\frac{z^3}{3}\right]_{-b/2}^{b/2} = \frac{hb^3}{12} \qquad (C.18)$$

A comparison of the results for the moments of inertia for the rectangle clearly indicates the importance of the location of the coordinate system relative to the area in question. As we may be required to determine the moment of inertia about axes with arbitrary locations, a method for accounting for shifting the position of coordinates is useful.

## C.6 THE PARALLEL AXIS THEOREM

Let's again consider an arbitrary area A positioned some distance from the z-axis as shown in Fig C.14. The centroidal axis of the area A is known and is identified with the z' axis. Assume that we have determined the second moment of the area with respect to the centroidal axis z'. Suppose we are required to compute the inertia $I_z$ with respect to an axis z that is parallel to the centroidal axis. We begin again with Eq. C.12 and write:

$$I_z = \int_A y^2 dA = \int_A (y_1 + d)^2 dA$$

$$(C.19)$$

$$I_z = \int_A y_1^2 dA + 2d\int_A y_1 dA + d^2\int_A dA$$

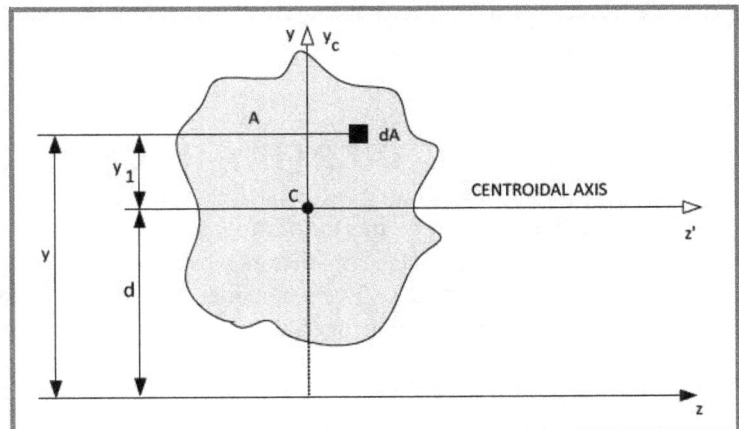

Fig. C.14 An arbitrary area A position a distance d from the z-axis.

The first term in Eq. (C.19) is the moment inertia $I_{z'}$ of the area about the centroidal axis. The second term $\int y_1\, dA$ is the first moment of the area about the centroidal axis, which is zero by definition of the centroid. The final term is simply $Ad^2$. Hence, the parallel axis theorem is given by:

$$I_z = I_{z'} + Ad^2 \qquad\qquad (C.20)$$

where $I_{z'}$ is the second moment of the area about the centroidal axis.

## EXAMPLE C.6

To demonstrate the use of Eq. (C.20), let's determine the moment of inertia $I_z$ of the rectangle shown in Fig C.15.

Fig. C.15 A rectangular area shifted by an amount $d = h$ relative to the z-axis.

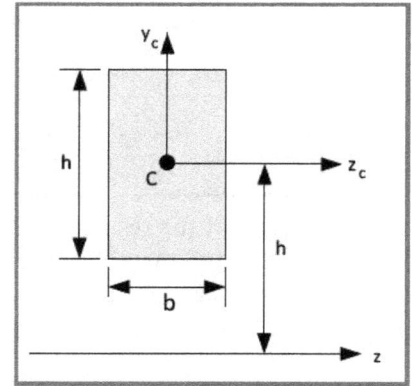

**Solution:**

To determine the moment of inertia $I_z$, we recall Eq. (C.20) and write:

$$I_z = I_{z'} + Ad^2$$

From Eq. (C.17) it is clear that $I_{z'} = bh^3/12$. Then we substitute into Eq. (C.20) to obtain:

$$I_z = bh^3/12 + (bh)(h^2) = (13/12)\, bh^3$$

This example illustrates two points. First, the parallel axis theorem is very helpful in determining the increase in the moment of inertia when the reference axis is some distance removed from the centroidal axis. Second, the moment of inertia is very sensitive to the movement of the reference axis relative to the centroidal axis. In this example, we moved the reference axis by an amount equal to the height h of the section and increased the inertia by a factor of 12.

## C.7 MOMENTS OF INERTIA OF COMPOSITE AREAS

To determine the moment of inertia of areas with complex shapes, we divide the area into subsections to form a composite area. Each subsection is a simple shape, such as a square, rectangle, circle or semi-circle with known properties. Let's consider the structural tee with a web and a flange as illustrated in Fig. C.16 and demonstrate the procedure for determining the moment of inertia.

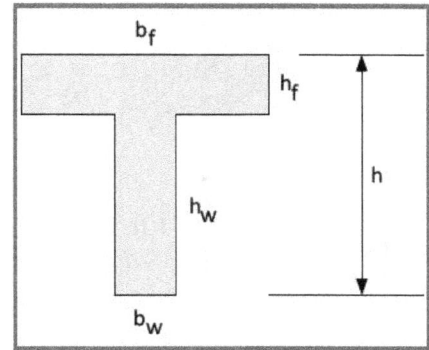

Fig. C.16 A structural tee is divided into two areas $A_{web}$ and $A_{flange}$.

The procedure for determining the properties of the composite area representing the structural tee involves the three steps listed below:

- Determine the location of the centroidal axis of the composite area.
- Determine the moment of inertia of each area of the composite section about its centroidal axis.
- Employ the parallel axis theorem to determine the moment of inertia of the total section relative to the centroidal axis.

## EXAMPLE C.6

Determine the moment of inertia $I_z$ of the structural tee shown in Fig. C.16 relative to its centroidal axis. The dimensions of the structural tee are given by:

$$h_f = 0.1\ h,\ h_w = 0.9\ h,\ b_w = 0.1\ b,\ \text{and}\ b = 150\ \text{mm and}\ h = 200\ \text{mm.}$$

**Solution:**

To determine the location of the centroidal axis of the composite area, subdivide the structural tee into two areas, namely $A_1$ and $A_2$ as shown in Fig. C.17. Then apply Eq. (C.10) to determine the location of the centroid relative to the z (reference) axis.

Fig. C.17 Subdivide the structural tee to form two rectangular areas and establish a convenient reference axis.

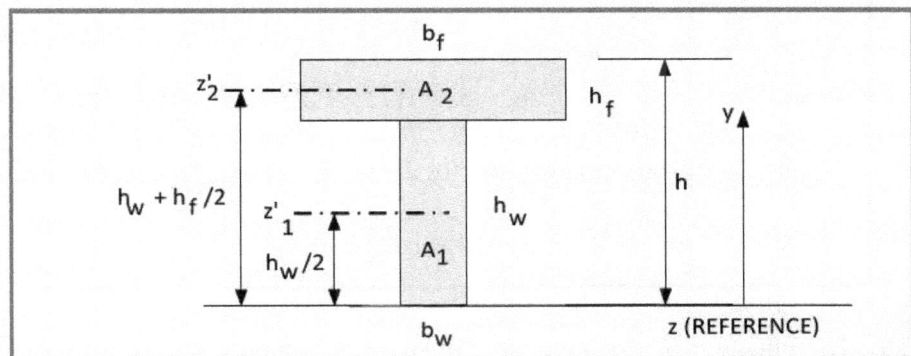

From Eq. (C.10), we write:

$$\Sigma Q_{z(REFERENCE)} = \overline{Y}\,A_t = \Sigma\,\overline{y}_n\,A_n = \overline{y}_1\,A_1 + \overline{y}_2\,A_2 \tag{a}$$

Substituting the dimensional quantities from Fig. C.17 into Eq. (a) yields:

$$\overline{Y} = \frac{A_1 \dfrac{h_w}{2} + A_2 (h_w + \dfrac{h_f}{2})}{A_1 + A_2} = \frac{\dfrac{b_w h_w^2}{2} + b_f h_f h_w + \dfrac{b_f h_f^2}{2}}{b_w h_w + b_f h_f} \tag{b}$$

Recall $h_f = 0.1$ h, $h_w = 0.9$ h, $b_w = 0.1$ b, $b_f = b$, b = 150 mm, and h = 200 mm. Substitute these quantities into Eq. (b) to obtain:

$$\overline{Y} = 0.713\, h = 0.713 \times 200 = 142.6 \text{ mm} \tag{c}$$

Next determine the moment for inertia of each area of the composite section about its centroidal axis. For $A_1$ (the web) with a width $b_w = 0.1$, b = 15 mm, and a height $h_w = 0.9$ h = 180 mm, we use Eq. (C.17) and write:

$$I_{z'1} = b_w\, h_w^{3} /12 = (15)(180)^{3}/12 = 7.29 \times 10^6 \text{ mm}^4 \tag{d}$$

For $A_2$ (the flange) with a width $b_f = 150$ mm, and a height $h_f = 0.1$ h = 20 mm, we write:

$$I_{z'2} = b_f\, h_f^{3} /12 = (150)(20)^{3}/12 = 0.100 \times 10^6 \text{ mm}^4 \tag{e}$$

Finally, employ the parallel axis theorem to determine the moment of inertia of the total area relative to the centroidal axis. Note for the composite area, we express the moment of inertia due to the two areas as:

$$I_z = I_{z1} + I_{z2} \tag{f}$$

We use Eq. (C.20) to expand Eq. (f) as:

$$I_z = I_{z1'} + A_1\, d_1^{2} + I_{z2'} + A_2\, d_2^{2} \tag{g}$$

The dimensions $d_1$ and $d_2$ are given by:

$$d_1 = \overline{Y} - h_w/2 = 142.6 - 90 = 52.6 \text{ mm} \tag{h}$$

$$d_2 = h_w + (h_f/2) - \overline{Y} = 180 + 10 - 142.6 = 47.4 \text{ mm} \tag{i}$$

Substituting numerical values for the terms in Eq. (g) yields the final result:

$$I_z = (7.290 \times 10^6) + (15)(180)(52.6)^2 + (0.1 \times 10^6) + (150)(20)(47.4)^2 = 21.60 \times 10^6 \text{ mm}^4$$

The determination of the moment of inertia for complex shapes is simple but tedious. Care must be exercised to avoid numerical errors in computing each of the quantities shown in Eq. (g). The moment of inertia is important in determining the bending stresses in beams because $I_z$ occurs in the denominator of the well know flexural formula $\sigma = - My/I_z$. The procedure for establishing the position of the centroidal axis is also important because it locates the neutral axis about which bending occurs. When the position of the neutral axis is known, we can establish $y_{max}$, $I_z$ and the maximum bending stress $\sigma$.

## EXAMPLE C.7

Determine the following quantities for the unsymmetrical wide flanged section shown in Fig. C.18.

1. The location of the centroid relative to a defined reference axis.
2. The moment of inertia of the web, top flange, and bottom flange relative to the centroidal axis.
3. The moment of inertia of the section relative to the centroidal axis.

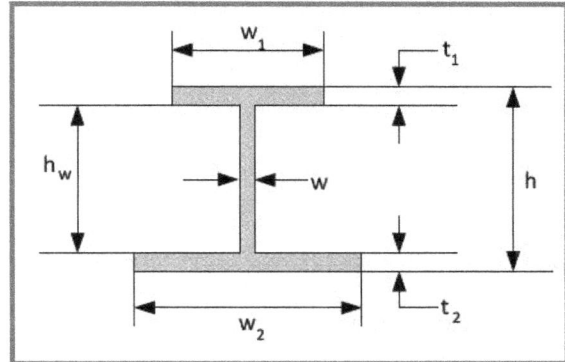

Fig. C.18 Dimensions of the unsymmetrical wide flanged section.

**Solution:**

To determine the location of the centroidal axis of the composite area, divide the unsymmetrical wide flanged section into three different areas consisting of the top flange, web and bottom flange as shown in Fig. C.19. Also establish a reference axis, z that is used as the datum for dimensioning the location of the centroid.

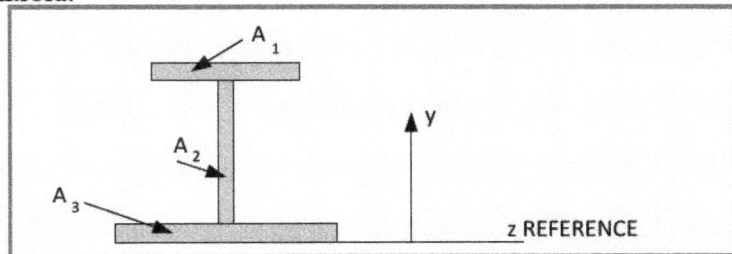

Fig. C.19 Dividing the unsymmetrical wide flanged section into three parts.

Let's apply Eq. (C.10) to determine the location of the centroid of this section relative to the z (reference) axis.

$$\Sigma Q_{z(REFERENCE)} = \overline{Y} A_t = \Sigma \, \overline{y}_n A_n = \overline{y}_1 A_1 + \overline{y}_2 A_2 + \overline{y}_3 A_3 \tag{a}$$

Substituting the symbols for the dimensions from Fig. C.18 into Eq. (a) yields:

$$\overline{Y} = \frac{A_1(h_w + t_2 + t_1/2) + A_2(t_2 + h_w/2) + A_3(t_2/2)}{A_1 + A_2 + A_3}$$

$$\overline{Y} = \frac{1}{2}\left[\frac{t_1 w_1(2h_w + 2t_2 + t_1) + h_w w(2t_2 + h_w) + w_2 t_2^2}{w_1 t_1 + h_w w + w_2 t_2}\right] \tag{b}$$

Suppose we assume the proportions of the cross section as:

$$w_1 = 10\ w;\ w_2 = 16\ w;\ t_1 = 0.15\ h;\ t_2 = 0.10\ h;\ \text{and}\ h_w = 0.75\ h \qquad \text{(c)}$$

Then Eq. (b) reduces to:

$$\overline{Y} = 0.4736\ h \qquad \text{(d)}$$

Finally if the height h = 300-mm and web thickness w = 15 mm, the location of the centroidal axis is given by:

$$\overline{Y} = 142.1\ \text{mm} \qquad \text{(e)}$$

To determine the moment of inertia of each area of the composite section about its centroidal axis, apply Eq. (C.17) to each of the three rectangular areas to obtain:

$$I_{z'1} = (1/12)w_1 t_1^3 = (1/12)(150)(45)^3 = 1.139 \times 10^6\ \text{mm}^4 \qquad \text{(f)}$$

$$I_{z'2} = (1/12)wh_w^3 = (1/12)(15)(225)^3 = 14.24 \times 10^6\ \text{mm}^4 \qquad \text{(g)}$$

$$I_{z'3} = (1/12)\ w_2 t_2^3 = (1/12)(240)(30)^3 = 0.540 \times 10^6\ \text{mm}^4 \qquad \text{(h)}$$

Next use the parallel axis theorem [Eq. (C.20)] to determine the moment of inertia of each of the three areas relative to the centroidal axis. For the top flange (Area 1):

$$I_{z1} = I_{z'1} + A_1 d_1^2 = I_{z'1} + (w_1\ t_1/4)(2h - t_1 - 2\overline{Y})^2$$

$$I_{z1} = 1.139 \times 10^6 + (150)(11.25)(600 - 45 - 284.2)^2$$

$$I_{z1} = 1.139 \times 10^6 + 123.7 \times 10^6 = 124.8 \times 10^6\ \text{mm}^4 \qquad \text{(i)}$$

For the web (Area 2):

$$I_{z2} = I_{z'2} + A_2 d_2^2 = I_{z'2} + (wh_w/4)(2t_2 + h_w - 2\overline{Y})^2$$

$$I_{z2} = 14.24 \times 10^6 + (15)(56.25)(60 + 225 - 284.2)^2$$

$$I_{z2} = 14.24 \times 10^6 + 0.0005 \times 10^6 = 14.24 \times 10^6\ \text{mm}^4 \qquad \text{(j)}$$

For the bottom flange (Area 3):

$$I_{z3} = I_{z'3} + A_3 d_3^2 = I_{z'3} + (w_2\ t_2/4)(2\ y - t_2)^2$$

$$I_{z3} = 0.540 \times 10^6 + (240)(7.5)(284.2 - 30)^2$$

$$I_{z3} = 0.540 \times 10^6 + 116.3 \times 10^6 = 116.8 \times 10^6\ \text{mm}^4 \qquad \text{(k)}$$

Before we complete the computation, examine the results listed in Eqs. (i – k). For both of the flanges, we note that the contribution of the $I_{z'}$ term was negligible compared to the $Ad^2$ term. For the web the $I_{z'}$ term was dominant and the $A_2 d_2^2$ term was negligible. To complete the solution for the moment of inertia of this composite area, we sum the moments of inertia due to the three areas as:

$$I_z = I_{z1} + I_{z2} + I_{z3} \qquad \text{(l)}$$

Substituting the results for Eqs. (i – l) into Eq. (l) yields:

$$I_z = (124.8 + 14.24 + 116.8) \times 10^6 = 255.8 \times 10^6 \text{ mm}^4 \qquad \text{(m)}$$

The contribution of the web to the moment of inertia about the centroidal axis of this unsymmetrical wide flange section was a small part of the total (less than 6%). This example shows that the main purpose of the web is to move the flanges outward from the neutral (centroidal) axis.

## PROBLEMS

C.1   Determine the area of a right triangle if its base is 30 mm and its height is 70 mm.

C.2   Determine the area of an isosceles triangle if its base is 30 mm and its height is 70 mm.

C.3   Determine the area of a right triangle if its base is 4in. and its height is 7 in.

C.4   Determine the area of an isosceles triangle if its base is 5 in. and its height is 9.5 in.

C.5   Determine the area of a circle with a diameter of 14 mm.

C.6   Determine the area of a circle with a diameter of 3.2 in.

C.7   Determine the area of a circle with a diameter of 88 mm.

C.8   Determine the area of a circle with a diameter of 3.6 in.

C.9   Determine the area of an ellipse if the major and minor axes are defined by $a = 2b = 16$ mm.

C.10  Determine the area of an ellipse if the major and minor axes are defined by $a = 3b = 6$ in.

C.11  Determine the area of an ellipse if the major and minor axes are defined by $a = 4b = 20$ mm.

C.12  Determine the area of an ellipse if the major and minor axes are defined by $a = (3.5)b = 7.0$ in.

C.13  Draw a right triangle with a base of 30 mm and height of 70 mm and then locate and dimension its centroid.

C.14  Draw an isosceles triangle with a base of 30 mm and height of 70 mm and then locate and dimension its centroid.

C.15  Draw a right triangle with a base of 4in. and height of 7 in. and then locate and dimension its centroid.

C.16  Draw an isosceles triangle with a base of 5 in. and height of 9.5 in. and then locate and dimension its centroid.

C.17  Verify the location of the centroid for the semicircular area defined in the figure to the right.

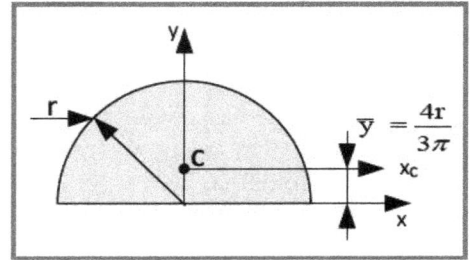

C.18  Verify the location of the centroid for the semi-elliptical area defined in the figure to the right.

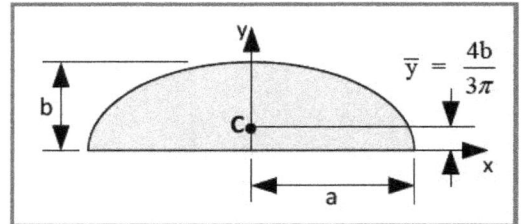

C.19  Verify the location of the centroid for the parabolic area defined in the figure to the left.

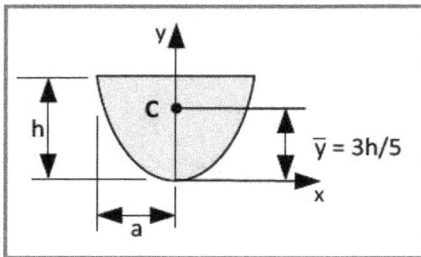

C.20  Verify the location of the centroid for the circular sector defined in the figure to the right.

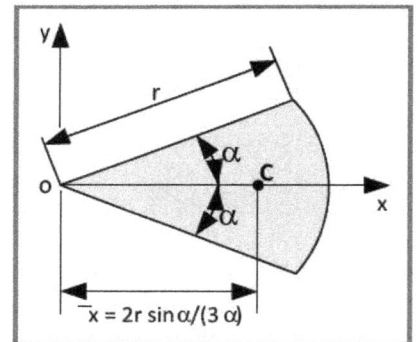

C.21  Verify the location of the centroid for the general spandrel defined in the figure to the left.

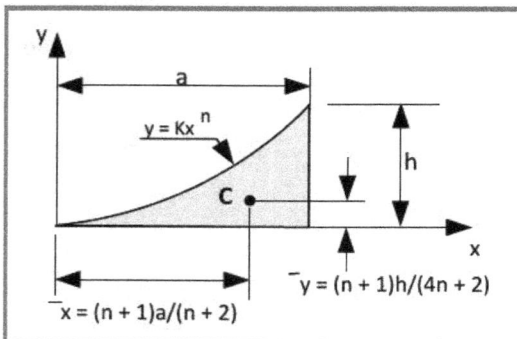

C.22 For the unusual plate shown in the figure to the right, derive an equation locating the centroid relative to point A. The elliptical hole is centered in the rectangular portion of the plate.

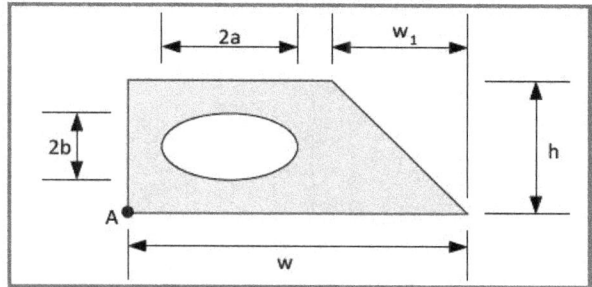

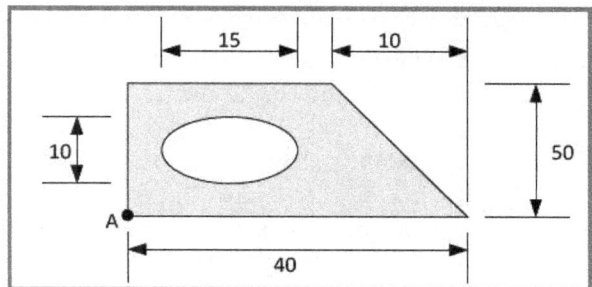

C.23 Determine the location of the centroid relative to point A for the plate defined in the figure to the right. All dimensions are given in mm, and the elliptical hole is centered in the rectangular portion of the plate.

C.24 Divide the plate, illustrated in the figure to the right, into three areas (an ellipse, rectangle and triangle). Determine the moment of inertia for each area about its centroidal axis.

C.25 Verify that the moment of inertia of an ellipse about its centroidal axis is given by:

$$I_x = \pi a b^3 /4 \qquad \text{and} \qquad I_y = \pi a^3 b/4$$

C.26 Determine the moment of inertia for the total area of the plate, shown in the figure to the left, about its centroidal axis.

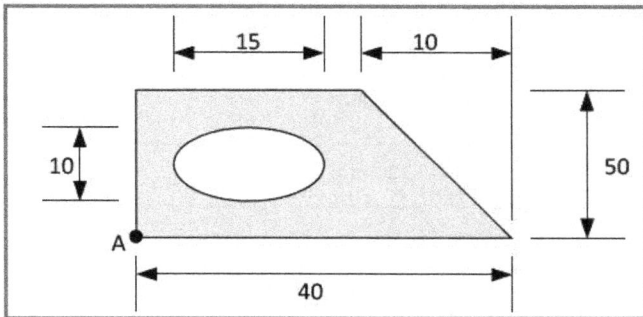

C.27 For the U shaped section, shown in the figure to the right, determine the location of the centroid $\bar{y}$ relative to an axis along its base. All dimensions are in inches.

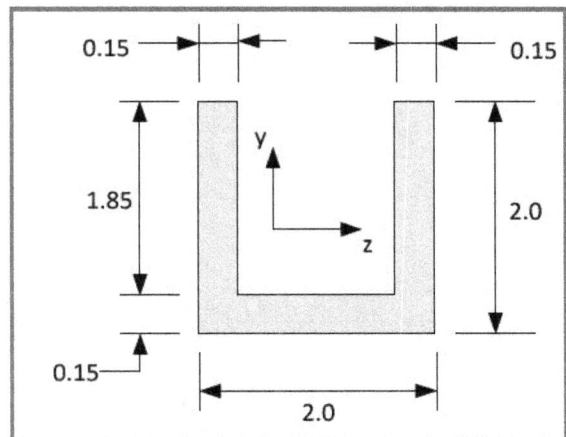

C.28 Determine the location of the centroid $\overline{y}$, relative to an axis along its base, for the U shaped section shown in the figure to the left. All dimensions are in inches.

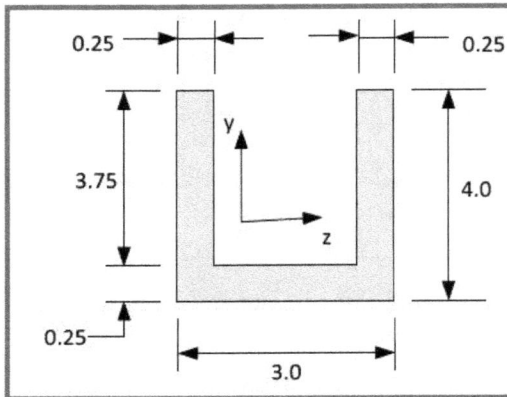

C.29 Determine the moment of inertia $I_z$ about the centroid of the U shaped member defined in (a) Problem C.27 and (b) Problem C.28.

C.30 Verify that the polar moment of inertia of a quarter circle area with a radius R is $J_o = \pi R^4/8$.

C.31 For the uncommon shape presented in the figure to the right, determine the centroid and the moment of inertia relative to the centroidal axis. The rectangular hole is centered in the elliptical area.

C.32 For the uncommon shape presented in the figure to the right, determine the moment of inertia relative to the z-axis.

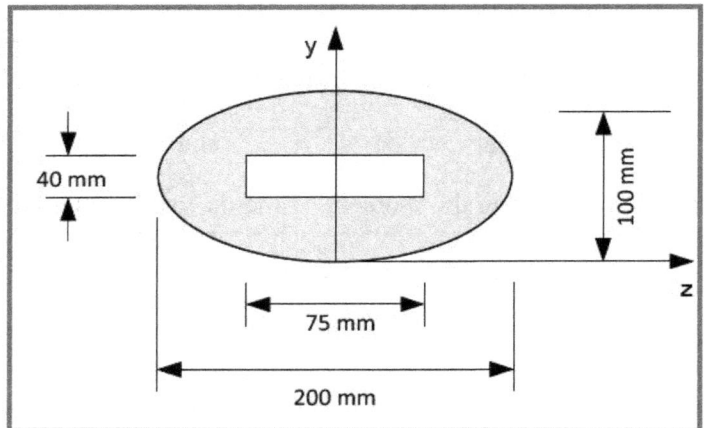

C.33 For the uncommon shape presented in the figure to the right, determine the moment of inertia relative to the y-axis.

C.34 For the irregular cross sectional area, described in the figure to the left, locate the centroid relative to both the x and y axes. The dimensions are given in mm.

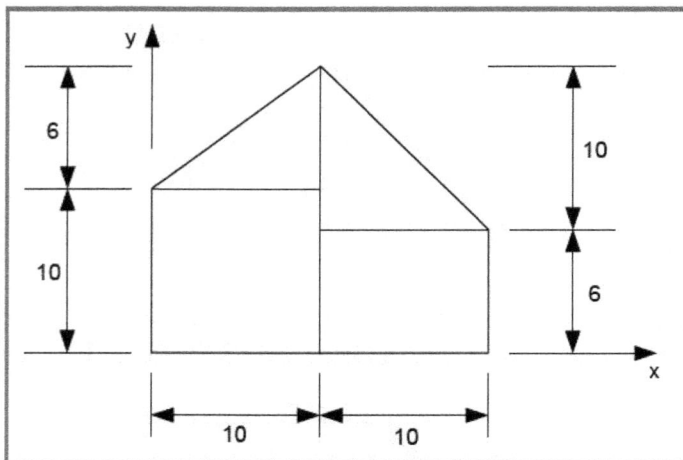

# INDEX

www.ingramcontent.com/pod-product-compliance
Lightning Source LLC
Chambersburg PA
CBHW081802200326
41597CB00023B/4112